TRANSFORMING CONSERVATION

Transforming Conservation

A Practical Guide to Evidence and Decision Making

Edited by William J. Sutherland

OpenBook Publishers

https://www.openbookpublishers.com

ISBN Paperback: 978-1-80064-856-2
ISBN Hardback: 978-1-80064-857-9
ISBN Digital (PDF): 978-1-80064-858-6
ISBN Digital ebook (EPUB): 978-1-80064-859-3
ISBN Digital ebook (AZW3): 978-1-80064-860-9
ISBN XML: 978-1-80064-861-6
ISBN HTML: 978-1-80064-862-3
DOI: 10.11647/OBP.0321

Cover image: Tiger Mountain / Issaquah, WA by Dave Hoefler.

Design by Katy Saunders

Contents

Preface

A Vision of Transformed Conservation Practice

The title of this book and preface are not exaggerations: we are serious about the need for delivering transformative change. We believe the processes outlined will deliver a better planet. Chapter 1 describes how conservation efforts are often far less effective than they could be. The consequences include wasting money, eroding public and political support, under-delivering the protection of species or habitats for a given budget, and deterring potential investments in conservation. The subsequent chapters describe how evidence can be used to make more effective decisions.

Fundamental strategic and cultural shifts are required. The strategic shift is to ensure evidence is available when and as needed and that decision makers have the necessary skills and tools to use evidence. The cultural shift is for it to be unacceptable to make decisions that ignore available evidence or effective decision-making processes.

Medicine has taken a strategic approach in which a doctor can access synthesised evidence or guidance in minutes. Similarly, if conservation science was to invest in synthesising evidence comprehensively, and embedding it in standards, policy, certifications, and guidelines, then decision makers could be assured that their actions were justified, based on the best available evidence. The cost of this is trivial in relation to the likely savings but it needs leadership to make the strategic investment.

The cultural shift refers to the need to be convinced and passionate about improving practice so that evidence use is expected and routinely delivered. Not using evidence would then be seen as being unprofessional and inefficient. Once transparent evidence use becomes the norm, many of the procedures described in this book would become routine.

The transition to evidence-based decision making will require new skills: finding and synthesising evidence; interpreting, evaluating and combining evidence; using experts efficiently; working appropriately with diverse communities of practitioners and stakeholders in co-assessing and co-designing practice; making effective decisions incorporating values and costs; embedding evidence in plans and guidance; and creating effective tests to improve the evidence base (Kadykalo 2021). Downey et al. (2021) provide some resources for training future generations to acquire those skills, but more is needed.

The good news is that this transition has begun. The fields of medicine and aviation safety have shown that such strategic and cultural transformation is possible and hugely beneficial. Increasingly, society expects decisions to be underpinned by evidence and some conservation organisations, such as the eight described in Chapter 11, have already placed evidence at their heart and are harnessing the benefits.

The measures in this book are not only likely to provide more cost effective delivery of policy and practice but they also provide confidence that the processes are responsible and the

https://doi.org/10.11647/OBP.0321.14

suggestions sensible. Improving credibility makes many aspects of project delivery easier, not least the attraction to funders.

We also anticipate that moves to evidence-based decisions will be self reinforcing. As practitioners, policy makers, businesses, and funders routinely ask why the suggested project or action is considered to be the best strategy, this will drive demand for the production, collation, and integration of evidence. To be most effective, evidence-based conservation needs to become the norm, where recipients expect to be asked to produce transparently collated evidence when applying for funds, funders expect to be shown the evidence and to contribute to funding it, and society at large demands the use of evidence in decision making and the improvement of the evidence base where it is lacking.

The most likely, and most effective, means of transforming conservation is for funders to ask whether their money is likely to be spent effectively (Chapter 9) and to support the embedding of learning within projects. Once multiple influential funders require the consideration of the evidence, the effect could be rapid and substantial.

Recently, philanthropists and funding bodies have made two major commitments to improving effectiveness. Twenty-five funders now ask their applicants to reflect on the evidence underpinning their applications (Parks et al., 2022). A range of funders has also agreed to ensure there is some funding for testing with a range of practitioner organisations committing to test an action annually (Tinsley-Marshall et al., 2022).

Most of the authors of this book are environmental scientists or practitioners and thus the examples we give relate to environmental practices. We are especially keen to continue the transformation of conservation practice so that it is more effective and so that conservation is more attractive to funders. However, the processes described here can be applied widely when deciding policies and practices. This book will examine these methods and their application across all fields, rather than discussing conservation specifically. The agenda described here equally apply to many other environmental areas, including agriculture, fisheries or energy production, as well as a wide range of other subjects that would gain from greater integration of science and policy, such as architecture, education, organisational management, overseas aid, philanthropy, policing, research funding, town planning or traffic management.

References

Downey, H., Amano, T., Cadotte, M., et al. 2021. Training future generations to deliver evidence-based conservation and ecosystem management. *Ecological Solutions and Evidence* 2(1): e121032, https://doi.org/10.1002/2688-8319.12032.

Kadykalo, A. N., Buxton, R. T., Morrison, P., et al. 2021. Bridging research and practice in conservation. *Conservation Biology* 35: 1725–37, https://doi.org/10.1111/cobi.13732.

Parks, D., Al-Fulaij, N., Brook, C., et al. 2022. Funding evidence-based conservation. Conservation Biology, 36, e13991. https://doi.org/10.1111/cobi.13991

Tinsley-Marshall, P., Downey, H., Adum, G., et al. 2022. Funding and delivering the routine testing of management interventions to improve conservation effectiveness. *Journal for Nature Conservation* 67: 126184, https://doi.org/10.1016/j.jnc.2022.1261

List of Authors

Tatsuya Amano, School of Biological Sciences, University of Queensland, Brisbane, 4072 Queensland, Australia

Longji Bako, The AP Leventis Ornithological Research Institute (APLORI), University of Jos Biological Conservatory, PO Box 13404, Laminga, Jos-East LGA, Plateau State, 930001, Nigeria

Marina Best, Parks Canada Agency, Indigenous Conservation, Indigenous Affairs and Cultural Heritage, 30 Rue Victoria, Gatineau, Quebec, Canada

Nicolas Boenisch, Foundations of Success Europe, 7211 AA Eefde, The Netherlands

Philipp Boersch-Supan, British Trust for Ornithology, The Nunnery, Thetford, Norfolk, UK; Department of Geography, University of Florida, Gainesville, FL 32611, USA

Des Browne, Houses of Parliament, Parliament Square, London SW1A 0PW; Biosecurity Research Initiative at St Catharine's College (BioRISC), St Catharine's College, Cambridge, UK

Yvonne Buckley, School of Natural Sciences, Trinity College, Dublin, Ireland

Mark Burgman, Centre for Environmental Policy, Imperial College London, London, UK

Marc W. Cadotte, Biological Sciences, University of Toronto-Scarborough, 1265 Military Trail, Toronto, ON, Canada

Stefano Canessa, Institute of Zoology, Zoological Society London, Regent's Park, London, UK

Samantha H. Cheng, Center for Biodiversity and Conservation, American Museum of Natural History, New York, NY 10026, USA; Global Science, World Wildlife Fund, Washington, DC 20037, USA

Alec P. Christie, Downing College, University of Cambridge, Regent Street, Cambridge, UK; Biosecurity Research Initiative at St Catharine's College (BioRISC), St Catharine's College, Cambridge, UK; Department of Zoology, University of Cambridge, The David Attenborough Building, Pembroke Street, Cambridge, UK

Geoffroy Citegetse, BirdLife International, Mermoz Pyrotechnie, Lot 23 Rue MZ 56, Dakar, Senegal

Carly N. Cook, School of Biological Sciences, Monash University, Clayton, VIC 3800, Australia

Steven J. Cooke, Carleton University, Ottawa, ON, Canada

Gemma Cranston, University of Cambridge Institute for Sustainability Leadership, The Entopia Building, 1 Regent Street, Cambridge CB2 1GG, UK

Angelita De la Luz, Beeflow, E Los Angeles Ave., Somis, California, USA 93066

Iain Dickson, Birdlife International, The David Attenborough Building, Pembroke Street, Cambridge, UK

Lynn V. Dicks, Department of Zoology, University of Cambridge, The David Attenborough Building, Pembroke Street, Cambridge, UK

Harriet Downey, Woodland Trust, Kempton Way, Grantham, Lincolnshire, UK

Johanna Eklund, Department of Geosciences and Geography, University of Helsinki, Finland

Nafeesa Esmail, The Wilder Institute, 1300 Zoo Road NE, Calgary, Alberta, Canada; Women for the Environment Africa (WE Africa) Fellow

Paul J. Ferraro, The Center for Behavioural & Experimental Agri-Environmental Research, Johns Hopkins University, Baltimore, Maryland, USA

Alison Field, Field and Forest, Land Management, Winchester, UK

Martin Fisher, Fauna & Flora International, The David Attenborough Building, Pembroke Street, Cambridge, UK

Robert Freckleton, School of Biosciences, University of Sheffield, Sheffield, UK

Winifred Frick, Bat Conservation International, 1012 14th Street NW, Suite 905, Washington, D.C. 20005, USA; Department of Ecology and Evolutionary Biology, University of California Santa Cruz, 130 McAllister Way, Santa Cruz, CA 95060, US

H. Charles J. Godfray, Oxford Martin School, University of Oxford, Oxford, UK

Matthew J Grainger, Norwegian Institute for Nature Research (NINA), PO Box 5685 Torgarden, 7485 Trondheim, Norway

Rhys Green, Department of Zoology, University of Cambridge, Downing Street, Cambridge, UK

Mark Hancock, RSPB Centre for Conservation Science, RSPB Scotland, North Scotland Office, Etive House, Beechwood Park, Inverness, UK; Cairngorms Connect, Achantoul, Aviemore, UK

Victoria Hemming, Department of Forest and Conservation Sciences, The University of British Columbia, Forest Sciences Centre, 3041–2424 Main Mall, Vancouver, British Columbia, V6T 1Z4 Canada

Jonathan Hughes, UN Environment World Conservation Monitoring Centre, 219 Huntingdon Road, Cambridge, UK

Robyn Irvine, Parks Canada Agency, Ecosystem Conservation, Protected Areas Establishment and Conservation, 30 Rue Victoria, Gatineau, Quebec, Canada

Hazel A. Jackson, Woodland Trust, Kempton Way, Grantham, Lincolnshire, UK

Kaylene E. Keller, US Fish and Wildlife Service, National Wildlife Refuge System 2800 Cottage Way, Suite W-2606, Sacramento, CA 95825, USA

Julia Koricheva, Department of Biological Sciences, Royal Holloway University of London, Egham, Surrey, UK

Charles Latrémouille, Partner, Conservation Measures Partnership (CMP), Montreal, Canada

Marissa McBride, Centre for Environmental Policy, Imperial College, Princes Gardens, South Kensington, London, UK

Angela R. McLean, Oxford Martin School, University of Oxford, Oxford, UK

Tom McPherson, Ingleby Farms and Forests, Slotsgade 1A, 4600 Køge, Denmark

William H. Morgan, Conservation Science Group, Department of Zoology, University of Cambridge, The David Attenborough Building, Pembroke Street, Cambridge, UK

Matt Muir, United States Fish and Wildlife Service International Affairs, 5275 Leesburg Pike, Falls Church, VA 22041, USA

Nibedita Mukherjee, Department of Social and Political Sciences, College of Business, Arts and Social Sciences, Brunel University, Uxbridge, London, UK

David O'Brien, NatureScot, Great Glen House, Leachkin Road, Inverness, UK

Nancy Ockendon, Endangered Landscapes Programme, Cambridge Conservation Initiative, The David Attenborough Building, Pembroke Street, Cambridge, UK

Danni Parks, The Whitley Fund for Nature, 110 Princedale Road, London, UK

Silviu Petrovan, Conservation Science Group, Department of Zoology, University of Cambridge, The David Attenborough Building, Pembroke Street, Cambridge, UK; Biosecurity Research Initiative at St Catharine's College (BioRISC), St Catharine's College, Cambridge, UK

Maud Quinzin, Ocean Discovery League, PO Box 182, Saunderstown, RI 02874, USA

Nicola P. Randall, Centre for Evidence Based Agriculture, Harper Adams University, Newport, Shropshire, UK

Ali Mohammad Rezaie, Department of Civil Engineering, University of Asia Pacific, Dhaka 1205, Bangladesh

Dilys Roe, School of Water, Energy and the Environment, Cranfield University, College Road, Cranfield, Bedford, UK

David Rose, School of Water, Energy and the Environment, Cranfield University, College Road, Cranfield, Bedford, UK

Libby Rumpff, School of BioSciences, University of Melbourne, Parkville 3052, Melbourne, Australia

Ullrika Sahlin, Centre for Environmental and Climate Science, Lund University, Lund, Sweden

Nick Salafsky, Foundations of Success, Bethesda, Maryland, USA

Stefan Schindler, Department Biodiversity & Nature Conservation, Environment Agency Austria (Umweltbundesamt), Vienna, Austria

Tom Sheldon, Science Media Centre, 183 Euston Road, London, UK

Ashley T. Simkins, Birdlife International, The David Attenborough Building, Pembroke Street, Cambridge, UK

Rebecca K. Smith, Conservation Science Group, Department of Zoology, University of Cambridge, The David Attenborough Building, Pembroke Street, Cambridge, UK; Biosecurity Research Initiative at St Catharine's College (BioRISC), St Catharine's College, Cambridge, UK

Jonathan Spencer, Field and Forest, Land Management, Winchester, UK

Ruth Stewart, Africa Centre for Evidence, University of Johannesburg, Bunting Road Campus, 2006, Johannesburg, South Africa

William J. Sutherland, Conservation Science Group, Department of Zoology, University of Cambridge, The David Attenborough Building, Pembroke Street, Cambridge, UK; Biosecurity Research Initiative at St Catharine's College (BioRISC), St Catharine's College, Cambridge, UK

Iroro Tanshi, Small Mammal Conservation Organisation SMACON, Nigeria

Nigel G. Taylor, Conservation Science Group, Department of Zoology, University of Cambridge, The David Attenborough Building, Pembroke Street, Cambridge, UK

Eleanor R. Tew, Forestry England, 620 Bristol Business Park, Coldharbour Lane, Bristol, UK

Des B.A. Thompson, NatureScot, Silvan House, 231 Corstorphine Road, Edinburgh, UK

Ann Thornton, Conservation Science Group, Department of Zoology, University of Cambridge, The David Attenborough Building, Pembroke Street, Cambridge, UK

Paul Tinsley-Marshall, Kent Wildlife Trust, Tyland Barn, Chatham Road, Sandling, Maidstone, UK; current address: Butterfly Conservation, Manor Yard, East Lulworth, Wareham, Dorset, UK

Thomas B. White, Conservation Science Group, Department of Zoology, University of Cambridge, The David Attenborough Building, Pembroke Street, Cambridge, UK

Kate Willott, Conservation Science Group, Department of Zoology, University of Cambridge, The David Attenborough Building, Pembroke Street, Cambridge, UK

Jeremy D. Wilson, RSPB Centre for Conservation Science, 2 Lochside View, Edinburgh, UK

Bonnie Wintle, School of Ecosystem and Forest Sciences, University of Melbourne, Victoria 3010 Australia

Thomas A. Worthington Conservation Science Group, Department of Zoology, University of Cambridge, The David Attenborough Building, Pembroke Street, Cambridge, UK

Hiromi Yamashita, Ristumeikan Asia Pacific University, Beppu, Oita, Japan; Nagoya University Graduate School of Environmental Studies, Nagoya, Aichi, Japan

Lindsay C. Young, Pacific Rim Conservation, PO Box 61827, Honolulu, Hawaii, USA 96839

Acknowledgements

For the last twenty years a global team has been working on the creation, collation, and adoption of evidence in policy and practice under the umbrella of the Conservation Evidence project. Over the past five years a number of teams have been busy creating a toolkit of approaches for embedding evidence into practice, whether writing evidence-based guidance or creating teaching material. Both of these have been a huge global effort with over 1,100 named collaborators (Smith, 2022). This book follows the completion of that toolkit.

This book has benefited enormously from those who have read and improved the text. Don Broom, Lisbet Rausing and Tim O'Riordan kindly provided detailed comments on the entire manuscript. The dream editorial team of Will Morgan, Silviu Petrovan, Rebecca Smith, Nigel Taylor, Ann Thornton, Tom White, Kate Willott and Hiromi Yamashita have done an astonishingly efficient task of writing, restructuring, editing the text, redrawing figures and sorting out the formatting.

The work on Conservation Evidence has received funding from the A.G. Leventis Foundation, Arcadia, British Ecological Society, Cambridge Conservation Initiative, Department for Environment, Food and Rural Affairs, Economic and Social Research Council, MAVA, Lord and Lady Moran, Natural England, Natural Environment Research Council, South West Water, Synchronicity Earth, The Nature Conservancy, and Waitrose and Partners. The Biosecurity Research Initiative at St Catharine's (BioRISC) is funded by The David and Claudia Harding Foundation.

Reference

Smith, R.K. 2022. The Conservation Evidence project — a truly collaborative and international effort. *Conservation Evidence Blog*, February 14 2022, https://conservationevidenceblog.wordpress.com/2022/02/14/the-conservation-evidence-project-a-truly-collaborative-and-international-effort

PART I
WHAT IS THE PROBLEM?

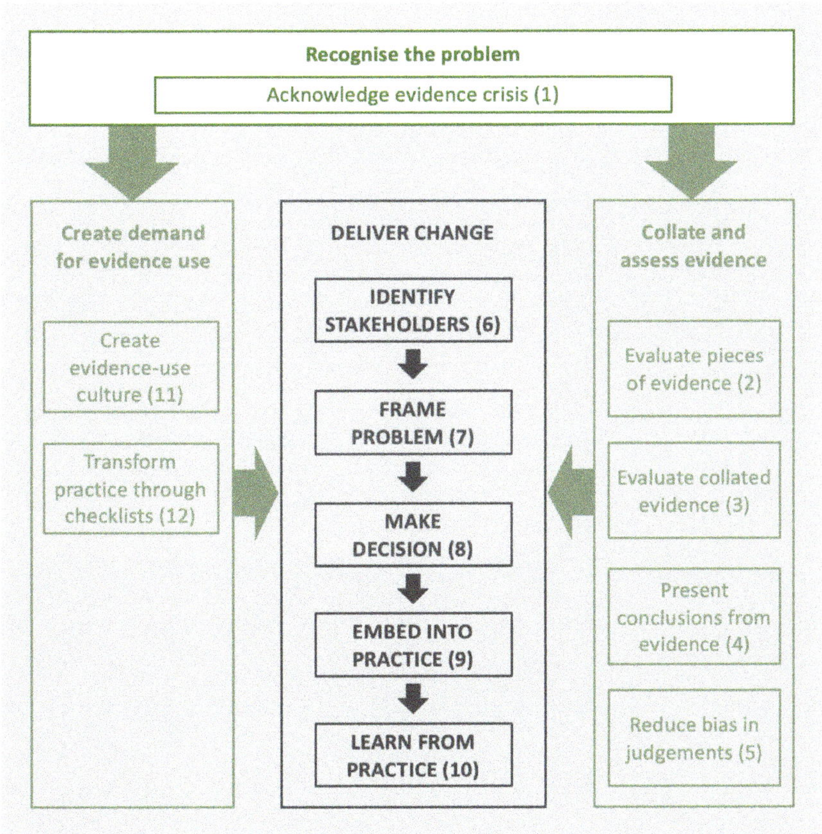

1. Introduction

The Evidence Crisis and the Evidence Revolution

William J. Sutherland[1,2]

Alongside many examples of highly effective practice, there are numerous studies showing ineffective practice. These studies suggest substantial improvements in efficiency are possible, meaning considerably more could be achieved for the same budget. Other fields, such as medicine and aviation, have shown how large improvements can be achieved through collecting and applying evidence effectively. A number of barriers hinder more effective practice, such as the challenges of accessing appropriate information and evidence complacency, where evidence is not sought despite being available. This chapter thus outlines the problems: the subsequent chapters provide the solutions.

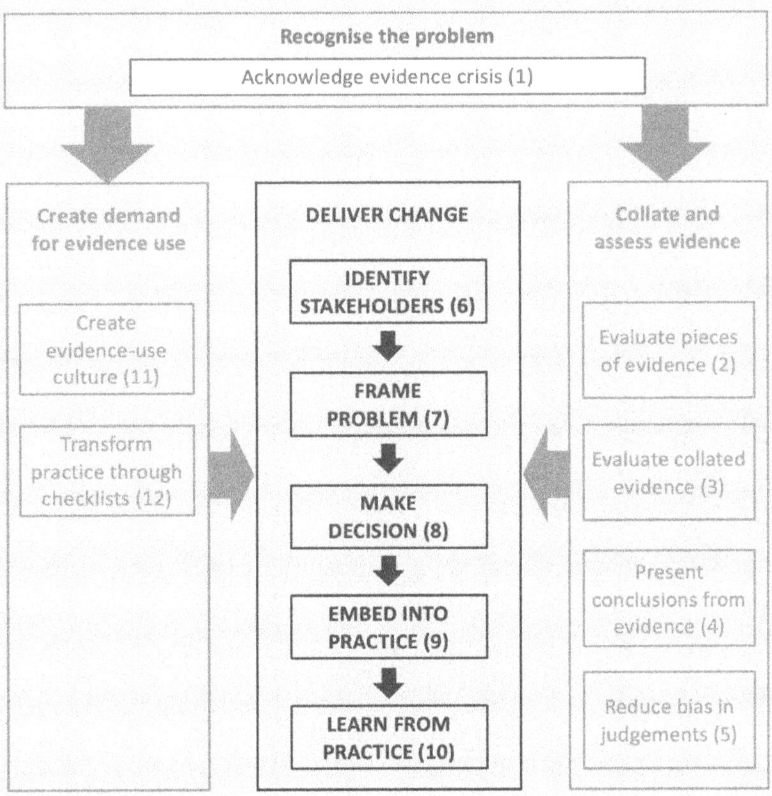

 https://doi.org/10.11647/OBP.0321.01

[1] Conservation Science Group, Department of Zoology, University of Cambridge, The David Attenborough Building, Pembroke Street, Cambridge, UK

[2] Biosecurity Research Initiative at St Catharine's (BioRISC), St Catharine's College, Cambridge, UK

Contents

1.1 The Aim of the Book

This book seeks to convince policy makers and practitioners that there are considerable problems with how decisions are usually made and that they need to take this seriously. It seeks to outline a range of straightforward practical solutions that can improve the use of evidence, improve the decision-making process and then embed evidence into practice. This increase in efficiency could either reduce expenditure or allow greater impact with the same budget. Furthermore, this increased efficiency could well make the field more attractive to fund. The overall aim is that better, and more, practice will result in a better planet.

1.2 The Evidence Crisis

There are innumerable success stories in which problems have been revealed and solved through a mix of science, policy and practice. These include many stunning protected areas (with Yellowstone in the USA as a pioneering example); improved species management meaning numerous species have been restored that would otherwise almost certainly have gone extinct (Bolam et al., 2021); numerous effective reintroduction projects (such as the beaver to numerous locations); reducing the problem of acid rain on forests and lakes by regulating to reduce sulphur dioxide and nitrogen oxide emissions; the Montreal Protocol restricting the use of various chemicals that cause the degradation of the ozone layer; the banning of dichlorodiphenyltrichloroethane (DDT) and related pesticides due to evidence of severe impacts of wildlife; the spread of many once severely persecuted species such as lynx due to reductions in persecution; and the recovery of the great whales from overexploitation.

Alongside these wonderful examples of effective use of evidence, there are many examples (Box 1.1) of poor evidence use and decision-making that lead to downright poor decisions to the detriment of both nature and people.

Although there are clearly substantial gains to be made by adopting effective practice, various studies (e.g. Pullin et al., 2004; Sutherland et al., 2004; Bayliss et al., 2011; Young and Van Aarde, 2011; Cvitanovic et al., 2014) have shown that conservation managers tend not to use scientific evidence to support their decision making. Instead they mainly rely on personal experience, anecdotes and the advice of colleagues. Walsh et al. (2015) asked 92 conservation managers, predominantly from Australia, New Zealand and the UK, whose work involved reducing predation on threatened birds, to provide opinions on 28 management techniques to reduce this problem. Managers were asked their opinions before and after giving them a summary of the literature about the interventions' effectiveness. On average, each survey participant changed their likelihood of implementing 45% of the interventions after reading the evidence of effectiveness. Unsurprisingly they switched to preferring actions shown to be effective and avoided the ineffective ones. This shows how enabling practitioners to access evidence can considerably change their preferred decisions toward more effective practices.

In his PhD, Andrea Santangeli (Santangeli, 2013) presented experimental tests of the effectiveness of ten conservation measures for protecting birds of prey on the resulting

Box 1.1 Examples of large-scale weak delivery in policy and practice

Problematic large-scale programmes

The Common Agricultural Policy (CAP) is one of the major expenditures of the European Union, with some successes such as reducing rural depopulation (Gray 2000; Garrone et al. 2019; Broom 2021). It is, however, widely acknowledged as environmentally problematic. A set of reforms have been implemented, intended to make the impact of agriculture less damaging to the environment with a series of agri-environmental measures. These measures have been expensive, with €3.23 billion spent on such schemes in 2012 alone. The ineffectiveness of some measures was identified by Kleijn et al. (2001), while a detailed review of agri-environment measures by Kleijn and Sutherland (2003) showed that the science was weak and many measures were already known to be ineffective. Indeed, only 54% showed clear benefits.

Following a review of agri-environment schemes (Kleijn and Sutherland, 2003), they advised policy makers that the 2003 reforms were unlikely to be effective. Following a subsequent review of the effectiveness of actions under the CAP, Dicks et al. (2014) concluded that the 2014 reforms would also be ineffective. This was shown to be the case (Pe'er et al., 2020). The EU's failure to effectively review the evidence before embarking on further reform in 2003 or 2014 resulted in ineffective measures being implemented across the EU. Despite this criticism, Batáry et al., (2015) showed no increase in the effectiveness of the measures over time.

An evaluation of large-scale tree planting programmes by Coleman et al. (2021) in Himachal Pradesh, Northern India, showed that tree plantings 'have not, on average, increased the proportion of forest canopy cover and have modestly shifted forest composition away from the broadleaf varieties valued by local people'. Hence, 'decades of expensive tree planting programmes in this region have not proved effective'. They conclude that planting will fail unless the underlying social and ecological processes that led to forest degradation are addressed.

As numerous studies have shown, many mangrove restoration schemes have also failed, due to a range of factors including using the wrong species, planting in the wrong location and a lack of community engagement (Brown et al., 2014; Kodikara et al., 2017; Lee et al., 2019).

Variation in delivery

Effective management can very work but depends on how it is delivered. Rigorously analysing the effectiveness of 1,506 protected areas on 27,055 waterbird populations showed that 27% of all populations were positively impacted once areas were classified as protected, 21% were negatively impacted, and 48% showed no detectable impact of protection (Wauchope et al., 2022). Analysis showed a strong signal that areas with further management for waterbirds or their habitat are more likely to benefit populations and a weak signal that larger areas are more beneficial than smaller ones.

Similarly, studies of marine protected areas show that their effectiveness varies greatly, with 29% not positively influencing fish populations. It turns out that staff capacity is key: those with adequate staff had ecological effects 2.9 times greater than those with inadequate capacity (Gill et al., 2017).

Another example showing that some actions are more cost-effective than others, leading to potential cost savings, is that around US$1.2 billion was invested between 2000 and 2019 into orangutan *Pongo* spp. conservation by governments, non-governmental organisations, companies, and communities (Santika 2022). Habitat protection was estimated to generate the highest return on investment overall, providing an average 12% improvement in orangutan distribution, followed by patrolling activities with a 9.2% improvement. However, rescue and rehabilitation, and translocation and reintroduction were considerably less cost effective.

Ineffective processes

To look at the evidence underpinning the proposed mitigation for the ecological damage caused by planning applications, Hunter et al. (2021) examined a sample of 50 UK housing applications. From this, they identified 446 recommendations comprising 65 different mitigation measures relating to eight taxa. Most (56%) measures were justified by citing published guidance. But their exploration of the literature underpinning this guidance revealed that empirical evaluations of the effectiveness of measures accounted for less than 10% of referenced texts. Over half of these measures had not been empirically evaluated at all.

production of fledglings. Six were beneficial, two made no difference and two were detrimental. In a subsequent paper, Santangeli and Sutherland (2017) calculated that a programme solely carrying out the effective measures would save about €78,854 (or 21.9% of the budget) annually. They then calculated the gain from this research: given the initial investment of about €156,211 for a PhD thesis, the financial return over a 10-year period ranges between 292% and 326%, depending on how costs are estimated. This shows that research on the effectiveness of actions can be hugely cost effective.

1.3 Why is Poor Decision Making So Common?

There are a host of practical reasons that hinder the use of evidence in decision making. The lack of availability of evidence can be a significant issue, due to (1) the subject being obscure or having not yet been researched resulting in a lack of relevant information, (2) the key information being behind a paywall, or (3) the evidence having not been compiled for a particular issue making it difficult to find or use. However, there is also an extensive range of further challenges to using evidence. Walsh et al. (2019) collated a comprehensive inventory of 230 factors that facilitate or limit the use of scientific evidence in conservation management decisions (Figure 1.1).

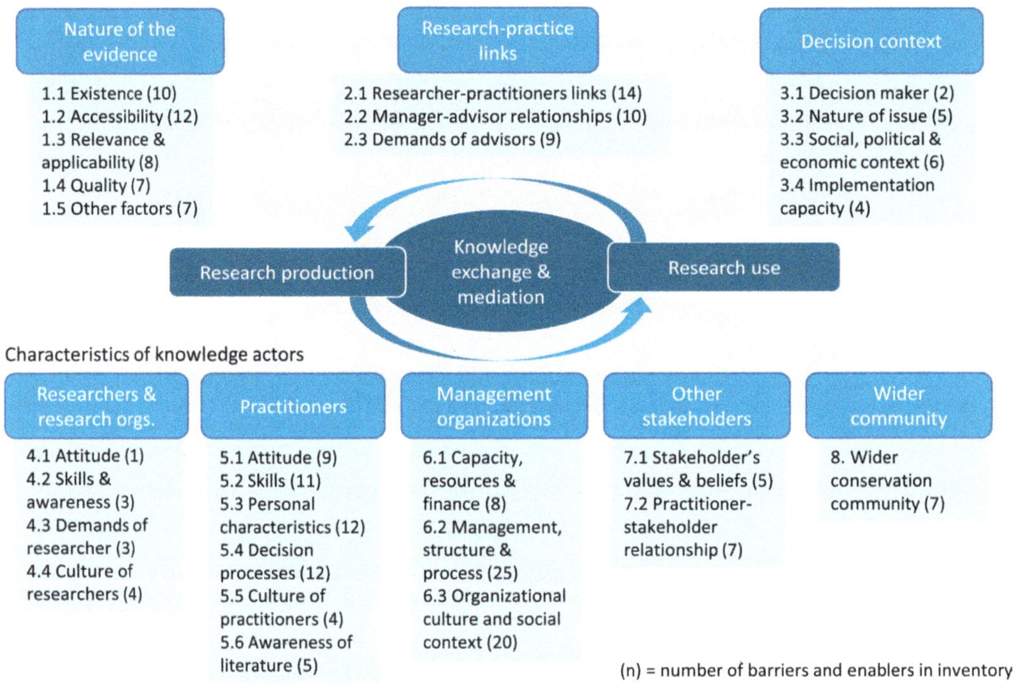

Figure 1.1 Taxonomy of 230 barriers and enablers to using scientific evidence in conservation management and planning decisions. (*Source:* Walsh et al., 2019, CC-BY-4.0)

However, alongside these many understandable reasons, there is also the widespread failure to use evidence even when it is easily available and practical to do so. There seem to be four main reasons for such poor decision making.

Conflict of interest

Conflict issues include cases of corruption because of financial or political gain. For example, the oil industry was aware of climate change yet undermined the crucial science (Oreskes and Conway, 2010). More commonly, personal values, interests and preferences (such as preferring a particular project or individual or disliking a particular outcome) can influence decisions; such motivational bias can lead to an unfavourable decision.

System 1 thinking

The Nobel Prize-winning psychologist Daniel Kahneman, in his book *Thinking Fast and Slow* (2011), made the distinction between two processes for thinking. System 1 is 'fast, instinctive and emotional', and well suited to the majority of quick, daily decisions. System 2 is 'slower, more deliberative and more logical', and suited to the occasional serious challenge. Evoking System 1 thinking for more complex decisions instead of slow, careful, analytical thinking (System 2) can lead to poor decisions.

Overconfidence effect

The overconfidence effect is an unwarranted confidence in ability. This can be illustrated by three phenomena: (1) people routinely overestimate their actual performance (e.g. the number of answers they got right); (2) people routinely overestimate their performance relative to others (e.g. most believing they are in the top 10%); (3) when asked to give an estimate range within which the answer lies, people are unjustifiably precise in their accuracy – so the correct answer regularly lies outside the stated range (Moore and Healy, 2008). These phenomena have been shown in repeated studies, including by those who are identified as experts (Burgman, 2015).

Evidence complacency

Almost 20 years ago, Sutherland et al. (2004) identified the serious challenges facing conservation practitioners wishing to read the scientific literature. These included journal paywalls, the lack of effective search engines and the challenges of extracting messages from scientific papers, resulting in practitioners making limited use of the available science. Since then, the power of search engines, open access papers and the development of evidence-based conservation has reduced these barriers, yet some of the challenges shown in Figure 1.1 remain. Although some conservation organisations produce and use excellent science, Sutherland and Wordley (2017) suggest that a culture of 'evidence complacency' remains in many areas of policy and practice. They used this term to describe a way of working in which, despite availability, evidence is not sought or used to make decisions, and the impact of actions is not tested.

As illustrated by the continuation of mistaken guidance to place babies on their fronts to sleep (see Section 1.4), ideas can become standard practices without consideration of the evidence or without a process of testing and evaluation. Another example is building bat bridges (placed to encourage bats to fly higher to reduce the risk of vehicle collisions), which were shown to be ineffective (Berthinussen and Altringham, 2012) yet continued to be installed with high financial cost. Such failures to test or search the literature wastes money and effort and also undermines confidence in expertise. Innovation is important, but novel ideas require testing before widespread adoption.

This book describes processes that can be adopted to reduce conflicts of interest, System 1 thinking, overconfidence and evidence complacency from undermining effective decision-making to the detriment of society.

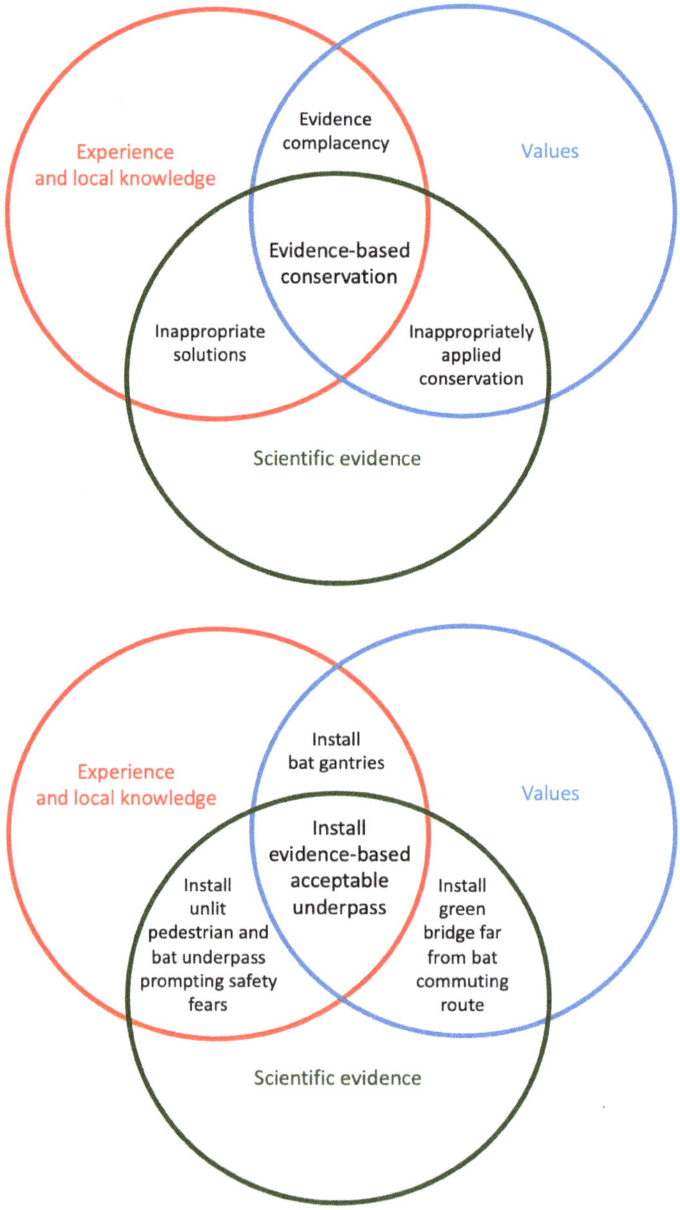

Figure 1.2 The role of evidence in evidence-based conservation, where values incorporate ethical, social, political and economic concerns. Top: how the components interact. Bottom: an example of how this might be applied to the problem of bats being killed when flying across a road (*Source*: author, with thanks to Claire Wordley).

1.4 The Evidence Revolution

We define evidence as 'relevant information used to assess one or more assumptions related to a question of interest' (modified from Salafsky et al., 2019). This includes published studies as well as reports, observations, citizen science and local knowledge.

Sackett et al. (1996) described evidence-based clinical decision making as combining research evidence with clinical expertise, alongside consideration of the preferences of the patient. Clinical expertise is the proficiency and judgement that individual clinicians acquire through clinical experience and clinical practice (Satterfield et al., 2009). Figure 1.2 similarly shows how scientific evidence, values, and experience and local knowledge can be combined and *how a combination of all three is usually essential.*

The idea of systematically using evidence has a long history. Box 1.2 outlines some components of this history of evidence use in medicine and conservation.

Medicine provides the classic example of policy and practice being transformed by evidence. Seventeen million parents owned and followed a copy of Spock's (1946) book, *The Common Sense Book of Baby and Child Care.* One key issue was whether to place babies on their front or back whilst sleeping. Spock's suggestion was to place the baby on its front, for the common sense reason of reducing the risk of choking following vomiting; sleeping on the front was routinely recommended in books up to 1988 (Gilbert et al., 2005). Gilbert and colleagues' 2005 analysis showed that by 1970 the published data indicated that there was a statistically significant increased risk of sudden infant death syndrome for front-sleeping babies. They concluded that had the evidence been reviewed and adopted in 1970, those results might have prevented over 10,000 infant deaths in the UK, with many more globally. Surely few things are more important than unnecessary baby deaths, yet the research was not converted into practice. A series of evidence reviews have also similarly shown fallacies in common sense treatments (enemas for women in labour [Cuervo et al., 1999], resting in bed after heart attacks [Sackett et al., 1996], giving corticosteroids to women about to have a premature baby to help the baby breathe [Liggins and Howie, 1972]). Medicine has been transformed with multiple mechanisms in place to reduce the chance of practising actions known to be ineffective or damaging.

But is evidence-based practice actually more effective? Emparanza et al. (2015) took advantage of the creation of an evidence-based practice unit in a hospital in Spain in 2003. They compared the fate of patients before and after the creation of the unit and in comparison with the rest of the hospital, which remained unchanged. Patients in the evidence-based unit had a significantly lower risk of death, 6.27% vs 7.75%, and a shorter length of stay, 6.01 vs 8.46 days, than those receiving standard practice. This is a 19% reduction in deaths and a 29% reduction in hospital stays.

Other fields show how improvements in decision making can be made. Aircraft flight safety has an impressive record: the total number of annual fatalities has tumbled 5–10 fold since the 1970s (Aviation Safety Network database https://aviation-safety.net/database) despite enormous increases in passengers carried over that time (Gössling and Humpe, 2020). Each aircraft disaster is followed by review, reflection and, if necessary, changes in practice. As

Box 1.2 A brief history of evidence use

The concept of hypotheses leading to experimental testing goes back at least to Hasan ibn al-Haythan (c965–1040AD) and his work on the nature of light and vision (Rosanna, 2003).

Scurvy, the awful disease and mass killer of sailors, precipitated thinking about evidence (Bown, 2003). James Lancaster, the captain of four ships sailing from England to India in 1601, was an early pioneer in the quest for an evidence-based solution. He organised that on one ship each sailor had three teaspoons of lemon juice per day while on the other three ships no one was given lemon juice. On the ship provided with lemon juice no one contracted scurvy, but 110 of 278 sailors on the other ships did. This is an impressive early controlled experiment, even if pseudoreplicated, with the results seeming to provide the correct answer.

Almost 150 years later in 1747, James Lind, the surgeon on the HMS *Salisbury*, decided to carry out an experiment in which twelve scurvy sufferers ('as similar as I could have them') were divided into six pairs who he treated similarly (the same diet and location) apart from the six experimental treatments ranging from two oranges and one lemon daily to 25 drops elixir of vitriol (i.e. sulphuric acid) taken three times daily.

The pair who took the citric fruits recovered; one became so well he could help run the experiment (and give the others their sulphuric acid!). The replication is minimal but this study is rightly applauded as one of the earliest replicated medical experiments.

Six years after revolutionising medicine, Lind published what can be considered the first attempt at anything resembling a systematic review. His *A Treatise of the Scurvy* (1753) concentrated on collecting the various observations 'upon attested facts and observations, without suffering the illusions of theory to influence and pervert the judgement' that had been made on the subject in an objective matter removing speculation 'before the subject could be set in clear and proper light, it was necessary to remove a great deal of rubbish.'

The popular story is that Lind was a visionary whose research was ignored by the bureaucratic Royal Navy and Merchant Navy, who only introduced citrus fruits in 1795 and 1865 respectively. However, Milne (2012) shows that Lind's advice was more confusing, although he was visionary in the methods he introduced.

The four messages of our book were discovered 250 years ago: (1) it is often straightforward to embed experiments into practice; (2) experiments are powerful in showing what works; (3) collating evidence (rather than speculation) is highly beneficial; (4) people can be bewilderingly reluctant to replace ineffective practices with those shown to work.

Modern evidence-based practice is usually attributed to Archie Cochrane, who questioned the lack of evidence for many common practices and, in 1972, published the classic tome *Effectiveness and Efficiency: Random Reflections on Health Services* in which he emphasised the need for randomised trials and literature reviews. The Cochrane Collaboration was established and named after him as a tribute to this pioneering work.

The expression 'evidence-based medicine' was invented by Guyatt et al. (1992) with the objective of altering the approach of medical decision making from 'intuition, unsystematic clinical experience, and pathophysiologic rationale' to using scientific, clinically relevant research.

Evidence-based practice has become routine in medicine and engineering. Elements of it have been adopted globally in, for example, education, policing and poverty alleviation, with numerous evidence centres and What Works Centres.

Evidence-based conservation

The term 'evidence-based conservation' was first used in the book *The Conservation Handbook* (Sutherland, 2000), which outlined the potential field. This was inspired by hearing Graham Bentham mention evidence-based medicine during an ineffective meeting to plan a university-wide research strategy.

Andrew Pullin, with his medic wife Teri Knight, had also been thinking about the use of evidence in conservation (Pullin and Knight, 2001). After the two teams discovered their joint interests at a British Ecological Society Policy Group field meeting, they wrote a collaborative paper to lay out the processes of evidence-based conservation (Sutherland et al., 2004).

Since then there has been a flourishing of interest associated with evidence-based conservation including various Centres of Environmental Evidence forming a global community creating and improving the practice of systematic reviews and systematic maps. The Conservation Evidence project, based in Cambridge, is working on an industrial-scale collation of evidence and collaborating with the global community to create new means of using evidence. However, there is yet to be a transformational change in the use of evidence in conservation.

an early example, after a series of crashes in the B-17 aircraft in 1942, Alphonso Chapanis interviewed pilots who had survived such crashes. He identified the problem of pilots muddling the adjacent, and similar-looking, landing gear and wing flap levers, which sometimes led to disastrous consequences (Syed, 2015). Adding wheel-shaped symbols to the landing gear lever and wedge-shaped symbols to the wing flap levers overcame the problem and became routine.

As another example, the manager of the low-budget Oakland Athletics baseball team used statistical analysis over the opinions of scouts in selecting team members. Following a 20-game winning streak, this evidence-based approach has become universal across baseball teams (Lewis, 2003).

Whilst the conservation community could review previous problematic programmes such as the Common Agricultural Policy, Indian tree-planting project, or the various unsuccessful mangrove planting schemes (Box 1.1) and incorporate the lessons in future plans, this feedback loop does not form a routine part of current processes, leading to the perpetuation of mistakes. Similar transformations to medicine are possible (and crucial) for other fields. As described in

Chapters 2 and 3, for conservation, evidence collation is being delivered at scale, and a number of conservation organisations have restructured themselves around the use of evidence in delivering action (Chapter 11).

The need for evidence is being increasingly recognised. Figure 1.3 is an attempt to bring together the challenges and options for change. The idea is that crises, such as evidence failures, drive demands for change. This book describes how to deliver tools and approaches that may change how practitioners, policy makers, businesses, and funders work, leading to an improved world.

1.5 The Case for Adopting Evidence Use

The main argument for increasing the use of evidence is that it will improve practice resulting in better outcomes and reduced costs. It is also likely that improving effectiveness will attract or secure further funding (Box 1.3).

Box 1.3 Advantages of including evidence

- Projects are more likely to be successful, reducing reputational risk.
- Projects are more cost effective.
- Organisations are less vulnerable to criticism if the practice is based on evidence.
- Decisions are more transparent.
- Learning from past successes and failures is embedded in the system.
- Less knowledge is lost within and between organisations.
- Funders have more confidence that their projects will deliver.
- Organisations may be seen as more professional and attract further funding.
- The field may be considered more professional so may attract additional funding.

1.5.1 The potential gains from transforming conservation

This chapter outlines what seems to be an overwhelming case for taking evidence use much more seriously. This includes billions of euros, dollars and pounds spent on practices known to be ineffective, and numerous studies showing that evidence is rarely used, alongside studies showing guidance is typically out of date and poorly evidence-based. Table 1.1 lists those studies that test gains from improving efficiency as reviewed in Section 1.2 showing the clear opportunity to improve following the good practice.

Subsequent chapters will show similar, equally serious problems in other stages of the decision-making process: as examples, it is rare to start by considering the full range of

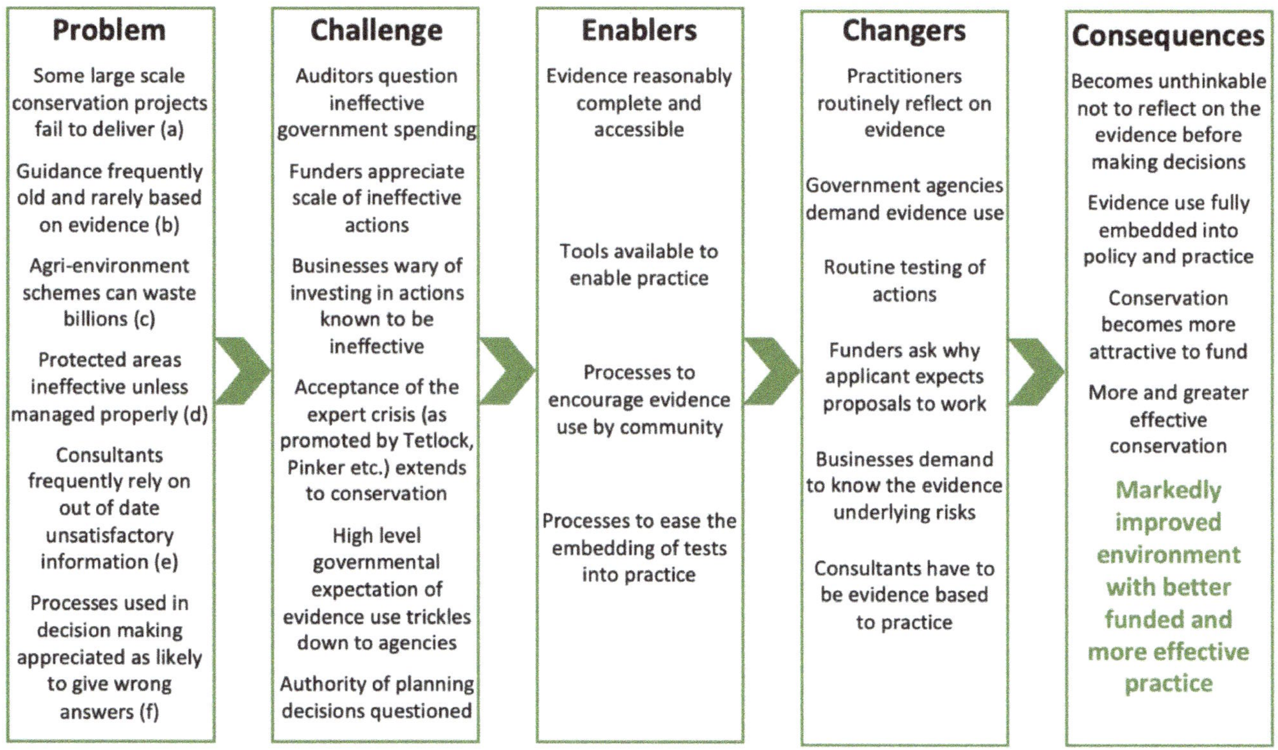

Figure 1.3 How the various evidence crises are likely to create demands for change that the enablers described in this book could help deliver, resulting in a series of improvements and a markedly better planet. References: (a) Coleman et al. (2021); (b) Downey et al. (2022); (c) Dicks et al. (2014); Pe'er et al. (2020); (d) European Court of Auditors (2020); Giakoumi et al. (2018); Wauchope et al. (2022); (e) Hunter et al. (2021); Laurance (2022); (f) Burgman (2015); Kahneman et al. (2021); Kahneman (2011); Pinker (2021); Tetlock (2009). (*Source:* author)

possible options; experts are usually used in ways that extensive research shows are likely to enhance bias and produce wrong answers; decision-making rarely follows processes known to be effective; costs are usually not presented in the manner that makes them comparable; other sources of knowledge are typically used haphazardly; there is ineffective learning from failures and there is rarely any effective learning.

Table 1.1 Summary of studies looking at inefficiencies or potential gains in investment from using evidence.

Action	Percentage change in effectiveness	Reference
Test of applying evidence-based medicine	19% reduction in deaths; 29% reduction in hospital stays	Emparanza et al. (2015)
Marine protected areas	29% not positively influencing fish populations	Gill et al. (2017)
The Common Agricultural Policy agri-environment measures	Compared to controls, 6% of studies showed decreases in species richness or abundance, 17% showed increases for some species and decreases for other species, 23% showed no change at all in response to agri-environment schemes (46% in total) and 54% showed increases.	Kleijn and Sutherland (2003)
	There has been no increase in the effectiveness of measures over time.	Batáry et al. (2015)
Experimental tests of the effectiveness of 10 conservation measures for protecting birds of prey	Just carrying out effective measures could achieve the same outcomes for 22% less expense	Santangeli and Sutherland (2017)
Estimated effectiveness orangutan *Pongo* spp. conservation measures in relation to investment (US$1.2 billion)	Some actions (habitat protection; patrolling activities) 300–400% more cost effective than others (habitat restoration, rescue and rehabilitation, and translocation)	Santika et al. (2022)
Review of the papers in *Conservation Evidence Journal* that test effectiveness of actions	Of those applied interventions that were tested 31% could be considered as unsuccessful	Spooner et al. (2015)

Action	Percentage change in effectiveness	Reference
Effectiveness of protected areas for waterbirds	27% of all populations positively impacted by protected areas; 21% negatively impacted; 48% no detectable impact	Wauchope et al. (2022)

1.6 The Inefficiency Paradox

The paradox described in this chapter is that society could relatively easily be more effective and save considerable amounts of money, but fails to be. Identifying solutions to this paradox is central to this book. Thus, the key chapters may seem to be those describing evidence assessment, expert elicitation or structured decision-making. Actually, they are the chapters describing how to achieve the cultural shift in which evidence becomes embedded in processes or organisational cultures.

Funders could achieve greater impact, governments could deliver more (or spend less), organisations could be more ambitious, consultants could be more effective, and businesses could reduce harm and increase benefits. Some pioneer organisations are taking evidence use seriously (see Chapter 11): they will lead the way in the necessary revolution.

1.7 Transforming Decision Making

It is conventional to illustrate decision making as a policy cycle with a logical flow of steps from identifying problems to determining the policy; it is, however, generally accepted that policy making rarely, if ever, proceeds this way (Owens, 2015). As shown in Figure 1.4, in practice, there is rarely a single assessment of each stage (i.e. as if going once round the outer loop). In reality, decisions may start at any of the stages and the process of making the decision means each stage becomes more focused spiralling inwards towards a final decision. Evidence is embedded in many of these stages.

With this framework, the decision-making process is complex and does not always start at the same stage. For example, it may start with considering a responsibility, such as what to do to protect a threatened species on your land, or by considering a threat and the range of options, or sometimes with someone promoting a solution and the need to consider whether this is appropriate. Table 1.2 shows how thinking usually progresses from the general to the specific as the decision becomes articulated.

Table 1.2 How the components of decision making shown in Figure 1.4 become more precise as thinking moves around and inwards around the hexagon towards making a decision.

Stage	Initial consideration	More specific consideration	Decision-making stage	Tools (and section numbers)
Identifying interests and responsibilities	Overall interests and responsibilities in this field	More specific consideration of priorities	Precise target of actions	Identifying stakeholders (6) Defining the scope of the project and its targets (7.2)
Identifying problems relating to these interests and responsibilities	General problems	More specific problems relating to interests and threats	Precise problem that proposed actions will deal with	Determining feature status (7.3.1) Assessing changes in feature status (7.3.2) Characterising current threats (7.4.1) Diagnosis: identifying the likely cause of declines (7.4.2) Horizon scanning to explore potential futures (7.4.4) Scenario planning for potential futures (7.4.5)
Identifying possible options	Broad range of options for dealing with threat	Consultation for other solutions	Fine tuning of options	Solution scanning (7.6.1) Innovating to devise novel solutions (7.6.2) Evaluating likely effects of actions (7.6.3) Theory of change (7.6.5) Developing questions and assumptions in situation models and theories of change (7.7.1) Identifying priority questions for policy and practice (7.7.2) Unpacking questions (7.7.3)

Stage	Initial consideration	More specific consideration	Decision-making stage	Tools (and section numbers)
Assessing likely effectiveness of options	General effectiveness of actions	Searching for further options and modifications	Specific consideration of likely effectiveness of exact options for precise conditions	A framework for assessing the weight of evidence (2.2) Weighing the evidence (2.3) Subjects of evidence (2.4) Sources of evidence (2.5) Types of evidence (2.6) Models (2.6.5)
Decision making	Broad consideration of likely options	Reducing list of options and refining those remaining	Final decision considering evidence and all factors	Deciding whether to invest in decision making (8.2) Decision sketching (8.3.1) Cost-benefit/cost-effectiveness analysis (8.3.2) Sketching means-ends networks (8.3.4) Consequences tables (8.5.1) Addressing trade-offs (8.5.7) Strategy tables (8.6) Decision trees (8.9) Creating models (8.10) Achieving consensus (8.11) Dealing with conflict (8.11.1)

Stage	Initial consideration	More specific consideration	Decision-making stage	Tools (and section numbers)
Agreeing policies or practice	Initial thoughts of policies	Fine-tuned policy	Specific detailed policy or practice	General principles for embedding evidence into practices (9.2) Evaluating evidence use (9.3) Evidence-based species and habitat management plans (9.4) Evidence-based guidance (9.5) Evidence-based policy (9.6) Evidence-based business decisions (9.7) Evidence-based writing and journalism (9.8) Evidence-based funding (9.9) Evidence-based decision-support tools (9.10) Evidence-based models (9.11)

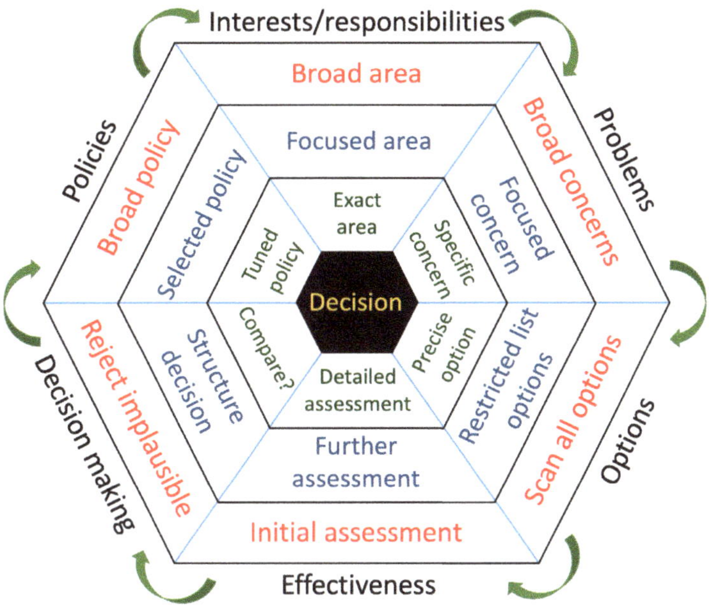

Figure 1.4 The Policy Hexagon. The logic is identifying responsibilities, identifying problems, finding options, considering effectiveness, and making a decision leading to the implementation of a policy or practice. (*Source*: author)

The approaches describe above seems to apply equally to other areas of practice. Table 1.3 lists a set of ten questions for identifying the extent of good practice.

Table 1.3 Suggested questions for determining the extent of good practice.

Question	Good practice	Weak practice
How have you determined the main issue that needs resolving?	Priorities identified, threats assessed and main problems diagnosed.	Assumed knowledge of what the main issue is (but cannot exclude other options).
How have you engaged with the community so they feel appropriately involved?	Identification of those who should be consulted, informed and involved in decision making followed by involvement.	Some discussions with local community.
Do you know how others deal with the problem (e.g. ask, 'how do they deal with the issue in Japan*?')?	An extensive range of options initially considered, for example as derived by a broad international group.	Consider the well-known options.
Has the published evidence been collated?	Reasonably comprehensive collation of evidence, in multiple languages, has been carried out by someone.	Uses literature that their experts are aware of.

Question	Good practice	Weak practice
How is the evidence used in making decisions?	Routinely include the process of assessing and reflecting on collated evidence.	Assume experts have knowledge of the evidence.
How has local knowledge been embedded in decisions?	Relevant knowledge, including on uses, values and beliefs, documented and used.	Informal consultation with locals.
How do experts contribute to decision making?	For key issues, methods are used to reduce bias.	Take views of an expert or the consensus of a group.
What is the process by which decisions are made?	Important decisions made using a structured decision-making approach.	Decision agreed by discussion.
Is the logic for the decision documented so it can be used by others?	Reasons for decisions documented so others can learn.	Lessons learnt and shared.
Is there ever any formal testing?	For important areas of uncertainty tests are routinely embedded into practice.	Occasionally try things out to see if they work and share verbally. Promote successes as case studies.

*or, 'how do they deal with the issue in the UK?' if in Japan

1.8 Structure of the Book

This book brings together the leading experts in the different elements of the process of improving decision making in conservation. The logic of the book follows two elements (presenting evidence and transforming culture) that feed into the decision-making process (Figure 1.5). The idea of using information and experts to make judgements is, of course, routine and unsurprising. The differences sought here are aiming to improve rigour, transparency and repeatability.

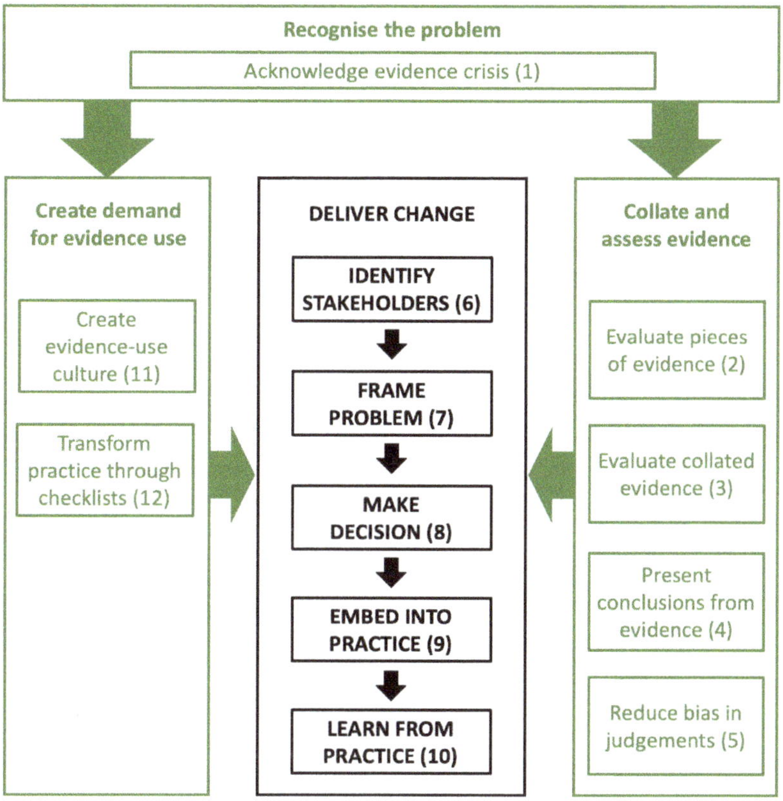

Figure 1.5 How the book sections and chapters (numbered) link together. The recognition of the evidence crisis leads to both a demand for evidence use (left) and processes for collating and assessing evidence and reducing bias in judgements (right). The central decision-making process running from identifying who should be involved to learning from the decision is driven by the demand for evidence and fed by evidence and judgements. (*Source:* author)

References

Amano, T., Székely, T., Sandel, B., et al. 2018. Successful conservation of global waterbird populations depends on effective governance. *Nature* 553: 199–202, https://doi.org/10.1038/nature25139.

Batáry, P., Dicks, L.V., Kleijn, D., et al. 2015. The role of agri-environment schemes in conservation and environmental management. *Conservation Biology* 29: 1006–16, https://doi.org/10.1111/cobi.12536.

Bayliss, H.R., Wilcox, A., Stewart, G.B., et al. 2011. Does research information meet the needs of stakeholders? Exploring evidence selection in the global management of invasive species. *Evidence and Policy* 8: 37–56, https://doi.org/10.1332/174426412X620128.

Berthinussen, A. and Altringham, J. 2012. Do bat gantries and underpasses help bats cross roads safely? *PLoS ONE* 7: e38775, https://doi.org/10.1371/journal.pone.0038775.

Bolam, F.C., Mair, L., Angelico, M., et al. 2021. How many bird and mammal extinctions has recent conservation action prevented? *Conservation Letters* 14: e12762, https://doi.org/10.1111/conl.12762.

Bown, S.R. 2003. *Scurvy: How a Surgeon, a Mariner, and a Gentleman Solved the Greatest Medical Mystery of the Age of Sail* (New York: Thomas Dunne Books, St. Martin's Press).

Broom, D.M. 2021. A method for assessing sustainability, with beef production as an example. *Biological Reviews* 96: 1836–53, https://doi.org/10.1111/brv.12726.

Brown, B., Fadillah, R., Nurdin, Y., et al. 2014. Case study: Community based ecological mangrove rehabilitation (CBEMR) in Indonesia. *SAPIENS* 7(2), http://journals.openedition.org/sapiens/1589.

Burgman, M.A. 2015. *Trusting Judgements: How to Get the Best out of Experts* (Cambridge: Cambridge University Press), https://doi.org/10.1017/CBO9781316282472.

Cochrane, A.L. 1972. *Effectiveness and Efficiency: Random Reflections on Health Services* (Nuffield Trust), https://www.nuffieldtrust.org.uk/research/effectiveness-and-efficiency-random-reflections-on-health-services.

Coleman, E.A., Schultz, B., Ramprasad, V., et al. 2021. Limited effects of tree planting on forest canopy cover and rural livelihoods in Northern India. *Nature Sustainability* 4: 997–1004, https://doi.org/10.1038/s41893-021-00761-z.

Cuervo, L.G., Rodriguez, M.N. and Delgado, M.B. 1999. Enemas during labour (Cochrane review). *Cochrane Database of Systematic Reviews* 4: Article Number CD000330, https://doi.org/10.1002/14651858.CD000330.

Cvitanovic, C., Fulton, C.J., Wilson, S.K., et al. 2014. Utility of primary scientific literature to environmental managers: An international case study on coral-dominated marine protected areas. *Ocean and Coastal Management* 102: 72–78, https://doi.org/10.1016/j.ocecoaman.2014.09.003.

Dicks, L.V., Hodge, I., Randall, N., et al. 2014. A transparent process for 'evidence-informed' policy making. *Conservation Letters* 7: 119–25, https://doi.org/10.1111/conl.12046.

Downey, H., Bretagnolle, V., Brick, C., et al. 2022. Principles for the production of evidence-based guidance for conservation actions. *Conservation Science and Practice* 4: e12663, https://doi.org/10.1111/csp2.12663.

Emparanza, J.I., Cabello, J.B. and Burls, A. 2015. Does evidence-based practice improve patient outcomes? *Journal of Evaluation in Clinical Practice* 21: 1059–65, https://doi.org/10.1111/jep.12460.

European Court of Auditors. 2020. *Special Report 26/2020. Marine Environment: EU Protection is Wide but not Deep* (Publications Office of the European Union), https://op.europa.eu/webpub/eca/special-reports/marine-environment-26-2020/en/.

Garrone, M., Emmers, D., Olper, A., et al. 2019. Jobs and agricultural policy: Impact of the common agricultural policy on EU agricultural employment. *Food Policy* 87: 101744, https://doi.org/10.1016/j.foodpol.2019.101744.

Giakoumi, S., McGowan, J., Mills, M., et al. 2018. Revisiting 'success' and 'failure' of marine protected areas: A conservation scientist perspective. *Frontiers in Marine Science* 5: Article 223, https://doi.org/10.3389/fmars.2018.00223.

Gilbert, R., Salanti, G., Harden, M., et al. 2005. Infant sleeping position and the sudden infant death syndrome: systematic review of observational studies and historical review of recommendations from 1940 to 2002. *International Journal of Epidemiology* 34: 874–87, https://doi.org/10.1093/ije/dyi088.

Gill, D., Mascia, M., Ahmadia, G., et al. 2017. Capacity shortfalls hinder the performance of marine protected areas globally. *Nature* 543: 665–69, https://doi.org/10.1038/nature21708.

Gössling, S. and Humpe, A. 2020. The global scale, distribution and growth of aviation: Implications for climate change. *Global Environmental Change* 65: 102194, https://doi.org/10.1016/j.gloenvcha.2020.102194.

Gray, J. 2000. The common agricultural policy and the re-invention of the rural in the European Community. *Sociologia Ruralis* 40: 30–52, https://doi.org/10.1111/1467-9523.00130.

Guyatt, G., Cairns, J., Churchill, D., et al. 1992. Evidence-based medicine: A new approach to teaching the practice of medicine. *JAMA* 268: 2420–25, https://doi.org/10.1001/jama.1992.03490170092032.

Hunter, S.B., zu Ermgassen, S.O.S.E., Downey, H., et al. 2021. Evidence shortfalls in the recommendations and guidance underpinning ecological mitigation for infrastructure developments. *Ecological Solutions and Evidence* 2: e12089, https://doi.org/10.1002/2688-8319.12089.

Kahneman, D. 2011. *Thinking, Fast and Slow* (New York: Macmillan).

Kahneman, D., Sibony, O. and Sunstein, C.R. 2021. *Noise. A Flaw in Human Judgement* (London: William Collins).

Kleijn, D., Berendse, F., Smit, R., et al. 2001. Agri-environment schemes do not effectively protect biodiversity in Dutch agricultural landscapes. *Nature* 413: 723–25, https://doi.org/10.1038/35099540.

Kleijn, D. and Sutherland, W.J. 2003. How effective are European agri-environment schemes in conserving and promoting biodiversity? *Journal of Applied Ecology* 40: 947–69, https://doi.org/10.1111/j.1365-2664.2003.00868.x.

Kodikara, K.A.S., Mukherjee, N., Jayatissa, L.P., et al. 2017. Have mangrove restoration projects worked? An in-depth study in Sri Lanka. *Restoration Ecology* 25: 705–16, https://doi.org/10.1111/rec.12492.

Laurance, W.F. 2022. Why environmental impact assessments often fail. *Therya* 13: 67–72, https://doi.org/10.12933/therya-22-1181.

Lee, S.Y., Hamilton, S., Barbier, E.B., et al. 2019. Better restoration policies are needed to conserve mangrove ecosystems. *Nature Ecology and Evolution* 3: 870–72, https://doi.org/10.1038/s41559-019-0861-y.

Lewis, M. 2003. *Moneyball. The Art of Winning an Unfair Game* (New York: W.W. Norton and Company).

Liggins, G.C. and Howie, R.N. 1972. A controlled trial of antepartum glucocorticoid treatment for prevention of the respiratory distress syndrome in premature infants. *Pediatrics* 50: 515–25, https://doi.org/10.1542/peds.50.4.515.

Lind, J. 1753. *A Treatise of the Scurvy. In Three Parts. Containing an Inquiry into the Nature, Causes, and Cure, of that Disease; Together with A Critical and Chronological View of what has been Published on the Subject*, First edition (Edinburgh: A. Kincaid and A. Donaldson).

Milne, I. 2012. Who was James Lind, and what exactly did he achieve? *The James Lind Library Bulletin: Commentaries on the history of treatment evaluation* 105: 503–08, https://www.jameslindlibrary.org/articles/who-was-james-lind-and-what-exactly-did-he-achieve/.

Moore, D.A. and Healy, P. J. 2008. The trouble with overconfidence. *Psychological Review* 115, 502–17, https://doi.org/10.1037/0033-295X.115.2.502.

Oreskes, N. and Conway, E.M. 2010. *Merchants of Doubt: How a Handful of Scientists Obscured the Truth on Issues from Tobacco Smoke to Global Warming* (New York: Bloomsbury Press).

Owens, S. 2015. *Knowledge, Policy, and Expertise: The UK Royal Commission on Environmental Pollution 1970–2011* (Oxford: Oxford University Press), https://doi.org/10.1093/acprof:oso/9780198294658.001.0001.

Pe'er, G., Bonn, A., Bruelheide, H., et al. 2020. Action needed for the EU Common Agricultural Policy to address sustainability challenges. *People and Nature* 2: 305–16, https://doi.org/10.1002/pan3.10080.

Pinker, S. 2021. *Rationality: What it is, Why it Seems Scarce, Why it Matters* (London: Allen Lane).

Pullin, A.S. and Knight, T.M. 2001. Effectiveness in conservation practice: Pointers from medicine and public health. *Conservation Biology* 15: 50–54, https://doi.org/10.1111/j.1523-1739.2001.99499.x.

Pullin, A.S., Knight, T.M., Stone, D.A., et al. 2004. Do conservation managers use scientific evidence to support their decision making? *Biological Conservation* 119: 245–52, https://doi.org/10.1016/j.biocon.2003.11.007.

Rosanna, G. 2003. Al-Haytham the man of experience. First steps in the science of vision. *Journal of the International Society for the History of Islamic Medicine* 2: 53–55, https://www.ishim.net/ishimj/4/10.pdf.

Sackett, D.L., Rosenberg, W.M. and Gray, J.A.M. 1996. Evidence based medicine: What it is and what it isn't. *BMJ* 312: 71–72, https://doi.org/10.1136/bmj.312.7023.71.

Salafsky, N., Boshoven, J., Burivalova, Z., et al. 2019. Defining and using evidence in conservation practice. *Conservation Science and Practice* 1: e27, https://doi.org/10.1111/csp2.27.

Santangeli, A. 2013. *Assessing the Effectiveness of Different Approaches to Species Conservation* (PhD thesis, Helsinki, Finland: University of Helsinki), https://helda.helsinki.fi/bitstream/handle/10138/38675/santangeli_dissertation.pdf?sequence=1&isAllowed=y.

Santangeli, A. and Sutherland, W.J. 2017. The financial return from measuring impact. *Conservation Letters* 10: 354–60, https://doi.org/10.1111/conl.12284.

Santika, T., Sherman, J., Voigt, M., et al. 2022. Effectiveness of 20 years of conservation investments in protecting orangutans. *Current Biology* 32: 1754–63, https://doi.org/10.1016/j.cub.2022.02.051.

Satterfield, J.M., Spring, B., Brownson, R.C., et al. 2009. Toward a transdisciplinary model of evidence-based practice. *The Milbank Quarterly* 87: 368–90, https://doi.org/10.1111/j.1468-0009.2009.00561.x.

Spock, B. 1946. *The Common Sense Book of Baby and Child Care* (New York: Duell, Sloan, and Pearce).

Spooner, F., Smith, R.K. and Sutherland, W.J. 2015. Trends, biases and effectiveness in reported conservation interventions. *Conservation Evidence* 12: 2–7, https://conservationevidencejournal.com/reference/pdf/5494.

Sutherland, W.J. 2000. *The Conservation Handbook: Research, Management and Policy*. (Oxford: Blackwell Scientific), https://doi.org/10.1002/9780470999356.

Sutherland, W.J., Pullin, A.S., Dolman, P.M., et al. 2004. The need for evidence-based conservation. *Trends in Ecology and Evolution* 19: 305–08, https://doi.org/10.1016/j.tree.2004.03.018.

Sutherland, W.J. and Wordley, C.F.R. 2017. Evidence complacency hampers conservation. *Nature Ecology and Evolution* 1: 1215–16, https://doi.org/10.1038/s41559-017-0244-1.

Syed, M. 2015. *Black Box Thinking* (New York: Penguin Random House).

Tetlock, P. 2009. *Expert Political Judgement: How Good Is It? How Can We Know?* (Princeton, New Jersey: Princeton University Press), https://doi.org/10.1515/9781400830312.

Walsh, J.C., Dicks, L.V., Raymond, C.M., et al. 2019. A typology of barriers and enablers of scientific evidence use in conservation practice. *Journal of Environmental Management* 250: 109481, https://doi.org/10.1016/j.jenvman.2019.109481.

Walsh, J.C., Dicks, L.V. and Sutherland, W.J. 2015. The effect of scientific evidence on conservation practitioners' management decisions. *Conservation Biology* 29: 88–98, https://doi.org/10.1111/cobi.12370.

Wauchope, H.S., Jones, J.P.G., Geldmann, J., et al. 2022. Protected areas have a mixed impact on waterbirds, but management helps. *Nature* 605: 103–07, https://doi.org/10.1038/s41586-022-04617-0.

Young, K.D. and Van Aarde, R.J. 2011. Science and elephant management decisions in South Africa. *Biological Conservation* 144: 876–85, https://doi.org/10.1016/j.biocon.2010.11.023.

PART II
OBTAINING, ASSESSING AND SUMMARISING EVIDENCE

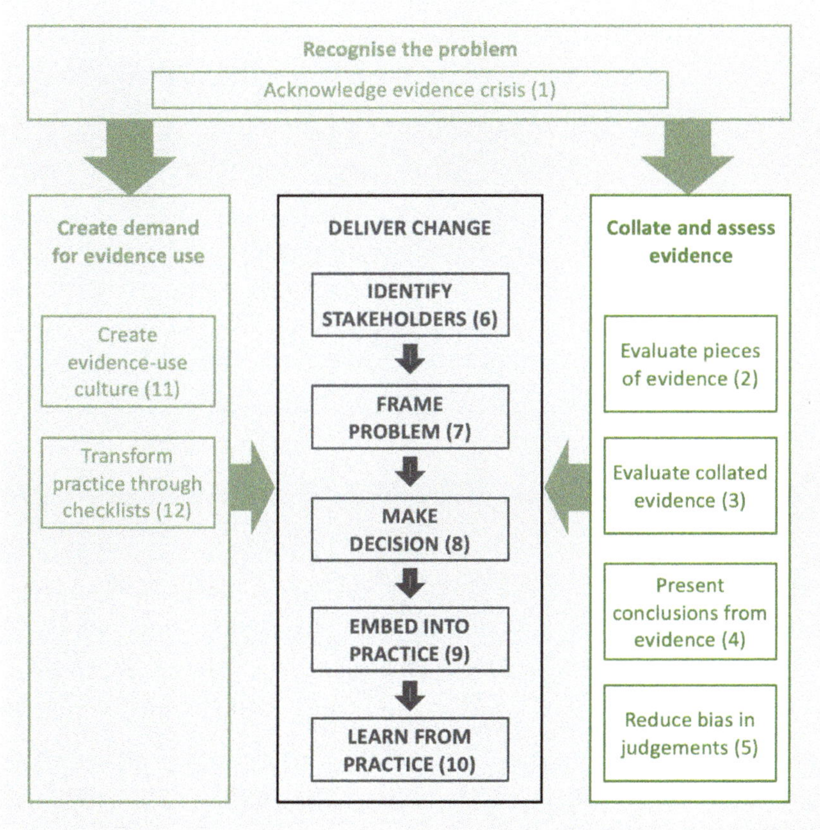

2. Gathering and Assessing Pieces of Evidence

Thomas B. White[1], Tatsuya Amano[2,3], Philipp Boersch-Supan[4,5], Alec Christie[1,6,7], Robert Freckleton[8], Maud C. Quinzin[9], Ali Mohammad Rezaie[10], William J. Sutherland[1,2], Hiromi Yamashita[11,12]

Finding and assessing evidence is core to making effective decisions. The three key elements of assessing any evidence are the rigour of the information, the trust in the reliability and objectivity of the source, and the relevance to the question under consideration. Evidence may originate from a range of sources including experiments, case studies, online information, expert knowledge (including local knowledge and Indigenous ways of knowing), or citizen science. This chapter considers how these different types of evidence can be assessed.

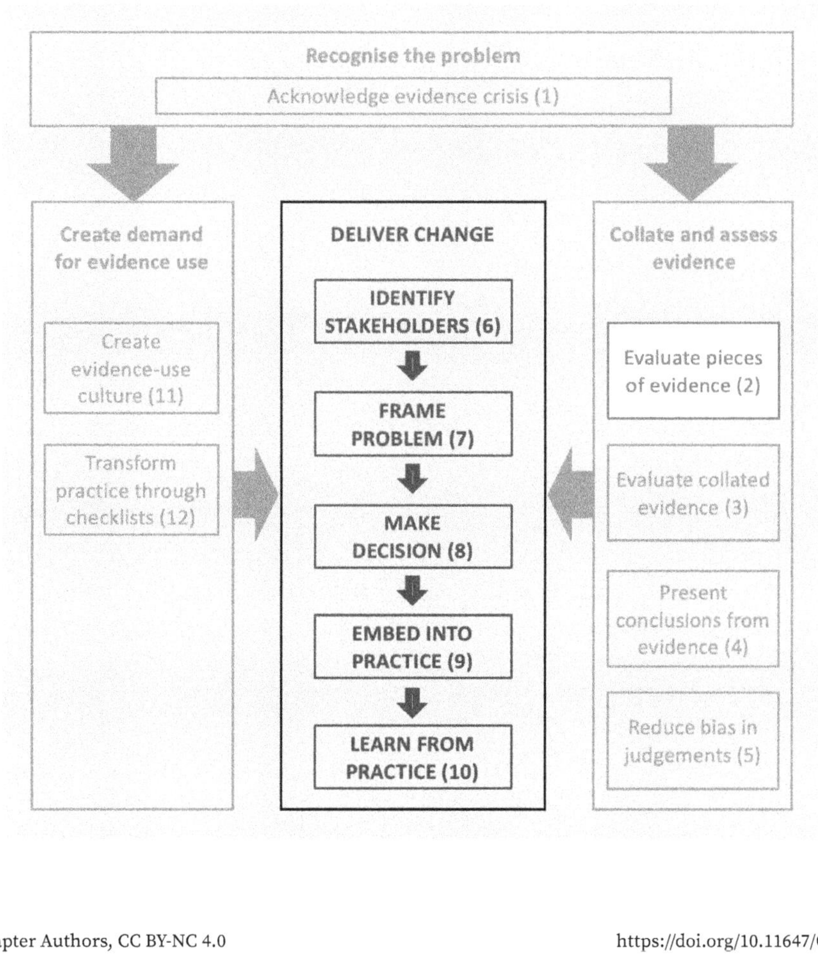

 https://doi.org/10.11647/OBP.0321.02

[1] Conservation Science Group, Department of Zoology, University of Cambridge, The David Attenborough Building, Pembroke Street, Cambridge, UK
[2] School of Biological Sciences, University of Queensland, Brisbane, 4072 Queensland, Australia
[3] Centre for Biodiversity and Conservation Science, The University of Queensland, Brisbane, 4072 Queensland, Australia
[4] British Trust for Ornithology, The Nunnery, Thetford, Norfolk, UK
[5] Department of Geography, University of Florida, Gainesville, FL 32611, USA
[6] Biosecurity Research Initiative at St Catharine's College (BioRISC), St Catharine's College, Cambridge, UK
[7] Downing College, University of Cambridge, Regent Street, Cambridge, UK
[8] School of Biosciences, University of Sheffield, Sheffield, UK
[9] Ocean Discovery League, PO Box 182, Saunderstown, RI 02874, USA
[10] Department of Civil Engineering, University of Asia Pacific, Dhaka 1205, Bangladesh
[11] Ritsumeikan Asia Pacific University, Beppu, Oita, Japan
[12] Nagoya University Graduate School of Environmental Studies, Nagoya, Aichi, Japan

Contents

2.1 What Counts as Evidence?

This book is about making more cost-effective decisions by underpinning them with evidence, defined as 'relevant information used to assess one or more assumptions related to a question of interest' (modified from Salafsky et al., 2019). This apparently straightforward approach easily becomes bewildering. For a decision about fisheries, the key evidence may relate to the markets, the communities, economics, legislation, fishing technologies, fish biology, and fishery models. This evidence is likely to be a mix of specific evidence applying to that community, such as the values of stakeholders or changes in fish catches, and generic evidence that applies widely, such as the size at which species start reproducing or the effectiveness of devices for reducing bycatch. The challenge is to make sense of this diverse evidence and marshal it so decisions can be made. This chapter is about assessing single pieces of evidence. The next chapter is about assessing collated evidence, such as a meta-analysis, followed by a third chapter describing the processes for converting evidence into conclusions that can then be the basis for decision making.

This process of embedding evidence into decision making is referred to by various, often interchangeable, terms. Evidence-based (e.g. Sackett et al., 2000) is a term coined in 1991 (Thomas and Eaves, 2015) that has become standard. It is generally used for the practice that aims to incorporate the best available information to guide decision making, often, but not exclusively, with an emphasis on scientific information. Evidence-informed (e.g. Nutley and Davies, 2000; Adams and Sandbrook, 2013) is similar but used to emphasise the importance of diverse types of evidence and contextual factors in decision-making, and used especially when evidence may not be central (Miles and Loughlin, 2011). Evidence-led (e.g. Sherman, 2003) is also overlapping but most often used by those stating an objective to be an evidence-led organisation to describe their aim to make evidence use central to practice.

Evidence can be embedded wherever a claim or assumption is made. For example, a project may make claims about the species present in the project site, the change in abundance of some key species, the spiritual significance of certain species, the threatening processes present, and the effectiveness of specific actions.

This chapter aims to provide a framework for considering how any piece of evidence can be evaluated and then explore what determines the reliability of different sources of evidence.

2.1.1 A taxonomy of the elements of evidence

There is a range of evidence types that serve different purposes. Table 2.1 provides a taxonomy of the key elements of different types of evidence, while Table 2.2 gives some examples of how this classification can be used.

Table 2.1 Some common distinguishing features that can be used to classify different types of evidence.

Feature		Evidence Type
Communication method		Physically documented (e.g. published, written, stored)
		Oral (e.g. statement, teaching, sayings, spoken stories)
		Performance (e.g. songs, dances, plays)
Generality		Specific (applying to local conditions)
		Generic (derived at a wider scale, where generality is sought to explain general patterns)
Source	Origin	Primary
		Secondary
		Unknown
	Type	Experiment
		Case studies
		Citizen science
		Statements, observations and conclusions
		Quantified
		Statement
		Tradition or culture
		Deduction
		Induction
		Summary
		Model
		Tacit (i.e. experience)
		Online open-source-investigations
		Artificial intelligence
	Location (if secondary)	Databases
		Publications
		Non-peer-reviewed literature
		Global data in multiple languages
		Practitioner knowledge
		Local and Indigenous knowledge
		Online material
Possible qualifier		Pattern
		Change
		Response

Feature	Evidence Type
Subject	Phenomenon
	Biological
	Social
	Status
	Threats
	Costs
	Uses
	Values
	Rules
	Beliefs

Table 2.2 Some examples of evidence and their suggested classification based on Table 2.1.

Example	Classification
Bird record submitted to eBird (an online database for bird sightings)	Physically documented, specific, primary, citizen science observation about status
Farmer says: 'snakes are more abundant near the river.'	Oral, specific, primary observation on a pattern relating to status
Practitioner's report presents data showing that there was a decrease in amphibian mortality when an underpass was installed	Physically documented, generic, primary, quantified response affecting status
A peer-reviewed scientific paper on experiment showing that nest protection for common redshank increased daily egg survival rates	Physically documented, generic, experiment, of the response affecting biology
Systematic review of the role of trees in reducing flood risk	Physically documented, generic, secondary summary on a pattern relating to threats
A local councillor says 'people will be more likely to support the construction of a tree nursery if local labourers and materials are used'	Oral, generic, tacit knowledge on response relating to values of society
Traditional dance describing how the community was saved by frigate birds	Performance, specific, traditional, pattern of beliefs

This framework provides the structure for the rest of the chapter, which, after describing a framework for assessing the weight of evidence, considers the different features of evidence, and how this framework can be applied.

2.2 A Framework for Assessing the Weight of Evidence

One of the earliest reports of framing evidence in terms of its weight comes from Greek mythology with Themis, the Greek goddess of justice, depicted carrying a pair of scales to represent the evidence for different sides of an argument. Assessing the weight of evidence is essential if the information from observations, studies or reviews are to be used to inform decisions as different pieces of evidence can vary in their strength, reliability and relevance (Gough, 2007). Such assessment is necessary even when considering formal methods, such as a meta-analysis, as it is important to evaluate the reliability of the meta-analysis, with its associated biases, and its relevance to the issue under consideration.

An approach to weighing the evidence is to consider each piece of evidence as a cuboid, as shown in Figure 2.1. The three axes of the cuboid are:

- *Information reliability (I)*: how much the information contained within a piece of evidence can be trusted, such as the rigour of the experimental design, or whether the statement is supported by information, such as photographs.
- *Source reliability (S):* how much trust can be placed in the source of the evidence, such as whether it is considered authoritative, honest, competent, and does not suffer from a conflict of interest, or bias.
- *Relevance (R)*: how closely the context in which the evidence was derived applies to the assumption being considered, such as whether it relates to a similar problem, action and situation.

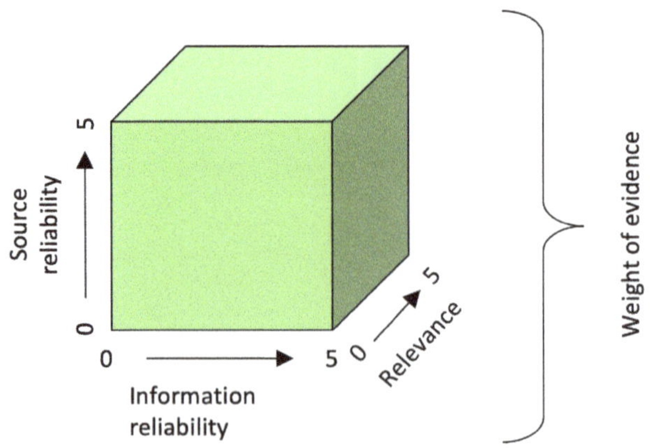

Figure 2.1 Assessing the weight of evidence according to information reliability, source reliability, and relevance (ISR). (*Source:* Christie et al., 2022, CC-BY-4.0)

Collectively these axes can be used to assess the weight of evidence with each piece of evidence given an information reliability, source reliability, and relevance (ISR) score (Christie et al., 2022). Table 2.3 gives a suggested set of criteria for assigning scores to assess the weight of evidence. If the evidence piece is completely irrelevant, or if there are considerable concerns in either the information reliability or source reliability, then the result is a score of zero and thus no total ISR weight. Conversely, evidence that has high relevance and reliability to a decision-making context would carry considerable weight. A score of two is given in each category for evidence about which little is known. For example, the claim 'otters have been seen nearby' provides little information to a nature reserve manager on what was seen, how reliable the observer is, or whether the observation was on the reserve or not.

Table 2.3 Criteria for classifying evidence weight scores, as shown in Figure 2.1.

Information reliability (I)		**Source reliability (S)**		**Relevance (R)**	
Very reliable approach	5	Considerable trust	5	Extremely relevant	5
Moderately reliable approach	4	Moderate trust	4	Very relevant	4
Weakly reliable approach	3	Some trust	3	Relevant	3
No knowledge of approach	2	No knowledge of source	2	Somewhat relevant	2
Some concerns over approach	1	Some concerns over reliability	1	Not very relevant	1
Considerable concerns over approach	0	Serious concerns over reliability	0	No relevance	0

Evidence is frequently quoted but without information on the original source; these should be treated carefully. In some cases the original source of a piece of physically documented evidence may not only be unknown but, when looked for, cannot be located and even may not exist. These are sometimes referred to as 'zombie studies' or 'zombie data'.

These ISR scores can be used for any piece of evidence, whether individual pieces (as described in this chapter) or collated evidence such as a meta-analysis (described in Chapter 3). The aim of the rest of this section is to consider the general principles for assessing the information reliability, source reliability, and relevance for different types of evidence.

2.2.1 Information reliability

Information reliability refers to how much confidence we have in the information contained within a piece of evidence, rather than its source. This could depend on the quality of the research design (see Section 2.6.1) or the support for a field observation (for example a statement, description, sketch, photo or DNA sample).

Questions to consider when assessing the reliability of the information provided by a piece of evidence could include:

- What is the basis for the claim?
- Are the methods used appropriate for the claim being made?
- Is the approach used likely to lead to bias?
- Is the material presented in its entirety or selectively presented?
- Is there supporting information, such as photographs or first-hand accounts?
- Are the conclusions appropriate given the information available?

Measurement error, sometimes called observational error, describes the difference between a true value and the measured value in the piece of evidence. For many types of evidence, this will be an important component of information reliability as this error can influence the precision and accuracy of evidence (see Freckleton et al., 2006).

Measurement error can be from multiple sources. For example, if we are counting the numbers of birds in a flock on an estuary, such counts are rarely exact and are estimates of the real number. Alternatively, we can exactly count the numbers of plants in a quadrat: however, any quadrat is a sample from a much larger population and there will consequently be error in our estimate. Similarly, many methods for censusing populations are indirect (e.g., scat counts, frass measurement, acoustic records or camera-trapping), so there is error translating these numbers into estimates of the size of the actual population.

The effect of measurement error will depend on whether it is *random* or *systematic*. Random errors affect the precision of an estimate (i.e. the variance from measurement to measurement of the same object), whereas systematic errors affect the accuracy of an estimate (i.e. how close the measured value is to the true value).

2.2.2 Source reliability

Source reliability refers to the reliability of the person, organisation, publication, website or social media providing information, including whether they can be considered authoritative or likely to be untrustworthy.

These scores are individual assessments and so different individuals will use different criteria. For example, for projects related to Pacific salmon conservation, stakeholders considered work reliable if the researchers had been seen to be involved in fieldwork, whereas government decision makers considered research reliable if it had been formally reviewed (Young et al., 2016).

Questions to consider when assessing the reliability of the source of evidence could include:

- Does the source have an interest in the evidence being used?
- Are the sources explicit about their positions, funding or agendas?
- What is the source of funding and could it influence outcomes?
- Is there evidence (real or perceived) of agendas or ulterior motives?

- What is the track record of the source in delivering reliable information?
- Is the source an expert in their field?
- Does the source have the appropriate experience for making this claim?
- If published, does it seem to be in a reliable unbiased publication?
- If published, was it peer reviewed?

2.2.3 Relevance to local conditions

Assessing relevance requires asking if a piece of evidence can be expected to apply to the issue being considered. As such it involves considering extrinsic factors, such as the similarities in location, climate, habitat, socio-cultural, and economic contexts. It also includes intrinsic factors about the problem or action being proposed, such as the type of action, specifics of implementation, or type of threat being considered.

Assessing the relevance of evidence is critical. When German forester practitioners were provided with some evidence-based guidance on forest management, the majority had concerns about the lack of specificity, were sceptical that the guidance would work across the board, wanted to know the location, forest type or soil type of the forest, and objected to a 'cookbook' approach (Gutzat and Dormann, 2020).

Salafsky and Margoluis (2022) make the important distinction between specific and generic evidence. For example if looking at the evidence for a species' ecology, information on the typical diet may apply widely but the actual fruit trees visited are location dependent. Similarly, a species may be threatened by overexploitation at a global scale, but in the local area, that specific threat may be unimportant. Figure 2.2 shows the interlinking relationships between these two types of information when considering the evidence of effectiveness for conservation actions. General conclusions combine specific studies as described in various approaches in Chapter 3.

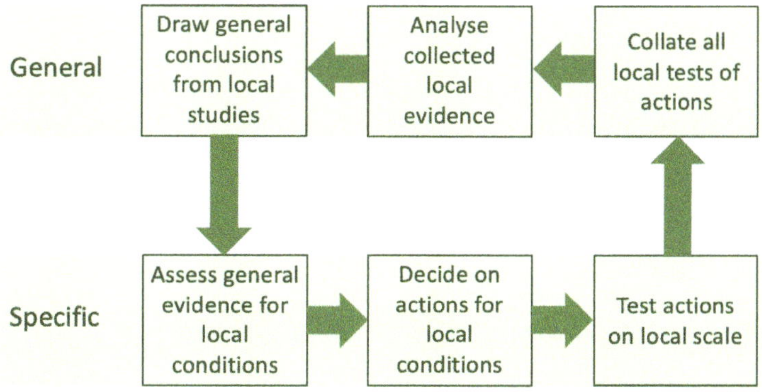

Figure 2.2 The links between general and specific information for conservation actions. Decision makers and practitioners have to interpret the general information for their specific conditions. They may generate specific evidence from their practice. The combined local studies generate general conclusions. (*Source:* authors)

Some questions to consider when assessing the relevance of a piece of evidence:

- Is the difference in the location and resulting differences in climate, community, etc., likely to affect the relevance?
- Do the species or ecological community differ and, if so, is this important?
- Are any sociocultural, economic, governance or regulatory variations likely to be important?
- Are there any differences in the season that could affect the relevance of the evidence?
- How much is the observed variation between the context and where the evidence was created likely to matter for the result observed? If an action is being considered, how similar is it to the action proposed?

2.2.4 Improbability of claims

Laplace's Principle states that 'the weight of evidence for an extraordinary claim must be proportioned to its strangeness' (Gillispie, 1999). This was reframed and popularised by Carl Sagan who, whilst discussing extra-terrestrial life in his TV series *Cosmos*, proclaimed, 'Extraordinary claims require extraordinary evidence'. Thus, a claim to have seen a common species will be readily accepted, whilst the equivalent claim to have seen a dodo would lead to calls for the observer to provide photos, videos, DNA, details of recreational drug use, etc. This improbability can be considered in terms of Bayesian statistics in which the prior belief is the likelihood of a claim or estimate, which can be updated based on additional evidence to give a posterior belief.

Improbability is an expression of the weight of evidence required to believe the claim. In many cases this is small. For example, being told that someone has reported that a tree has fallen blocking a road (i.e. an unsupported statement by a stranger) is likely as the event is unambiguous and it is hard to imagine why someone would report it inaccurately. Improbability also includes the likelihood of an alternative explanation, for example, that a rare species reported on a nature reserve is confused with a common similar species.

The factors influencing improbability include:

- How likely is it?
- Does it fall outside current knowledge?
- Does it sound plausible?
- Is there an alternative, more obvious (and parsimonious), explanation?
- Are there reasons why improbable claims might be made?
- Is this part of a pattern of claims, such as from those holding an agenda?

2.2.5 Presenting information reliability, source reliability, relevance (ISR) scores

Each piece of evidence can then be presented followed by the ISR score as assessed by the user (see Box 2.1).

Box 2.1 Examples of evidence with associated ISR scores

These fictitious examples show how ISR (information reliability, source reliability, relevance) scores could be presented after evidence assessment in relation to considering whether mowing road verges in Japan will be effective for biodiversity and feasible to implement.

- Species richness is higher on roadsides mowed once or twice per year with hay removal than on unmown roadsides. [I = 4, S = 5, R = 2] *A well carried out independent systematic review, although with limited studies (Jakobsson et al., 2018) with meta-analysis of studies largely in Europe and North America.*
- Last time the verge was left uncut for the summer there was a petition of dog walkers complaining to the council. [2, 2, 5] *Told by the shop owner who could not remember the source.*
- There were far more butterflies when the verge was cut regularly. [3, 1, 5] *Statement from the spokesperson from the dog walking group, who has a clear agenda, based on walking past most mornings.*
- Cutting twice a year rather than every two weeks has saved the neighbouring village 25,000 Yen a year. [5, 5, 5] *From a village clerk who checked accounts.*

2.3 Weighing the Evidence

We take all the different weighted pieces of evidence for a particular claim or assumption and balance these to arrive at a decision. Figure 2.3 shows how the evidence for an assumption can be visualised (Salafsky et al., 2019; Christie et al., 2022). The combined pieces of evidence (represented by green blocks) can fall on either side of the scales to refute (left-hand side) or support (right-hand side) and must outweigh the improbability of the assumption (the red block). This represents a situation where an assumption can either be supported or refuted in a binary manner and can be used to indicate the weight of evidence required for us to have confidence that an assumption holds.

This conceptual model helps to explain why assessing different types of evidence is so critical, which is the basis for the remainder of the chapter. If pieces of evidence are deemed highly reliable, and relevant, then there may be a strong evidence base backing up a claim or assumption. However, if there is limited evidence, or if the existing evidence has poor reliability and relevance to the context, then the evidence base may not tip the scales of the balance, and we cannot be confident in the assumption being made. Chapter 4 expands on this model to consider cases where the strength of support for the claim matters (for example there may be strong evidence for the effect of an action, or only evidence of a weak effect).

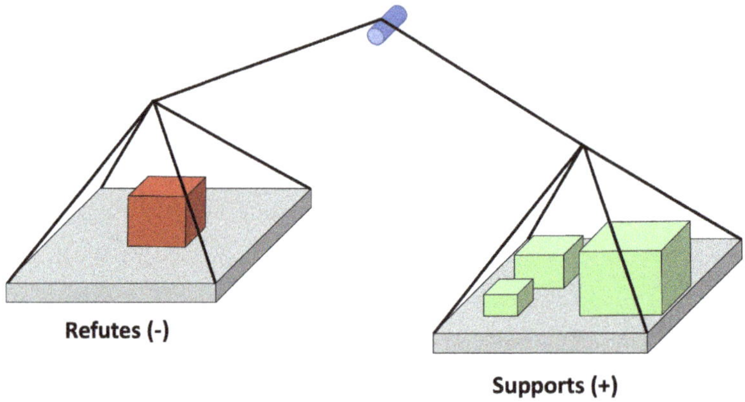

Figure 2.3 A means of visualising the evidence behind an assumption. The pieces of evidence are placed on either side of the scale, with evidence supporting the assumption falling on the right-hand side (green cuboids), and evidence refuting the assumption and the improbability (red cuboid) falling on the left-hand side. The volume of each cuboid represents its weight. The greater the tilt the higher the confidence in accepting (or rejecting) an assumption. (*Source:* adapted from Christie et al., 2022, following an idea of Salafsky et al., 2019, CC-BY-4.0)

2.4 Subjects of Evidence

2.4.1 Patterns or changes in status

A conservation project will often define features of interest (e.g. the focal conservation targets). These can include species, habitats, broader ecosystems or even cultural features. Thus, when designing a project an important subject of evidence is the patterns of these features and changes in that status. For example, what species are present at a site? What is the population size, and how has it changed over time? Similar questions could be asked for habitats: what is the extent and quality of habitat at the project site? How has the quality of habitat changed over time? Further details are provided in Chapter 7.

2.4.2 Patterns or changes in threats and pressures

Understanding the patterns and changes in the direct threats to biodiversity, and the indirect threats (i.e. underlying drivers) is an important subject for which evidence is required when designing a conservation project. For example, what are the key threats in a project area? What are the processes driving these threats? And how are these threats likely to change in future? Direct threats may include, but not be limited to, habitat loss and degradation, overexploitation, pollution, invasive species, or climate change. Further details are provided in Chapter 7.

2.4.3 Responses to actions

Another important subject of evidence in the design of conservation projects are the responses of features to actions that are implemented. This could be the effect of actions on biodiversity directly, the effect of actions on direct threatening processes, the effects on indirect threats (i.e. the underlying drivers of decline) or the effects of actions on other environmental and social variables (e.g. impacts on carbon storage, water flows, community health). Further details are provided in Chapter 7.

2.4.4 Financial costs and benefits

Costs are an important element of decisions, making evidence of costs an important topic. Conservationists lack the funds to implement all required interventions (Duetz et al., 2020) with Barbier (2022) calculating that nature protection may be underfunded by about $880 billion annually. This means conservationists need to prioritise actions based on their financial costs — whether that be deciding where to focus effort geographically, deciding which species and habitats to concentrate on, or which specific management action to implement. Financial costs and benefits are therefore other important types of evidence in decision making.

In this section we focus on financial and economic costs to deliver an action, but note that non-financial costs are also an important consideration in decision making; these include other environmental costs (e.g. changes in carbon storage, water quality, soil retention) and social costs (e.g. loss of access, cultural impacts).

Financial/accounting costs

Despite their importance, open and transparent information on costs is often surprisingly sparse (Iacona et al., 2018; White et al., 2022a). A study of the peer-reviewed literature of conservation actions identified that only 8.8% of studies reported the costs of the actions they were testing. Even when costs were reported there was little detail as to what types of costs were incorporated and rarely any breakdown of the constituent costs (White et al., 2022a). This detail is essential for determining the relevance of available cost estimates to a particular problem. For example, knowing that planting 1 ha of wildflower meadow in Bulgaria costs 5,000 Lev in 2021 is difficult to apply without knowing what equipment and consumables were paid for, if labour costs were included, if consultant fees were paid, or whether there were any overheads.

Table 2.4 lists the types of costs and benefits that should be considered when thinking about the costs of proposed actions. Costs can be accessed from databases of costs, project reports or budgets (although these are often not freely available) or from online catalogues of prices and cost data.

Table 2.4 Types of financial costs and benefits of conservation interventions. (*Source:* White et al., 2022b)

Type of financial costs	Benefits of conservation interventions
1) Ongoing Costs	**Costs incurred independently of the project or intervention under consideration.**
Central Administration	i.e. HR costs, construction/maintenance of buildings (e.g. office blocks, HQ, rents), project design. These costs are incurred regardless of the intervention being implemented.
Training and Skill Development	Costs of training and skill development of staff. These costs are incurred regardless of the intervention being implemented.
2) Costs of Implementing an Intervention	**Various financial costs incurred as a result of implementing the intervention.**
Direct Implementation	
Labour	Cost and amount of labour required to implement the intervention.
Capital	Capital required to implement the intervention (e.g. vehicles, extra office space, machinery).
Consumables	Consumables required to implement the intervention (e.g. equipment, supplementary food etc.).
Access	Cost required to access the intervention (e.g. transport costs, services that are not available on site). Access costs can sometimes be considered consumables.
Transaction	Cost associated with designing and planning the intervention or programme.
Joint Costs/Overheads	Overhead costs shared between multiple interventions where only a proportion of the cost can be assigned to the specific intervention or project being studied (e.g. project planning, electricity bills, administration staff etc.). These are distinguished from ongoing costs as the project being implemented does contribute to a portion of these costs.
Future Management	Future management costs that would not otherwise have been incurred (e.g. monitoring, replacement, reoccurring management actions).
3) Opportunity Costs	**Future costs related to the outcomes of an action (e.g. loss of income, future management costs).**

Type of financial costs	Benefits of conservation interventions
Opportunity Costs/Benefits Foregone	Market valuation — Financial income foregone as a result of an intervention (i.e. lower income crop harvests, reduced hunting revenues, excess burden of tax at societal scale).
	Non-market valuation — Unrealised benefits as a result of an intervention due to foregone ecosystem service provision in the alternative scenario (e.g. a loss of carbon storage potential, or aesthetic value due to the conservation action).
4) The Future Economic Benefits	**Economic benefits that may occur due to the outcomes of an action.**
Explicit Benefits / Extra Benefits from Enhanced Environment	Market valuation — Financial income generated as a result of the enhanced environment (e.g. ecotourism).
	Non-market valuation — Economic gains associated with greater ecosystem service provision (e.g. flood protection, carbon sequestration, water purification).
Avoided Costs / Costs Foregone	Market valuation — Financial costs avoided as a result of the intervention (e.g. fines, costs of human-wildlife conflicts).
	Non-market valuation — Averted economic loss associated with the gained ecosystem services in the intervention scenario (e.g. flood protection, carbon sequestration, water purification).

Financial benefits

Sometimes there may be financial benefits associated with a project that warrant consideration (see Table 2.4). Some conservation outcomes are easily valued, for example, there may be ecotourism revenue or the sale of sustainably harvested timber products. However, many benefits of conservation actions may be difficult to value financially (e.g. the value of ecosystem services, which are traditionally not accounted for, or seen as an externality to the economic system). It is worth considering these hidden, or unaccounted-for, benefits when assessing different conservation actions.

Economic costs and benefits

However, financial costs do not always represent the true cost of an action, and in some situations, it may be useful to estimate wider economic costs and benefits of actions. Economic costs are distinguished from accounting costs in that they represent the costs of choosing to implement a given action compared to the most likely alternative scenario, including implicit costs and benefits (e.g. opportunity costs, avoided costs), which are not seen directly on a balance sheet. For example, although the cost of setting up a protected area may include employing rangers, fencing and constructing buildings, the true cost to local communities may also include the lost agricultural income that can no longer be obtained due to the protected area placing limits on activities. Similarly, a road construction project may be installing a wildlife bridge over the road to mitigate impacts to mammal species; the cost of the bridge may be substantial, but if the bridge had not been built then the project may have been substantially delayed, or the constructor fined, due to impacts on those mammals, meaning these costs could be avoided with the construction of the bridge. Calculating economic cost would include consideration of these avoided and opportunity costs, which provide important evidence of the financial implications of a given intervention.

Distribution of costs and benefits

It is important to note that the costs and benefits of a given intervention will not be distributed evenly amongst all relevant stakeholder groups. A conservation project may be paid for by a non-governmental organisation, with financial benefits received by local government agencies, and substantial opportunity costs only felt by local communities. It is important therefore to consider who will win and lose as a result of an intervention to make sure the overall costs and benefits of the action are equitable and fair.

2.4.5 Values and norms

Decision making in conservation often requires understanding the values placed upon different scenarios, outcomes or behaviours by those stakeholders who may be influenced by the decisions. For example, the proposed reintroduction of white-tailed sea eagles to parts of the UK faced substantial opposition from farming groups, whilst gaining broad support from conservation charities. Having buy-in and involvement from local stakeholders is a vital component of ensuring conservation projects are successful, and not doing so can be a common cause of project failure (Dickson et al., 2022). Therefore, gaining evidence on the values held by different stakeholder groups, and the acceptability of different actions, is vital for designing effective and equitable conservation projects.

Just as costs and effectiveness are important for designing effective conservation actions, it is also important that decision makers consider the values placed on different outcomes, behaviours or actions by all relevant stakeholders to ensure that proposed actions are acceptable, equitable, and likely to be effective in local contexts (e.g. Gavin et al., 2018; Gregory, 2002; Whyte et al., 2016).

Values can be motivated by the want for enjoyment (e.g. a comfortable life, happiness), security (e.g. survival, health), achievement, self-direction (e.g. to experience for yourself and learn), social power (e.g. status), conformity, or for the welfare of others (Schwartz and Bilsky, 1987). Stakeholders in conservation may have values placed upon different components of the environment, desirable outcomes from a conservation project, how species and habitats should be used, cultural or religious values placed on different areas of habitats, or values on who should be implementing conservation projects. Values can either be concepts or beliefs of what is desirable.

Where it is thought that something is true, beliefs are often held by different individuals or groups. For example, a local community may highly value a particular area of habitat due to its religious significance. Understanding such beliefs, and integrating them into decision-making, is important for designing conservation projects that are likely to be successful in local contexts. These beliefs may be embedded into stories.

Where values are held widely by many individuals or groups, these can be termed social norms, which describe the shared standards of acceptable behaviour or outcomes by groups or widely by society. These can also be termed 'collectivist' values (Schwartz and Bilsky, 1987).

Whilst some information on values may be documented and widely available, often obtaining this information will require close consultation with different local stakeholders and communities (see Chapter 6). Such approaches can also be used to obtain evidence on status, threats, effects, and costs that may be known by different stakeholder groups.

2.5 Sources of Evidence

In the section above, we described the range of subjects of evidence that may be required when designing a conservation project. But this information can come from a range of different sources and localities. For example, evidence can be contained within peer-reviewed literature, NGO reports, online databases, or embodied within people's knowledge and experience. Evidence can come from primary sources, secondary or tertiary sources, or where the original source of the information may not be known. Lastly, evidence can be available from sources in multiple languages.

2.5.1 Primary, secondary or unknown sources

Evidence can come from a range of localities (see below), but there are broad levels of traceability to the sources. Evidence can be from a primary source, where the evidence is available from its original locality. For example, the original research paper in which the data was published, or an original datapoint recorded by the individual who made the observation. Evidence can also be from secondary sources, which refer back to the original source of the evidence. For example, this could be in a review of many studies, in databases that reference the original source, or from experts and individuals when talking about specific studies or observations made by others. Lastly, there may be some instances where the source of evidence is unknown. This may be the case for example where unreferenced claims are made within published literature, or when experts know something based on experience but cannot pinpoint a source of the information. Unknown sources make it difficult to determine the reliability of a piece of evidence.

2.5.2 Databases

Databases are a major source of data on species and habitat status, threats, management actions, the effectiveness of actions, and the costs of actions. Stephenson and Stengel (2020) give a list of 145 conservation databases. This will typically need to be supplemented with national and local sources of documented evidence along with experience and local knowledge.

2.5.3 Peer-reviewed publications

A major source of relevant evidence to conservationists is in peer-reviewed literature, where there are a wealth of publications detailing information on species status, threats, effects of actions, as well as stakeholder values. Publications are peer-reviewed, meaning that (in theory) they are vetted by experts in the field before publication, to ensure the quality of the study and its results. Studies will be published in a range of different journals which vary in their scope and aims (e.g. what types of studies they will publish). This process helps to improve the quality of the published evidence base and is an important part of the scientific process.

Published studies vary in their quality. The main means of assessing study quality should be through evaluating the study i.e. how it was conducted and the methodology used (see Section 2.6.1 for discussion of this issue for experiments). Very good, rigorous studies can be published in little known journals and awful ones accepted in prestigious ones, but the quality of a journal is often used to give some broad indication as to the quality of the research. The following issues should however be considered when thinking about journal quality:

- **Level of peer review** — Does the journal have a rigorous peer-review process? For example, the rigour of the peer-review process at some journals may be poor, meaning there is less vetting of research before publication.
- **Impact measures** — Much is made of journal quality and especially evaluating them by measures such as impact factors (the frequency with which papers are quoted by others) although these measures have serious flaws (Kokko and Sutherland, 1999). There are also means of manipulating impact factors.
- **Predatory journals** — There is also the issue of predatory journals that have the appearance of others but publish with minimal assessment (as illustrated by the acceptance and publication of joke papers submitted to ridicule them). There is a list of probable predatory journals (Beall's list, https://beallslist.net).
- **Ethics** — The publishing system has many ethical barriers that can disadvantage people accessing the evidence base and researchers wanting to publish. A common model is that journals charge authors to publish open-access (e.g. article processing charges, open-access fees), or charge readers fees to access articles. This creates barriers to the access of evidence and limits who can publish. A recent study found that only 5% of conservation journals met all of the Fair Open Access Principles (Veríssimo et al., 2020); principles developed to help move publishing towards sustainable, ethical practice.

- **Publication delays** — There can be large delays in the publishing of studies in the peer-reviewed literature due to delays with submission, and the publication process. For example, Christie et al. (2021) show that the average delay in publications providing evidence for the effectiveness of actions between the end of data collection and final publication in the literature was 3.2 years. This delay in the evidence base prevents timely access to information.

As a result of some of the issues above, there have been calls to move away from some aspects of the system. For example, Stern and O'Shea (2019) call for moves towards a publish first, curate second model of publishing where authors publish work online, peer-review happens transparently, and journals then choose which papers they wish to include. Peer Community In is an initiative that offers free, transparent peer reviews of preprints.

Databases exist to help access the scientific literature. For example, Scopus, Web of Science and Google Scholar are commonly used databases to help find relevant research literature articles based on topic or keyword searches.

2.5.4 Grey literature

Many studies of conservation actions are published in reports, conference proceedings, theses, etc. (also known as grey literature, defined as not controlled by commercial publishers), rather than as papers in academic journals. The main criticisms of grey literature are that it usually has not been peer reviewed, sometimes is not available online and even if online can sometimes be difficult to locate and search. Another is that some grey literature has a clear agenda and may be less neutral. However, there is nothing inherently unreliable about grey literature, and academic journals cannot be assumed to publish high-quality research just because they have a peer review process. There will be many high-quality pieces of evidence published in the grey literature.

Increasingly, preprint servers are hosting papers that have not yet been peer reviewed or are in the process of peer review. Although there are many models, the most common involves the preprint server simply being a repository. However, there are a growing number of preprint servers that enable readers to make comments on the documents that parallel what would happen in traditional peer review. As with any source of material, the reader must beware of variable quality (see Hoy, 2020) with potential for evidence misuse leading to trust crises, as has been observed with some preprints during the COVID-19 pandemic (Fleerachers et al., 2022). Nonetheless, such servers provide rapid access to information that may be particularly salient to conservation given that there is inherent urgency when, say, trying to recover an imperilled species (Cooke et al., 2016).

Databases also exist to help access the grey literature. Applied Ecology Resources provide a database of grey literature that can help support and improve biodiversity management (https://www.britishecologicalsociety.org/applied-ecology-resources/). The Conservation Evidence website (https://www.conservationevidence.com/) has a catalogue of open-access online reports that have been searched by them and their collaborators, now including 25 report series and identifying 278 reports that tested actions.

2.5.5 Global evidence in multiple languages

Much scientific evidence is published in languages other than English, although it is often ignored at the international level (Lynch et al., 2021). Recent research shows that up to 36% of the conservation literature is published in non-English languages (Amano et al., 2016a). Further, the number of non-English-language conservation articles published annually has been increasing over the past 39 years, at a rate similar to English-language articles (Chowdhury et al., 2022).

Ignoring scientific evidence provided in non-English-language literature could cause severe biases and gaps in our understanding of global biodiversity and its conservation. For example, using the same selection criteria as those used by Conservation Evidence (Sutherland et al., 2019), the translatE project screened 419,679 papers in 16 languages and identified 1,234 relevant papers that describe tests of conservation actions, especially in areas and for species with little or even no relevant English-language evidence (Amano et al., 2021a). Incorporating non-English-language evidence can expand the geographical coverage of English-language evidence by 12% to 25%, especially in biodiverse regions (e.g. Latin America), and taxonomic coverage by 5% to 32%, although non-English-language papers tend to adopt less robust study designs (Amano et al., 2021a). Konno et al. (2020) showed that incorporating Japanese-language studies into English-language meta-analyses caused considerable changes in the magnitude, and even direction, of overall mean effect sizes. These findings indicate that incorporating scientific literature published in non-English languages is important for synthesising global evidence in an unbiased way and deriving robust conclusions.

Finding evidence, for example in the form of literature or data, typically involves the following four stages:

1. Developing search strategies
2. Conducting searches
3. Screening evidence based on eligibility criteria
4. Extracting relevant information

Finding evidence in non-English languages could be challenging as it would require sufficient skills in the relevant languages at all stages, for example, for developing search strings (Stage 1), using language-specific search systems (Stage 2), reading full texts for screening and assessing validity (Stage 3), and extracting specific information from eligible sources (Stage 4).

One obvious solution to securing relevant language skills at all stages is to develop collaboration with native speakers of the languages, who should also be familiar with the ecology and conservation of local species and ecosystems. As conservation science has increasingly been globalised, it is now relatively easy to find experts on a specific topic who are native speakers of different languages. For example, the translatE project worked with 62 collaborators who, collectively, are native speakers of 17 languages, for screening non-English-language papers (Amano et al., 2021a). Such collaborators should be involved in as many stages of finding evidence as possible and given appropriate credit (e.g. in the form of co-authorship).

In healthcare, Cochrane Task Exchange (https://taskexchange.cochrane.org/) provides an online platform where you could post requests for help with aspects of a literature review, such as screening, translation or data extraction.

The quality of machine translation (e.g. Google Translate and DeepL) has been improving, aiding some of the stages in finding evidence (e.g. for reading full texts [Stages 3 and 4], Zulfiqar et al., 2018; Steigerwald et al., 2022). However, even a few critical errors, for instance, when translating search strings (Stage 1) and extracting information (Stage 4) could have major consequences. Therefore, we still need robust tests to assess the validity of machine translation at each of the stages in finding relevant non-English-language evidence. For now, we should still try to find collaborators with relevant language skills and use machine translation with caution, i.e. only when a native speaker of the language is available for double-checking the translation output.

When developing search strategies (Stage 1), identifying appropriate sources of non-English-language evidence (e.g. bibliographic databases) is key, as few international sources index non-English-language evidence (Chowdhury et al., 2022). For example, non-English-language literature is searchable on Google Scholar (https://scholar.google.com) using non-English-language keywords. Searches on Google Scholar should be restricted only to pages written in the relevant language (from Settings), apart from languages where this search option is not available (Amano et al., 2016a), as otherwise Google Scholar's algorithm is known to make non-English-language literature almost invisible (Rovira et al., 2021). Another effective approach is to use language-specific literature search systems, such as SciELO (https://scielo.org) for Spanish and Portuguese, J-STAGE (https://www.jstage.jst.go.jp) for Japanese, KoreaScience (https://www.koreascience.or.kr) for Korean, and CNKI (https://cnki.net) for simplified Chinese. Again, it is important to involve native speakers of relevant languages who are familiar with language-specific sources of evidence.

It is also recommended to seek input from native speakers of the languages when developing appropriate search strings in non-English languages (Stage 1). It is often difficult to find the most appropriate non-English translations for scientific terms (Amano et al., 2021b). For example, 'biodiversity' in German can be any of 'Biodiversität', 'biologische Vielfalt' and 'Artenvielfalt'. In such a case, using all translations for the search helps make it as comprehensive as possible. When searching with species names, including common names in the relevant language in a search string can be effective; bird species names in multiple languages, for example, are available in the IOC World Bird List (Gill et al., 2022) and Avibase (2022).

The translatE project provides useful tools and databases that aid multilingual evidence searches. This includes a list of 466 peer-reviewed journals in ecology and conservation in 19 languages (translatE Project, 2020), a list of language-specific literature search systems for 13 languages (available in Table 1 in Chowdhury et al., 2022), a list of 1,234 non-English-language studies testing the effectiveness of conservation actions (available as S2 Data in Amano et al., 2021a), and more general tips for overcoming language barriers in science (Amano et al., 2021b) including resources and opportunities for non-native-English speakers (Amano, 2022).

2.5.6 Practitioner knowledge and expertise

Another important source of evidence is the knowledge held by practitioners and other stakeholders, which may not necessarily be documented in the peer-reviewed or non-peer-reviewed literature. The type of individual, or group of individuals, who could hold useful evidence will vary substantially depending on the context but could include: local communities and landowners, business owners, local and regional government, NGOs, Indigenous groups, scientists/researchers or policy makers. For example, it could be the knowledge of species and habitats held by local conservationists or landowners, or it could be the knowledge held by members of a scientific advisory board for NGOs or government agencies.

Often there is an overlap between this source of evidence, and other sources, as knowledge held by individuals can sometimes be traced back to other sources. For example, a conservationist may offer some evidence of the presence of a given species on a site. This knowledge may be from a primary observation and not documented, or it may have been learnt by the conservationist from a secondary source in the documented literature.

This source of evidence is particularly useful for local information in a conservation context where practitioners may have a detailed understanding of project sites and the status of species and habitats, threats present within them, and the suitability of sites for particular interventions. This is also an important source for gathering evidence on stakeholder values that can be used to help ensure actions taken are equitable and acceptable to different stakeholder groups.

The same groups, or individuals, who act as a source of evidence in this context, can also help judge the evidence to make better decisions. Chapter 5 expands upon this.

2.5.7 Indigenous and local knowledge

One particular source of expert knowledge and expertise is that held within Indigenous and local knowledge systems. Indigenous and local knowledge systems are dynamic bodies of integrated, holistic, social, cultural and ecological knowledge, practices and beliefs pertaining to the relationships of people and other living beings with one another and with their environments (IPBES, 2017). As described in Chapter 6, there is a wide range of reasons for using this knowledge and ethical ways of using it with protocols developed locally.

The ways of knowing for Indigenous and local knowledge systems include sense perception, reason, emotion, faith, imagination, intuition, memory and language (Berkes, 2017). Berkes discusses four layers of Indigenous and local knowledge: empirical (e.g. knowledge over animals, plants, soils and landscape); resource management (e.g. ecological, medicinal, scientific, and technical knowledge and practice); institutions of knowledge (e.g. the process of social memory, creativity and learning); and overarching cosmologies (e.g. underpinning knowledge-holders' understanding of the world). Under Salafsky et al.'s (2019) list of seven types of 'evidence' to be used in conservation, Indigenous and local knowledge is especially strong in providing 'direct and circumstantial evidence' (through long-term observation); 'specific evidence' (local information about a specific hypothesis in a particular situation); and 'observational and experimental evidence' (experience from long-term trial and error). Tengö et al. (2017) point out that 'Indigenous and local knowledge systems, and the holders of such knowledge, carry insights that are complementary to science, in terms of scope and content, and also in ways of knowing and governing social-ecological systems during turbulent times and articulating alternative ways forward'.

The collation, use and sharing of evidence may involve a range of ethical issues. One infamous example is the collection of 70,000 rubber seeds from Brazil by Henry Wickham in 1876 to take to Kew Gardens for germination, which became the source of rubber plantations in South-East Asia that replaced the South American market (Musgrave and Musgrave, 2010). Another example is 'teff', a gluten-free cereal that is high in protein, iron and fibre, which has been cultivated in Ethiopia for more than 2,000 years. In 2003, a dozen varieties of teff seeds were sent to a Dutch agronomist through a partnership with the Ethiopian Institute of Biodiversity Conservation for research and development. Four years later, the European Patent Office granted a patent for teff flour and related products (including the Ethiopian national staple pancake, ingela) to his Dutch company. The Ethiopian government and the Convention on Biological Diversity (CBD) tried to employ the Access and Benefit Sharing (ABS) mechanism to compensate for the loss to Ethiopians, but it was unsuccessful (Andersen and Winge, 2012).

Ensuring participatory approaches are used is key to working with indigenous and local knowledge; Verschuuren et al. (2021) set out the guidance on the approach to recognising and working with Indigenous and local knowledge on biodiversity and ecosystem services. Yet some, such as Krug et al. (2020) and Reyes-Garcia et al. (2022), suggest IPBES should improve its own processes to appropriately engage Indigenous peoples and local communities and increase linguistic diversity in its ecosystem assessments (Lynch et al., 2021). Local initiatives such as Kūlana Noi'i in Hawai'i (https://seagrant.soest.hawaii.edu/wp-content/uploads/2021/09/Kulana-Noii-2.0_LowRes.pdf) can offer guidelines along local language and value, even though engagement will look different with different communities, as discussed in Chapter 6.

Indigenous knowledge is often embedded in Indigenous languages, many of which are threatened and going extinct (Nettle and Romaine, 2000; Maffi, 2005; Amano et al., 2014). Where appropriate, respecting knowledge sovereignty and, where mutually beneficial, there is a need for storing and incorporating Indigenous and local knowledge. There are three main approaches for accessing and using knowledge: 'knowledge co-assessment', 'knowledge co-production', and 'knowledge co-evolution'.

Knowledge co-assessment is the process in which those involved in the decision reflect and assess the different knowledge prior to making a decision (Sutherland et al., 2014). This is the most cost-effective solution and, therefore, the one that is carried out at scale.

Knowledge co-production with local communities in which research priorities are identified by the community and the research planned collectively. There are a number of good examples — the systematic review of Zurba et al. (2022) identified 102 studies. This makes sense after an inclusive process of co-assessment to check whether the knowledge already exists and if there is the capacity for the considerable work involved.

Knowledge co-evolution includes the objectives of capacity building, empowerment, and self-determination alongside knowledge co-production (Chapman and Schott, 2020).

A new role for the scientists and environmental conservationists would be to restructure and systemise a diverse knowledge system through the eyes of stakeholders, and to work with the stakeholders to utilise it to solve local environmental issues (Satō et al., 2018). In the context of local and Indigenous knowledge, the information providers are also users of this knowledge. This context-based knowledge holds multiple perspectives that are necessary to solve local issues, through the actions in which diverse stakeholders influence each other, and is constantly evolving. The sharing of this tacit knowledge mainly occurs through social interactions, informal networking, observation and listening. Therefore, an outsider's ability to fully understand Indigenous and local knowledge may be compromised without strong trusting relationships built over time. The transfer of this vast and diverse local and Indigenous knowledge is often intergenerational. They are passed from the elders or first people to the youth through oral storytelling, talking circles or place-based education (Ross, 2016). A couple of recent projects/studies are also holding workshops, using digital storytelling and producing a local seasonal calendar to share or transfer the Indigenous and local knowledge to youth groups (Hausknechta et al., 2021; McNamara and Westoby, 2016).

Ideally, if the targeted evidence is to be used for decisions regarding land that Indigenous Peoples still manage or have tenure rights over (Indigenous land), Indigenous governance should be respected and strengthened (e.g. Artelle et al., 2019). In such cases, Indigenous agency and leadership rule the decision making while external logistical and/or technical support should be offered instead of trying to impose a western governance or decision-making system. This is supported by the fact that Indigenous lands host a consequent proportion of global biodiversity as well as protected areas and can be efficient as protected areas (Garnett et al., 2018; IPBES, 2019). If strictly local governance is not possible or the evidence is to be collected for use elsewhere than on the landscape/seascape where the knowledge is from, then the three above-mentioned approaches to access and use local and Indigenous knowledge can be adopted.

As described in Section 2.2, evidence can be assessed using the criteria of information reliability, source reliability and relevance. This can potentially be applied to any source of information. At the same time, different communities may have different ways of assessing these axes. For example, some languages embed evidence with different words, a process called 'evidentiality': languages may thus have specific words for different ways of knowing (Aikhenvald, 2003).

2.5.8 Online material

For information outside some of the sources listed above it is likely that an internet search will be used to obtain further information. This is especially appropriate for straightforward facts (What is the most common frog in the region? What size in the National Park? What does this species feed on?) and where the information required is not critical to the success of the project. Such sources could include Wikipedia articles, blog posts, news articles and organisation websites. Again, the same principles for considering information and source reliability apply to information online as to other sources of evidence. For example, does it seem to be an acceptable source or linked to an authoritative source? Does it seem authoritative? Does it provide sensible material on other issues you know about? Does it seem to have an agenda? Is the topic controversial?

Possible checks include:

- Does the URL have a name related to the material provided?
- Does the mailing address look legitimate in a search engine or street views?
- Is the evidence is provided by respected sources or by commercial interests?
- Is it clear where the material comes from? Does the material link back to a primary or secondary source?
- Does it seem up to date and, if not, does that matter?
- When was the source was last updated (there are a range of ways of checking this)?
- Who was involved in creating this evidence creation and do they seem credibile?
- Which other sites link to that source — found by typing the website name as link: http://www.[WEBSITE].com?
- Is there any criticism of the source?

Although probably not a major issue for most conservation evidence at present there is an increasing quantity of entirely fake material online. In fact, some conservationists have documented and warned of the rise of 'extinction denial' where there is denial of the evidence documenting biodiversity loss, often in media coverage (Lees et al., 2020). This may be more common in online sources of evidence, but also pervade other information sources. At a more extreme level, powerful images and videos can be reused claiming to show something different (reverse image search can be used to identify where the image was originally presented) and apparent mass support or criticism of a statement may be created by bots.

2.6 Types of Evidence

2.6.1 Experiments and quasi-experimental studies

Research studies provide data on status, change, correlations, impacts of threats and outcomes of actions. Pimm et al. (2019) state 'measuring resilience is essential to understand it'; this principle of measurement applies widely. A number of factors indicate the accuracy of research studies. Checklist 2.1 outlines a range of issues to consider when evaluating study quality.

Figure 2.4 illustrates six broad types of study designs. There are four key aspects of study design that can improve the reliability and accuracy of results and inferences derived from them: (1) Randomisation; (2) Controls; (3) Data sampled before and after an intervention or impact has occurred; (4) Temporal and spatial replication.

Christie et al. (2019, 2020b) demonstrated that study designs that incorporate randomisation, controls, and before–after sampling (e.g. randomised control–impact [RCI or RCT] and

2.1 Checklist for assessing study quality

- ☐ Can the result be attributed to measurement error?
- ☐ Is the result likely to be due to chance variation?
- ☐ Is the result likely to be due to bias?
- ☐ Is the sample size sufficient?
- ☐ Is correlation being confused with causation?
- ☐ Might the result be due to regression to the mean (e.g. a measure considered effective after being introduced after an atypical spike of occurrences followed by reduction to usual level)?
- ☐ Are any controls suitable? They should be equivalent to the treatment, and just missing the factor being studied.
- ☐ Is randomisation used to allocate individuals or groups to interventions?
- ☐ Is there replication and not pseudoreplication (e.g. taking multiple samples from a field and treating them as different data points)?
- ☐ Are the results statistically significant?
- ☐ Is no effect confused with non-significance?
- ☐ How large is the effect size?
- ☐ Do the presented results appear to be selected (cherry-picked) from a larger set of results?
- ☐ Are a few extreme measures critical to the analysis?

(*Source:* modified from Sutherland et al., 2013)

This checklist can be downloaded from https://doi.org/10.11647/OBP.0321#resources. They can be modified and tailored for specific uses.

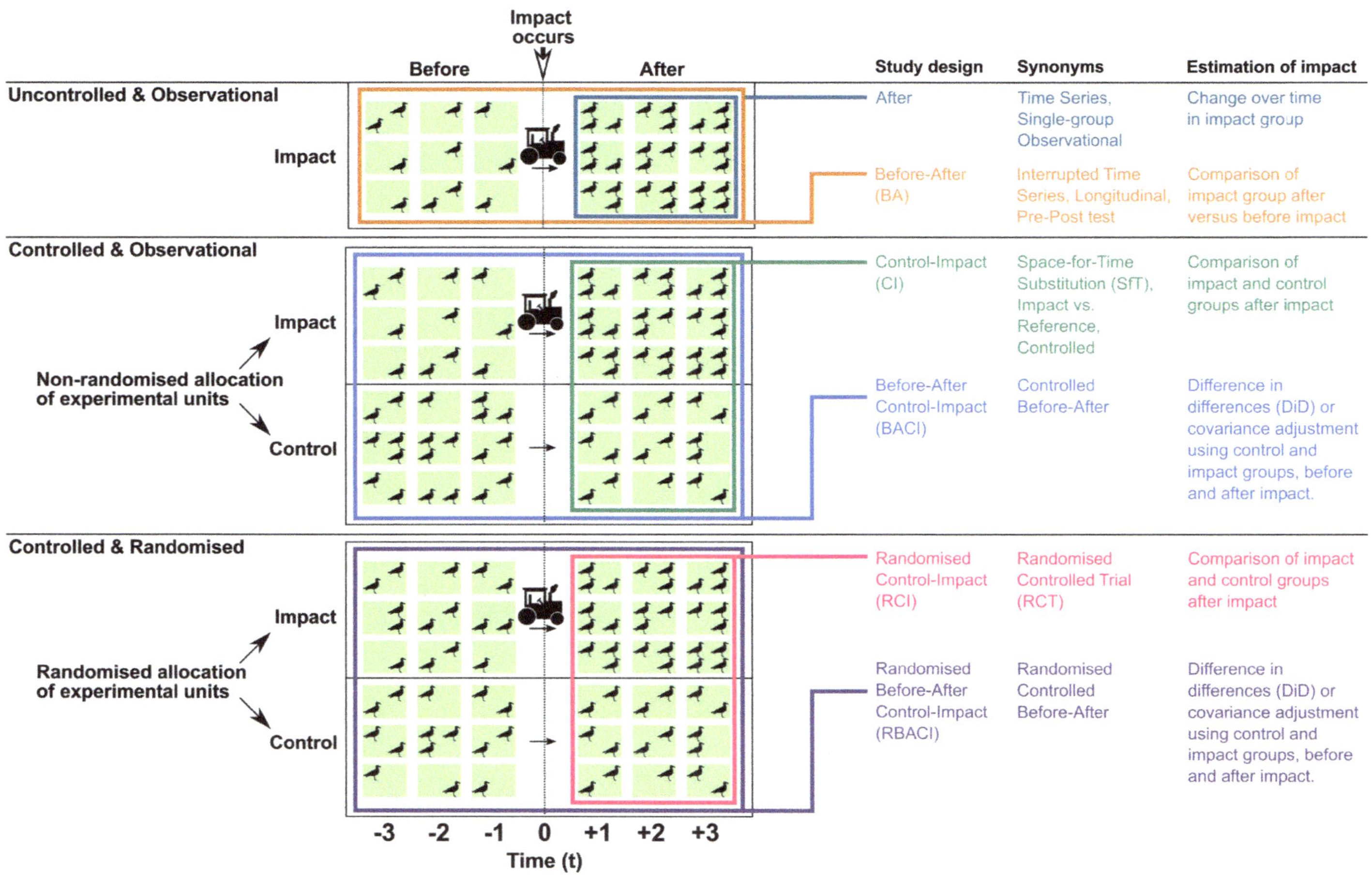

Figure 2.4 A summary of the main six broad types of study designs. Designs with an R preceding their names feature randomisation (RCI and RBACI) whilst all other designs do not. (*Source:* Christie et al., 2020a, CC-BY-4.0)

randomised or non-randomised before–after control–impact designs [RBACI and BACI]) are typically more reliable and accurate at estimating the magnitude and direction of an impact or intervention than simpler designs that lack these features (control–impact, before–after, and after designs). In turn, control–impact and before–after designs are more accurate than after designs. Increasing replication for the more reliable, complex study designs led to increases in both accuracy and precision, but increasing replication for the simpler study designs only led to increases in precision (a more precise but still largely inaccurate figure). Accuracy can thus only be improved by changing the fundamental aspects of a study design (e.g. randomisation, controls, before–after sampling) and not simply by increasing the replication (Table 2.5).

Table 2.5 Comparison of the effectiveness of six experimental and quasi-experimental methods. These give the percentage of estimates that correctly estimated the true effect's direction (based on the point estimate), magnitude to within 30% and direction (based on point estimate and 95% Confidence Intervals), and the increase in accuracy with greater replication (improvements in performance with increasing replication from two control and two impact sites to 50 control and 50 impact sites). (*Source:* Christie, 2021)

Design type	Design name	Effect's direction	Improvement of direction with greater replication	Effect's direction and magnitude (within 30%)	Improvement of direction and magnitude with greater replication
Experimental	Randomised Controlled Trials (RCTs)	91.2%	+0.9%	46.7%	+9.0%
	Randomised Before-After Control-Impact (R-BACI)	88.4%	+1.0%	32.2%	+11.4%
Quasi-experimental	Before–After Control–Impact (BACI)	85.9%	1.5%	29.9%	+9.2%
	Control–Impact (CI)	75.2%	+0.2%	17.6%	+2.4%
	Before–After (BA)	73.9%	+0.5%	19.0%	+2.5%
	After	49.5%	+0.3%	6.4%	+2.7%

However, in practice, simpler study designs are commonly used in ecology and conservation (Christie et al., 2020a) due to a whole host of logistical, resource-based, knowledge and awareness-based constraints — for example, having the knowledge and resources to survey before and after an impact (if it is planned), the awareness and knowledge of more powerful statistical study designs, and the ability to conduct randomised experiments (due to ethical or practical barriers).

The main message is that study quality really matters and that some designs, although widely used, are unlikely to give reliable estimates of the effects of impacts and interventions. Some useful quotes to bear in mind are: 'You can't fix by analysis what you bungled by design...' (Light et al., 1990), 'Study design is to study, as foundation is to building' (Christie, 2021). Thus it pays to use more reliable study designs whenever possible. This may not necessarily mean using randomised experiments or before-after control-impact designs (due to various constraints), but could involve using matching counterfactuals, synthetic controls, or regression discontinuity designs that are widely used in fields such as economics to evaluate policy implementations (Ferraro, 2009; Christie et al., 2020a).

In addition, any cost-based feasibility assessments of implementing study designs should carefully consider the social, environmental, and political costs of Type I and Type II errors associated with different designs (Mapstone, 1995). Researchers should always explicitly acknowledge the limitations and assumptions behind designs used and make appropriate, cautious conclusions.

2.6.2 Case studies

Case studies are a detailed examination of a particular process or situation in conservation. They are produced frequently in conservation and can provide evidence of the socio-economic context and drivers of loss in specific cases, or the implementation or outcome of a particular project or action. Case studies may be published in several sources including databases, reports or the scientific literature.

Case studies often delve into the detail of the implementation of projects, and investigate why and how outcomes occurred, specifics of the socio-economic contexts or identify challenges that hindered progress. If done well, they provide a highly relevant and detailed account of a particular situation, which can be used to help design more effective programmes in that or similar contexts. Unless they embed a test, they tend to be weaker in providing evidence for specific actions. Case studies can have considerable value in showing how projects can be created, providing potential vision, and identifying challenges and practicalities of programme implementation. The list of online material (Chapter 13) provides a range of websites giving evidence including case studies.

Case studies provide low accuracy of general effectiveness, which is better identified through stronger study designs, with baseline data, controls and replication (Christie et al., 2020a). Secondly, there is a particularly high risk of publication bias if case studies of successful projects tend to be presented but unsuccessful ones are disproportionately underreported. This can also impact the generality of findings from multiple case studies.

2.6.3 Citizen science

Citizen science (also called 'community science') is an umbrella term for a large variety of data collection approaches that rely on the active engagement of the general public (Haklay et al., 2021). The scope of citizen science data relevant to conservation science includes a broad suite of data types, especially direct species observations, but also acoustic recordings (Newson et al., 2015; Rowley et al., 2019), environmental DNA samples (eDNA; Buxton et al., 2018), or species records extracted from camera trap photographs (Jones et al., 2020; Swanson et al., 2016).

Long-term structured surveys, which use randomly selected sites and survey methods that are standardised over time and space, are the gold standard for robust status and change assessments. Such structured surveys require large and long-term commitments and can be costly to organise and coordinate (Schmeller et al., 2009) and therefore such monitoring programmes (Sauer et al., 2017; Greenwood et al., 1995; Lee et al., 2022; van Swaay et al., 2019; Pescott et al., 2015) have a long history of leveraging the efforts of amateur naturalists, and thus form one end of the citizen science spectrum.

However, the rapid growth in citizen science biodiversity data is predominantly the result of so-called unstructured or semi-structured projects where data entry is often facilitated through the use of digital technologies (smartphone apps, online portals), but without formal survey designs or standardised protocols, and with less stringent requirements for observer knowledge or long-term observer commitment (Pocock et al., 2017).

Such projects may have a primary goal other than population monitoring, e.g. raising awareness about focal taxa or facilitating personal record keeping for amateur naturalists, but the vast amounts of data collected in this way can potentially contribute substantially to biodiversity monitoring, particularly in parts of the world with little or no formal data collection (Amano et al., 2016b; Bayraktarov et al., 2019).

Assessing status and change from such data is challenging because of known biases in site selection, visit timing, survey effort, and/or to surveyor skill (Boersch-Supan et al., 2019; Isaac and Pocock 2015; Johnston et al., 2018, 2021, 2022). Thus there is usually a trade-off between collecting a large amount of relatively heterogeneous (i.e. lower quality) data or a smaller amount of higher quality data conforming to a defined common structure. The consequences of this quantity versus quality trade-off are an active topic of statistical research, but careful statistical accounting for observer heterogeneity and preferential sampling can turn citizen science data into a powerful tool for the sustainable monitoring of biodiversity.

There is a growing set of modelling approaches to address the challenges of unstructured data sets and/or leverage to the strengths of both structured and unstructured data sources. However, the less structured a data source is, the higher the analytical costs are in terms of strong modelling assumptions, increased model complexity, and computational demands (Fithian et al., 2015; Robinson et al., 2018; Johnston et al., 2021). Furthermore, validating complex statistical models for citizen science remains a challenge given independent validation data are often lacking.

Despite ongoing statistical developments it is therefore crucial to recognise that improvements in data quality often have much greater benefits for robust inferences than increases in data quantity (Gorleri et al., 2021). For existing citizen science data sets, it is crucial to appraise likely sources of bias and error (Dobson et al., 2020) and whether or not the data contain information that allows these biases to be accounted for. Sampling quality will be determined by factors such as observer expertise, observer motivation, spatial and temporal sampling structure (randomness of site and sampling time selection, evenness of coverage across space and/or relevant to ecological gradients) and the sample size, whereas attempts to account for bias require metadata on observer identity, effort measures, the precision of the location and time metadata, and possibly other predictors that may inform reliability of the records. For data sources with heterogeneous metadata availability and/or very uneven to sampling across space and time, stringent quality control or filtering before analysis can greatly improve the quality of inferences (Johnston et al., 2021; Gorleri et al., 2021). For ongoing schemes, it is important to add relevant metadata during data collection into the scheme design, and to educate scheme participants about the value of specific metadata, e.g. by encouraging the collection of presence-absence data or complete lists over presence-only data, or steering observer efforts to achieve more balanced sampling across space and time (Callaghan et al., 2019).

2.6.4 Statements, observations and conclusions

Much of the evidence used in decision making often involves statements and observations. If knowledge is assessed using the aforementioned criteria of information reliability, source reliability, and relevance, then when documenting knowledge we want to ensure that statements elaborate on these three criteria. Thus the claim 'otters have been seen near the reserve' is likely to score poorly on all these axes. The statement 'Eric Jones, a keen local fisherman says in early spring 2022 that he watched an otter within about 10 m viewing distance for fifteen minutes, on the river just after it exits the nature reserve. He says he has seen mink regularly and he was certain it was larger' is likely to attract a higher score. Table 2.6 lists elements to include in documenting knowledge.

The table lists the elements that can underpin most statements. The more comprehensive the statement is, the more reliable it is likely to be considered. Relevance then needs to be considered case by case in relation to the context and the claim being assessed. The aim is not to show the statement is true but to specify the evidence so that it can be assessed by others.

The acceptance of such information depends on a range of elements, such as those outlined in Table 2.6. One objective is to encourage the creation of statements that can be evaluated and be more likely to be accepted.

Table 2.6 A classification, with examples, of the elements of most statements.

Evidence source

What is the origin of the claim

- *Own work (I have seen... My study showed...)*
- *Experienced (I read Smith's paper, which showed... Marie told me she had seen...)*
- *Reported (Carlos said Smith's paper said... Erica said Marie had seen...)*
- *Unattributed (I heard a study had shown... I heard people had seen...)*

Authority of source

Why should this be believed

- *Experience in subject or site*
- *Indicators of reliability*

Knowledge source

Where did the claim come from

- *Published*
- *Grey*
- *Own experience*
- *Verbal*
- *Cultural*

Knowledge type

What sort of information is this

- *Experiment*
- *Documented*
- *Observation*
- *Stories*

Verification

Anything that establishes its veracity

- *Peer reviewed*
- *Verified through a process of critical appraisal of the research*
- *Verified by named expert*
- *Object, such as sample or photo, provided*
- *Acceptance by the community*
- *Status within the community and how other community members see it — e.g. Matauranga framework*

Substance

For example

- *Status*
- *Change*
- *Interaction*
- *Use*
- *Values*
- *Response*

Relevance

The aim here is to consider information that would be needed to judge the relevance of the statement to another context. For example:

- *Where was the evidence from?*
- *When was the evidence from?*
- *If looking at the evidence for an action, what were the specifics of the action tested?*

Two examples of this process are shown below.

> Shin So-jung said that she has noticed, based on monitoring water levels on the river, which she has visited weekly for over 30 years, that the sediment load was higher than she had ever seen before both along the coast and the river where our project is.

> 'I have fished the lake most days for the past thirty years. After the papyrus swamp was cleared three years ago my observed fish catch is a quarter of what it was a few years ago. All other fishers state the same'. Meshack Nyongesa

2.6.5 Models

We are all familiar with the use of models in everyday decision making through the use of weather forecasts. Weather forecasts are based on the predictions of models that are complex computer simulations, fed with extensive remotely sensed data. The outputs from these models are disseminated in several ways. TV weather forecasts, for instance, make predictions of what the weather will be in the form, 'it will be sunny tomorrow with cloudy intervals, the temperature will be 12 °C…'. Other forecasts are presented in a probabilistic way, e.g. a 20% probability of rain.

In neither case is the presentation of the output fully satisfactory. The first forecast presents the outcome without expression of uncertainty (e.g. what is the likelihood that the forecast is wrong and that it rains instead?). The second forecast expresses a probability, which seems a better expression of uncertainty. However, the stated probability will be conditional on several constituent uncertainties and the interpretation of the statement '20% probability of rain' depends on these.

Whether the output of a model will be useful or not is difficult to predict. What is critically important is the purpose for which a model is to be used compared with that for which it was originally developed and being clear about the limits of the model in the predictions made.

Evaluating model quality

It may seem obvious that a poor model will generate low-quality output, however, the degree to which this is true depends on what the model is being used for. It is useful to consider four facets of model quality: biological understanding, data quality, assumptions, and quality control.

Biological understanding

For the outputs of a model to be robust, arguably they need to be based on some degree of understanding of the underlying system. The degree of model detail can vary widely, however, and the level of biological understanding required for a model to be useful depends on the situation in which the model is being used. For example, the simple model $N_{(t+1)}=rN_t$ allows us to forecast the change in numbers from one year to the next as a function of the intrinsic rate of increase (r). This model is based on minimal biological understanding, beyond an estimate of the rate of population growth.

Although an extremely simple model such as this can potentially be useful under some circumstances (e.g. low population size, unlimited resources), they lack the ecological reality to be useful under others (e.g. large population sizes, limited resources). Consequently, models of increasing levels of sophistication have been developed, incorporating layers of ecological complexity. At the extreme, complex simulations include many details, in some cases down to the behaviour of individual organisms.

Biologically realistic models and simulations have the advantage that they better represent the true system and can be used to perform more sophisticated analyses. This is especially advantageous when we use the models to consider novel conditions or interventions. In conservation, the utility of a model is often in forecasting the likely outcome of different hypothetical options.

Complex models can have disadvantages, however. The chief limitation is that data or parameters are required to generate each component of the model. In the absence of information, it is impossible to include important details. For example, if we do not know basic information, such as the clutch size of a bird, we cannot model nesting. As the level of granularity increases, the number of parameters explodes. A second major limitation that follows from this is that parameters contain errors. These errors propagate through the model and, if there are many uncertain parameters then the outcome can be very uncertain. Alternatively, if errors are not estimated, then the model output can appear unrealistically certain.

Many useful models include no biological mechanisms whatsoever and are based on statistical analysis, basically driven by correlation. Data-driven statistical models have been used extensively in a range of applications, foremost among which are Species Distribution Models (SDMs) and Ecological Niche Models (ENMs). These are models that predict species occupancy (presence/absence) as a function of environmental and habitat variables. There are several advantages to this approach: first, it is possible to harness large amounts of existing data across many species (e.g. atlas data); second, there exists an extensive range of readily available environmental and land-use datasets; and, third, the predictions of these models can extend over regional, national or even global scales.

Of course, these models are not completely lacking in biological understanding. The choice of predictor variables will be driven by an understanding of the ecology of the species being modelled. However, there is no mechanistic modelling based on ecological or biological processes. A consequence of this is that it may not be possible to robustly predict outside of the range of conditions under which the model was fitted or to consider new variables that are not already included within the model.

Data quality

As models increase in complexity or realism, more data are required to parameterise them. The availability of data can be a major constraint in the modelling process. This is particularly a limitation on the development of models that are applied in a site-specific manner. The finer the resolution (spatial and temporal), the greater the data requirements.

If data are lacking, shortcuts can be taken in the parameterisation of models. For example, if we are modelling a particular species for which data are lacking, data may be available on similar species that can be used. This is a crude form of data imputation, the statistical methods for which are now well developed. Imputation uses the correlations between measured variables to predict values of missing data and can be used to fill in holes in datasets. Importantly, data are frequently not missing at random: for example, in conservation data are typically missing from species/populations that are low density, difficult to assess or elusive. Under such circumstances, missing data can represent a non-random subset and ignoring this could lead to biases in the dataset.

Ideally, the process of data collection and model development should be integrated: as populations are managed, monitoring will generate more data that can be used to improve the precision of models. This idea has been formalised in a couple of ways. Adaptive Management (Section 10.1) is a process of iterative improvement of decisions through feedback between interventions and models. Alternatively, the Value of Information (Section 10.5) is a quantitative measure of how much new information adds to the quality of decision making.

Assumptions

Any model is based on a set of assumptions. For a biological process-based model, these will include assumptions about the structure of the system and the various processes that drive it. Ideally, these should be clearly stated and justified with respect to previous studies and literature. Where alternative assumptions could be made, there should be a clear rationale for the decision taken.

Statistical assumptions are also important whether these are made in the model parameterisation or in the formulation of purely statistical models. All statistical methods have explicitly stated assumptions, and these can be tested through, for example, diagnostic plots. The presentation of these is important to reassure that the parameters of the models are estimated accurately. Failure of statistical assumptions requires that alternative methods of analysis are employed.

Quality control

Models are made available to end-users in a range of formats, which include software, tools (e.g. spreadsheets, web interfaces) and summaries of outputs (e.g. scientific outputs and reports). An obvious question is whether the implementation provided is reliable enough to be trusted.

The scientific peer-review process is intended to ensure that underpinning scientific logic is justified and, ideally, models will have been subjected to this peer review. The peer-review process assesses quality in terms of the assumptions and the formulation of the model. Peer review may not include rigorous testing of the model or evaluation of any implementation of the model, such as a piece of software.

If the model is presented as a piece of software, or is based on complex code, it is difficult to be sure that the model does not contain bugs or errors. Rigorous testing can reveal any issues, however this can take time. Problems can be avoided if model developers employ robust programming practices. These include unit testing of model components, and open workflows (e.g. using platforms such as GitHub) so that code can be inspected and tested. Professional programmers are familiar with such approaches, and scientific journals increasingly require such quality control for complex models.

Will a model be useful?

There are no fixed rules for judging whether a model is likely to be useful or not. A model can be very useful in one context, but hopeless in another. The key to judging whether a model is useful and if the outputs can be trusted is to have clear questions and be able to assess the degree to which the model output can address these objectives. In assessing the model, the following questions might be helpful to consider:

- What are the desired outputs of the model? For example, if we ask, 'Will grazing increase or decrease the populations of breeding birds?' then a simple model might be helpful. On the other hand, the answer to the question 'by how much do we need to reduce grazing to increase breeding bird population sizes by 30%' would require a model with sufficient sophistication to generate outputs that are numerically accurate enough.
- Has the model been validated? Ideally, a model should be validated by comparing its outputs with independent data. However, this is frequently difficult not least because models are often developed to predict novel management interventions that have not been attempted before. Nonetheless, it may be possible to compare the predictions of the model with data collected under baseline conditions. Note, it is not enough to just compare the model outputs with the input data, as this is likely to be a biased test of 'predictive' ability.
- Does the model output include an estimate of uncertainty? As outlined above, the parameters of a model are subject to uncertainty owing to statistical error as well as inherent temporal and spatial variability. These uncertainties propagate through the model and into the model outputs. As in the weather forecast example above, there is a need to express this uncertainty in the model predictions. If models do not include uncertainties in their predictions, there is a danger of undue faith in their outputs.
- How relevant is the model to my site? It is unlikely that a model will have been generated for any specific site that is being considered and, in any case, inter-annual variability can be considerable. If a model has been developed at another site, the relevance of the model will depend on a suite of factors including weather, habitat, soil conditions etc. In some cases, models will have been created by collating data across several studies/sites. If this is true, then an assessment will need to be made of the range of conditions under which the model outputs are valid.

- Are statistical models robust enough to make predictions? Statistical models make efficient use of data and can be used to make predictions. However, using such models requires care: do the conditions under which the data were collected remain valid? Do the predictions of the model involve extrapolation and is this valid?
- Are alternative models available and how do these compare? Sometimes there are several independent models for the same species or system. More commonly, the creators of a model will build a suite of models, typically of increasing levels of complexity from simplest to most sophisticated. Where possible these alternatives should be explored and the implications of different formulations considered.

2.6.6 Expertise

It is common for expert knowledge to be used as a form of evidence. We consider an expert to be anyone knowledgeable or well-informed about a specific context or topic. We consider expert knowledge as providing an assessment from study and accumulated experience. This thus differs from the statements described in the previous section that can be linked to specific sources and thus evaluated. Many experts have remarkable knowledge and can assess, for example, the management needed to achieve an objective.

However, care should be taken to avoid bias and inaccurate information in expert knowledge. As described in Chapter 5, experts may not be as authoritative as they seem and approaches are described that can be used to: i) elicit more reliable judgements from experts when collating evidence, ii) see if an expert makes statements that turn out to be true, and iii) check the basis of assertions. Ideally, when using expertise as a type of evidence, do not rely on a single expert, but rather collate independent advice from a diverse group of experts.

2.6.7 Online open source investigations

Another type of evidence is that from online open-source investigations, also called open source intelligence or open source research. In much of journalism the model comprises trusted anonymous sources so that the reader has to decide whether to trust both the journalist and the journalist's unknown source. By contrast, open-source investigations comprises the transparent assessment of facts through extensive online communities. The underlying principles are to Identify, Verify and Amplify. The most famous adopter is Bellingcat, an organisation that played a key role in many security issues, including the shooting down of the Malaysian Aircraft MH17 and identifying those who carried out the Salisbury novichok poisoning (Higgins, 2022). They also actively seek to expose fake stories, images and videos and have been used for environmental issues such as identifying participants in illegal wildlife trade or illegal deforestation. The Bellingcat website provides training materials.

2.6.8 Artificial intelligence

It seems likely that in the future more evidence will be provided by artificial intelligence. As examples, literature searches are becoming increasingly automated allowing for 'living' reviews, species' detection and classification algorithms are becoming increasingly accurate and the automated answering of questions is improving in quality and feasibility. Artificial intelligence does, of course, have its own range of errors and biases and is also prone to manipulation.

2.7 Acknowledgements

We thank Jonathan Campbell-James for initiating our thinking on assessing evidence and Tim Lucas for a tip leading to evidentiality.

References

Adams, W. and Sandbrook, C. 2013. Conservation, evidence and policy. *Oryx* 47: 329–35, https://doi.org/10.1017/S0030605312001470.

Aikhenvald, A.Y. 2003. Evidentiality in typological perspective. *Typological Studies in Language* 54: 1–32, https://doi.org/10.1075/tsl.54.04aik.

Amano, T. 2022. translatE. OSF, https://doi.org/10.17605/OSF.IO/Y94ZT.

Amano, T., Berdejo-Espinola, V., Christie, A.P., et al. 2021a. Tapping into non-English-language science for the conservation of global biodiversity. *PLoS Biology* 19: e3001296, https://doi.org/10.1371/journal.pbio.3001296.

Amano, T., González-Varo, J.P. and Sutherland, W.J. 2016a. Languages are still a major barrier to global science. *PLoS Biology* 14: e2000933, https://doi.org/10.1371/journal.pbio.2000933.

Amano, T., Lamming, J.D. and Sutherland, W.J. 2016b. Spatial gaps in global biodiversity information and the role of citizen science. *Bioscience* 66: 393–400, https://doi.org/10.1093/biosci/biw022.

Amano, T., Rios Rojas, C., Boum II, Y., et al. 2021b. Ten tips for overcoming language barriers in science. *Nature Human Behaviour* 5: 1119–22, https://doi.org/10.1038/s41562-021-01137-1.

Amano, T., Sandel, B., Eager, H., et al. 2014. Global distribution and drivers of language extinction risk. *Proceedings of the Royal Society B*. 281: 20141574, https://doi.org/10.1098/rspb.2014.1574.

Andersen, R. and Winge, T. 2012. *The Access and Benefit-Sharing Agreement on Teff Genetic Resources: Facts and Lessons*, FNI Report 6/2012 (Laysaker, Norway: Fridtjof Nansen Institute), https://www.fni.no/getfile.php/131843-1469869194/Filer/Publikasjoner/FNI-R0612.pdf.

Artelle, K.A., Zurba, M., Bhattacharyya, J., et al. 2019. Supporting resurgent Indigenous-led governance: A nascent mechanism for just and effective conservation. *Biological Conservation* 240: 108284, https://doi.org/10.1016/j.biocon.2019.108284.

Avibase. 2022. *Avibase — The World Bird Database*, https://avibase.bsc-eoc.org/.

Barbier, E.B. 2022. The policy implications of the Dasgupta Review: Land use change and biodiversity. *Environmental and Resource Economics* Special Issue, https://doi.org/10.1007/s10640-022-00658-1.

Bayraktarov, E., Ehmke, G., O'Connor, J., et al. 2019. Do big unstructured biodiversity data mean more knowledge? *Frontiers in Ecology and Evolution* 6: 239, https://doi.org/10.3389/fevo.2018.00239.

Berkes, F. 2017. *Sacred Ecology*, Fourth edition (New York: Routledge), https://doi.org/10.4324/9781315114644.

Boersch-Supan, P.H., Trask, A.E. and Baillie, S.R. 2019. Robustness of simple avian population trend models for semi-structured citizen science data is species-dependent. *Biological Conservation* 240: 108286, https://doi.org/10.1016/j.biocon.2019.108286.

Buxton, A., Groombridge, J. and Griffiths, R. 2018. Comparison of two citizen scientist methods for collecting pond water samples for environmental DNA studies. *Citizen Science: Theory and Practice* 3(2): Article 2, https://doi.org/10.5334/cstp.151.

Callaghan, C.T., Poore, A.G., Major, R.E., et al. 2019. Optimizing future biodiversity sampling by citizen scientists. *Proceedings of the Royal Society B*. 286: 20191487, https://doi.org/10.1098/rspb.2019.1487.

Chapman, J.M. and Schott, S. 2020. Knowledge coevolution: Generating new understanding through bridging and strengthening distinct knowledge systems and empowering local knowledge holders. *Sustainability Science* 15: 931–43, https://doi.org/10.1007/s11625-020-00781-2.

Chowdhury, S., Gonzalez, K., Aytekin, M.Ç.K., et al. 2022. Growth of non-English-language literature on biodiversity conservation. *Conservation Biology* 36, e13883, https://doi.org/10.1111/cobi.13883.

Christie, A. 2021. *Determining the Biases and Consistencies in the Evidence for Conservation* (PhD thesis, Cambridge: University of Cambridge), https://doi.org/10.17863/CAM.68583.

Christie, A.P., Abecasis, D., Adjeroud, M., et al. 2020a. Quantifying and addressing the prevalence and bias of study designs in the environmental and social sciences. *Nature Communications* 11: 6377, https://doi.org/10.1038/s41467-020-20142-y.

Christie, A.P., Amano, T., Martin, P.A., et al. 2019. Simple study designs in ecology produce inaccurate estimates of biodiversity responses. *Journal of Applied Ecology* 56: 2742–54, https://doi.org/10.1111/1365-2664.13499.

Christie, A.P., Amano, T., Martin, P.A., et al., 2020b. Poor availability of context-specific evidence hampers decision-making in conservation. *Biological Conservation* 248: 108666, https://doi.org/10.1016/j.biocon.2020.108666.

Christie, A.P., Chiaravalloti, R.M., Irvine, R., et al. 2022. Combining and assessing global and local evidence to improve decision-making. OSF, https://osf.io/wpa4t/?view_only=05e59759bba74302b52f9d8ef41c04b1.

Christie, A.P., White, T.B., Martin, P.A., et al., 2021. Reducing publication delay to improve the efficiency and impact of conservation science. *PeerJ* 9: e12245, https://doi.org/10.7717/peerj.12245.

Cooke, S.J., Nguyen, V.M., Wilson, A.D.M., et al. 2016. The need for speed in a crisis discipline: Perspectives on peer review duration and implications for conservation science. *Endangered Species Research* 30: 11–18, https://doi.org/10.3354/esr00721.

Deutz, A., Heal, G.M., Niu, R., et al. 2020. *Financing Nature: Closing the Global Biodiversity Financing Gap* (The Paulson Institute, The Nature Conservancy, and the Cornell Atkinson Center for Sustainability), https://www.paulsoninstitute.org/conservation/financing-nature-report/.

Dickson, I., Butchart, S.H., Catalano, A., et al. 2022. Introducing a common taxonomy to support learning from failure in conservation. *Conservation Biology* Accepted Author Manuscript, https://doi.org/10.1111/cobi.13967.

Dobson, A.D.M., Milner-Gulland, E.J., Aebischer, N.J., et al. 2020. Making messy data work for conservation. *One Earth* 2: 455–65, https://doi.org/10.1016/j.oneear.2020.04.012.

Ferraro, P.J. 2009. Counterfactual thinking and impact evaluation in environmental policy. *New Directions for Evaluation* Special Issue Summer 2009: 75–84, https://doi.org/10.1002/ev.297.

Fithian, W., Elith, J., Hastie, T., et al. 2015. Bias correction in species distribution models: Pooling survey and collection data for multiple species. *Methods in Ecology and Evolution* 6: 424–38, https://doi.org/10.1111/2041-210X.12242.

Fleerackers, A., Riedlinger, M., Moorhead, L., et al. 2022. Communicating scientific uncertainty in an age of COVID-19: An investigation into the use of preprints by digital media outlets. *Health Communication* 37: 726–38, https://doi.org/10.1080/10410236.2020.1864892.

Freckleton, R.P., Watkinson, A.R., Green, R.E. et al. 2006. Census error and the detection of density dependence. *Journal of Animal Ecology* 75: 837–851, https://doi.org/10.1111/j.1365-2656.2006.01121.x

Garnett, S.T., Burgess, N.D., Fa, J.E., et al. 2018. A spatial overview of the global importance of Indigenous lands for conservation. *Nature Sustainability* 1: 369–74, https://doi.org/10.1038/s41893-018-0100-6.

Gavin, M.C., McCarter, J., Berkes, F., et al. 2018. Effective biodiversity conservation requires dynamic, pluralistic, partnership-based approaches. *Sustainability* 10: 1846, https://doi.org/10.3390/su10061846.

Gill, F., Donsker, D. and Rasmussen, P. (eds.) 2022. IOC World Bird List (v12.1), available at https://www.worldbirdnames.org/new/.

Gillispie, C.C., Fox, R. and Gratton-Guinness, I. 1999. *Pierre Simon Laplace, 1749–1827: A Life in Exact Science* (Princeton, NJ: Princeton University Press).

Gough, D. 2007. Weight of evidence: A framework for the appraisal of the quality and relevance of evidence. *Research Papers in Education* 22: 213–28, https://doi.org/10.1080/02671520701296189.

Gorleri, F.C., Hochachka, W.M. and Areta, J.I. 2021. Distribution models using semi-structured community science data outperform unstructured-data models for a data-poor species, the Plain Tyrannulet. *Ornithological Applications* 123: duab038, https://doi.org/10.1093/ornithapp/duab038.

Greenwood, J.J., Baillie, S.R., Gregory, R.D., et al. 1995. Some new approaches to conservation monitoring of British breeding birds. *Ibis* 137: S16–S28, https://doi.org/10.1111/j.1474-919X.1995.tb08437.x.

Gregory, R.S. 2002. Incorporating value trade-offs into community-based environmental risk decisions. *Environmental Values* 11: 461–88, https://doi.org/10.3197/096327102129341181.

Gutzat, F. and Dormann, C. 2020. Exploration of concerns about the evidence-based guideline approach in conservation management: Hints from medical practice. *Environmental Management* 66: 435–49, https://doi.org/10.1007/s00267-020-01312-6.

Haklay, M., Dörler, D., Heigl, F., et al. 2021. What is citizen science? The challenges of definition. In: *The Science of Citizen Science*, ed. by K. Vohland, et al. (Cham, Switzerland: Springer), https://doi.org/10.1007/978-3-030-58278-4_2.

Hausknechta, S., Freeman, S., Martinc, J., et al. 2021. Sharing Indigenous knowledge through intergenerational digital storytelling: Design of a workshop engaging elders and youth, *Educational Gerontology* 47: 285–96, https://doi.org/10.1080/03601277.2021.1927484.

Higgins, E. 2022. *We Are Bellingcat. An Intelligence Agency for the People* (London: Bloomsbury).

Hoy, M.B. 2020. Rise of the Rxivs: How preprint servers are changing the publishing process. *Medical Reference Services Quarterly* 39: 84–89, https://doi.org/10.1080/02763869.2020.1704597.

Iacona, G.D., Sutherland, W.J. and Mappin, B., 2018. Standardized reporting of the costs of management interventions for biodiversity conservation. *Conservation Biology* 32: 979–88, https://doi.org/10.1111/cobi.13195.

IPBES. 2017. Annex II to decision IPBES-5/1: Approach to recognizing and working with Indigenous and local knowledge in the Intergovernmental Science-Policy Platform on Biodiversity and Ecosystem Services. April 2017, https://ipbes.net/sites/default/files/inline/files/ipbes_ilkapproach_ipbes-5-15.pdf.

IPBES. 2019. Summary for policymakers of the global assessment report on biodiversity and ecosystem services of the Intergovernmental Science-Policy Platform on Biodiversity and Ecosystem Services, ed. by. S. Díaz, et al. (Bonn, Germany; IPBES secretariat), https://doi.org/10.1111/padr.12283.

Isaac, N.J. and Pocock, M.J. 2015. Bias and information in biological records. *Biological Journal of the Linnean Society* 115: 522–31, https://doi.org/10.1111/bij.12532.

Jakobsson, S., Bernes, C., Bullock, J.M. et al. 2018. How does roadside vegetation management affect the diversity of vascular plants and invertebrates? A systematic review. *Environmental Evidence* 7: Article 17, https://doi.org/10.1186/s13750-018-0129-z.

Johnston, A., Fink, D., Hochachka, W.M., et al. 2018. Estimates of observer expertise improve species distributions from citizen science data. *Methods in Ecology and Evolution* 9: 88–97, https://doi.org/10.1111/2041-210X.12838.

Johnston, A., Hochachka, W.M., Strimas-Mackey, M.E., et al. 2021. Analytical guidelines to increase the value of community science data: An example using eBird data to estimate species distributions. *Diversity and Distributions* 27: 1265–77, https://doi.org/10.1111/ddi.13271.

Johnston, A., Matechou, E. and Dennis, E.B. 2022. Outstanding challenges and future directions for biodiversity monitoring using citizen science data. *Methods in Ecology and Evolution* 00: 1–14 (early view online), https://doi.org/10.1111/2041-210X.13834.

Jones, F.M., Arteta, C., Zisserman, A. et al. 2020. Processing citizen science- and machine-annotated time-lapse imagery for biologically meaningful metrics. *Scientific Data* 7: Article 102, https://doi.org/10.1038/s41597-020-0442-6.

Kokko, H. and Sutherland, W.J. 1999. What do impact factors tell us? *Trends in Ecology and Evolution* 14: 382–84, https://doi.org/10.1016/S0169-5347(99)01711-5.

Konno, K., Akasaka, M., Koshida, C., et al. 2020. Ignoring non-English-language studies may bias ecological meta-analyses. *Ecological Evolution* 10: 6373–84, https://doi.org/10.1002/ece3.6368.

Krug, C.B., Sterling, E. and Cadman, T. 2020. Stakeholder participation in IPBES: Connecting local environmental work with global decision making. *Ecosystems and People* 16: 197–211, https://doi.org/10.1080/26395916.2020.1788643.

Lee, A.T., Brooks, M. and Underhill, L.G. 2022. The SABAP2 legacy: A review of the history and use of data generated by a long-running citizen science project. *South African Journal of Science* 118: Article 12030, https://doi.org/10.17159/sajs.2022/12030.

Lees, A.C., Attwood, S., Barlow, J., et al. 2020. Biodiversity scientists must fight the creeping rise of extinction denial. *Nature Ecology & Evolution* 4: 1440–43, https://doi.org/10.1038/s41559-020-01285-z.

Light, R.J., Singer, J.D. and Willett, J.B. 1990. *By Design: Planning Research on Higher Education* (Cambridge, MA: Harvard University Press).

Lynch, A.J., Fernández-Llamazares, Á., Palomo, I., et al. 2021. Culturally diverse expert teams have yet to bring comprehensive linguistic diversity to intergovernmental ecosystem assessments. *One Earth* 4: 269–78, https://doi.org/10.1016/j.oneear.2021.01.002.

Maffi, L. 2005. Linguistic, cultural, and biological diversity. *Annual Review of Anthropology* 34: 599–617, https://doi.org/10.1146/annurev.anthro.34.081804.120437.

Mapstone, B.D. 1995. Scalable decision rules for environmental impact studies: Effect size, type I, and type II errors. *Ecological Applications* 5: 401–10, https://doi.org/10.2307/1942031.

McNamara K.E. and Westoby, R. 2016. Intergenerational sharing of Indigenous environmental knowledge in the Torres Strait. In: *Geographies of Global Issues: Change and Threat. Geographies of Children and Young People, vol 8*, ed. by N. Ansell, et al. (Singapore: Springer), https://doi.org/10.1007/978-981-4585-54-5_6.

Miles, A. and Loughlin, M. 2011. Models in the balance: Evidence-based medicine versus evidence-informed individualized care. *Journal of Evaluation in Clinical Practice* 17: 531–53, https://doi.org/10.1111/j.1365-2753.2011.01713.x.

Musgrave, W. and Musgrave, T. 2010. *An Empire of Plants: People and Plants that Changed the World* (London: Cassell Illustrated).

Nettle, D. and Romaine, S. 2000. *Vanishing Voices: The Extinction of the World's Languages* (New York: Oxford University Press), https://vdoc.pub/download/vanishing-voices-the-extinction-of-the-worlds-languages-6n5a4e0qa5o0.

Newson, S.E., Evans, H.E. and Gillings, S. 2015. A novel citizen science approach for large-scale standardised monitoring of bat activity and distribution, evaluated in eastern England. *Biological Conservation* 191: 38–49, https://doi.org/10.1016/j.biocon.2015.06.009.

Nutley, S.M. and Davies, H.T.O. 2000. Getting research into practice: Making a reality of evidence based practice: Some lessons from the diffusion of innovations. *Public Money and Management* 20: 35–42, https://doi.org/10.1111/1467-9302.00234.

Pescott, O.L., Walker, K.J., Pocock, M.J.O., et al. 2015. Ecological monitoring with citizen science: The design and implementation of schemes for plants in Britain and Ireland. *Biological Journal of the Linnean Society* 115: 505–21, https://doi.org/10.1111/bij.12581.

Pimm, S.L., Donohue, I., Montoya, J.M. et al. 2019. Measuring resilience is essential to understand it. *Nature Sustainability* 2: 895–97, https://doi.org/10.1038/s41893-019-0399-7.

Pocock, M.J.O., Tweddle, J.C., Savage, J., et al. 2017. The diversity and evolution of ecological and environmental citizen science. *PLoS ONE* 12: e0172579, https://doi.org/10.1371/journal.pone.0172579.

Reyes-García, V., Fernández-Llamazares, Á., Aumeeruddy-Thomas, Y., et al. 2022. Recognizing Indigenous peoples' and local communities' rights and agency in the post-2020 Biodiversity Agenda. *Ambio* 51: 84–92, https://doi.org/10.1007/s13280-021-01561-7.

Robinson, O.J., Ruiz-Gutierrez, V. and Fink, D. 2018. Correcting for bias in distribution modelling for rare species using citizen science data. *Diversity and Distributions* 24: 460–72, https://doi.org/10.1111/ddi.12698.

Rovira, C., Codina, L. and Lopezosa, C. 2021. Language bias in the Google Scholar ranking algorithm. *Future Internet* 13: Article 31, https://doi.org/10.3390/fi13020031.

Ross, J.B. 2016. Indigenous intergenerational teachings: The transfer of culture, language, and knowledge in an intergenerational summer camp. *American Indian Quarterly* 40: 216–50, https://doi.org/10.5250/amerindiquar.40.3.0216.

Rowley, J.J.L., Callaghan, C.T. and Cutajar, T. 2019. FrogID: Citizen scientists provide validated biodiversity data on frogs of Australia. *Herpetological Conservation and Biology* 14: 155–70, http://www.herpconbio.org/Volume_14/Issue_1/Rowley_etal_2019.pdf.

Sackett, D.L., Straus, S.E., Richardson, W.S., et al. 2000. *Evidence-Based Medicine: How to Practice and Teach EBM,* Second edition (Edinburgh: Churchill Livingstone).

Salafsky, N., Boshoven, J., Burivalova, Z., et al. 2019. Defining and using evidence in conservation practice. *Conservation Science and Practice* 1: e27, https://doi.org/10.1111/csp2.27.

Salafsky, N. and Margoluis, R. 2022. *Pathways to Success: Taking Conservation to Scale in Complex Systems* (Washington: Island Press).

Satō T., Chabay, I. and Helgeson, J. (eds.). 2018. *Transformations of Social-Ecological Systems: Studies in Co-creating Integrated Knowledge Toward Sustainable Futures* (Ecological Research Monographs, Springer), https://doi.org/10.1007/978-981-13-2327-0.

Sauer, J.R., Pardieck, K.L., Ziolkowski, D.J., et al. 2017. The first 50 years of the North American Breeding Bird Survey. *The Condor* 119: 576–93, https://doi.org/10.1650/CONDOR-17-83.1.

Savaresi, A. 2016. Doing the right thing with traditional knowledge in international law: Lessons for the climate regime. BENELEX Working Paper No.8, Edinburgh School of Law Research Paper 2016/16, https://doi.org/10.2139/ssrn.2780332.

Schmeller, D.S., Henry, P.-Y., Julliard, R., et al. 2009. Advantages of volunteer-based biodiversity monitoring in Europe. *Conservation Biology* 23: 307–16, https://doi.org/10.1111/j.1523-1739.2008.01125.x.

Schwartz, S.H. and Bilsky, W. 1987. Toward a universal psychological structure of human values. *Journal of Personality and Social Psychology* 53: 550–62, https://doi.org/10.1037/0022-3514.53.3.550.

Sherman, L.W. 2003. Misleading evidence and evidence-led policy: Making social science more experimental. *The Annals of the American Academy of Political and Social Science* 589: 6–19, https://doi.org/10.1177/0002716203256266.

Steigerwald, E., Ramírez-Castañeda, V., Brandt, D.Y.C., et al. 2022. Overcoming language barriers in academia: Machine translation tools and a vision for a multilingual future. *BioScience* biac062, https://doi.org/10.1093/biosci/biac062.

Stephenson, P.J. and Stengel, C. 2020. An inventory of biodiversity data sources for conservation monitoring. *PLoS ONE* 15: e0242923, https://doi.org/10.1371/journal.pone.0242923.

Stern, B.M. and O'Shea, E.K. 2019. A proposal for the future of scientific publishing in the life sciences. *PLoS Biology* 17: e3000116, https://doi.org/10.1371/journal.pbio.3000116.

Sutherland, W.J., Spiegelhalter, D. and Burgman, M. 2013. Twenty tips for interpreting scientific claims. *Nature* 503: 335–37, https://doi.org/10.1038/503335a.

Sutherland, W.J., Gardner, T.A., Haider, J., et al. 2014. How can local and traditional knowledge be effectively incorporated into international assessments? *Oryx* 48: 1–2, https://doi.org/10.1017/S0030605313001543.

Sutherland, W.J., Taylor, N.G., MacFarlane, D., et al. 2019. Building a tool to overcome barriers in the research-implementation space: the Conservation Evidence database. *Biological Conservation* 238: 108199, https://doi.org/10.1016/j.biocon.2019.108199.

van Swaay, C.A., Dennis, E.B., Schmucki, R., et al. 2019. *The EU Butterfly Indicator for Grassland Species: 1990–2017: Technical Report* (Butterfly Conservation Europe; ABLE/eBMS), https://butterfly-monitoring.net/.

Swanson, A., Kosmala, M., Lintott, C., et al. 2016. A generalized approach for producing, quantifying, and validating citizen science data from wildlife images. *Conservation Biology* 30: 520–31, https://doi.org/10.1111/cobi.12695.

Tengö, M., Hill, R., Malmer, P., et al. 2017. Weaving knowledge systems in IPBES, CBD and beyond—lessons learned for sustainability. *Current Opinion in Environmental Sustainability* 26–27: 17–25, https://doi.org/10.1016/j.cosust.2016.12.005.

Thomas, A. and Eaves, F.F., 3rd. 2015. A brief history of evidence-based medicine (EBM) and the contributions of Dr David Sackett. *Aesthetic Surgery Journal* 35: NP261–63, https://doi.org/10.1093/asj/sjv130.

translatE Project. 2020. List of non-English-language journals in ecology and conservation. Version 1, https://translatesciences.com/resources/#journals.

Veríssimo, D., Pienkowski, T., Arias, M., et al. 2020. Ethical publishing in biodiversity conservation science. *Conservation and Society* 18: 220–25, https://www.biology.ox.ac.uk/publication/1127544/hyrax.

Verschuuren, B., Mallarach, J.M., Bernbaum, E., et al. 2021. *Cultural and Spiritual Significance of Nature: Guidance for Protected and Conserved Area Governance and Management*. Best Practice Protected Area Guidelines Series (Gland, Switzerland: IUCN), https://doi.org/10.2305/IUCN.CH.2021.PAG.32.en.

White, T.B, Petrovan, S.O, Booth, H., et al. 2022b. Determining the economic costs and benefits of conservation actions: A framework to support decision making. *Conservation Science & Practice.* https://doi.org/10.1111/csp2.12840.

White, T.B., Petrovan, S.O., Christie, A.P., et al. 2022a. What is the price of conservation? A review of the status quo and recommendations for improving cost reporting. *BioScience* 72: 461–71, https://doi.org/10.1093/biosci/biac007.

Whyte, K.P., Brewer, J.P. and Johnson, J.T. 2016. Weaving Indigenous science, protocols and sustainability science. *Sustainability Science* 11: 25–32, https://doi.org/10.1007/s11625-015-0296-6.

Young, N., Corriveau, M., Nguyen, V.M., et al. 2016. How do potential knowledge users evaluate new claims about a contested resource? Problems of power and politics in knowledge exchange and mobilization. *Journal of Environmental Management* 184: 380–88, https://doi.org/10.1016/j.jenvman.2016.10.006.

Zulfiqar, S., Wahab, M.F., Sarwar, M.I., et al. 2018. Is machine translation a reliable tool for reading German scientific databases and research articles? *Journal of Chemical Information and Modeling* 58: 2214–23, https://doi.org/10.1021/acs.jcim.8b00534.

Zurba, M., Petriello, M.A., Madge, C., et al. 2022. Learning from knowledge co-production research and practice in the twenty-first century: Global lessons and what they mean for collaborative research in Nunatsiavut. *Sustainability Science* 17: 449–67, https://doi.org/10.1007/s11625-021-00996-x.

3. Assessing Collated and Synthesised Evidence

William J. Sutherland[1,2], Samantha H. Cheng[3,4], Alec P. Christie[1,2,5], Steven J. Cooke[6], Julia Koricheva[7], Nicola Randall[8], Rebecca K. Smith[1,2], Ruth Stewart[9]

Multiple pieces of evidence can be brought together in a variety of ways including systematic maps, subject-wide evidence synthesis, systematic reviews, rapid evidence assessment, meta-analysis and collated open access effect sizes. Each has different uses. This chapter describes each along with suggestions for how to interpret the results and assess the evidence.

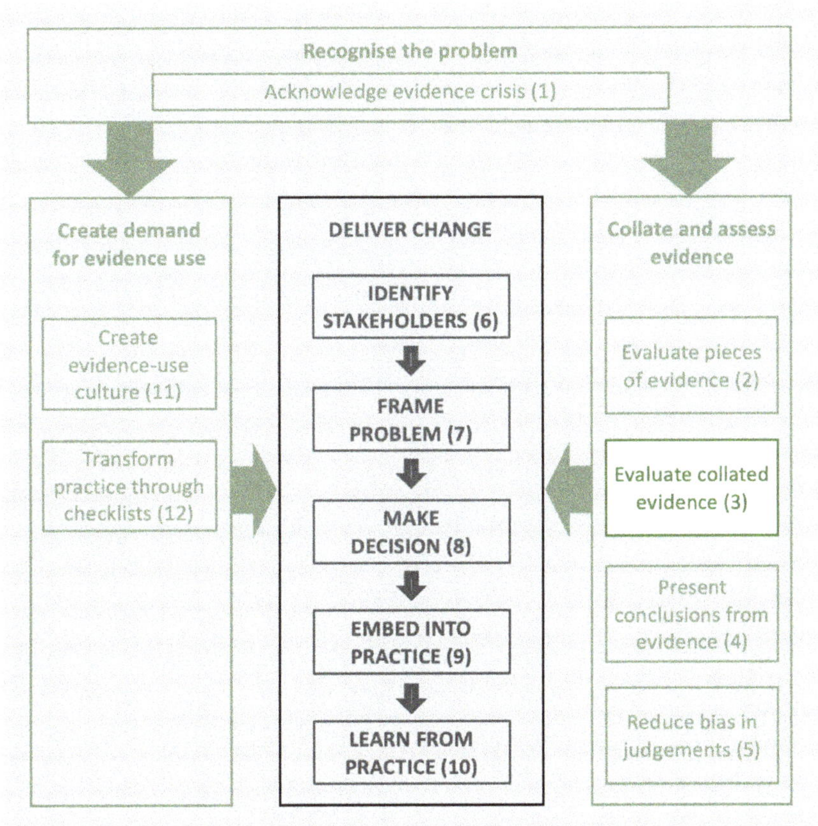

 https://doi.org/10.11647/OBP.0321.03

[1] Conservation Science Group, Department of Zoology, University of Cambridge, The David Attenborough Building, Pembroke Street, Cambridge, UK
[2] Biosecurity Research Initiative at St Catharine's College (BioRISC), St Catharine's College, Cambridge, UK.
[3] Center for Biodiversity and Conservation, American Museum of Natural History, New York, NY 10026, USA
[4] Global Science, World Wildlife Fund, Washington, D.C. 20037, USA
[5] Downing College, Regent Street, Cambridge, UK
[6] Canadian Centre for Evidence-Based Conservation, Institute of Environmental and Interdisciplinary Science, Carleton University, Ottawa, ON, K1S 5B6, Canada
[7] Department of Biological Sciences, Royal Holloway University of London, Egham, Surrey, UK
[8] Centre for Evidence-Based Agriculture, Harper Adams University, Newport, Shropshire, UK
[9] Africa Centre for Evidence, University of Johannesburg, Bunting Road Campus, 2006, Johannesburg, South Africa

Contents

3.1 Collating the Evidence

The previous chapter considered how to assess single pieces of evidence. However, there are often multiple pieces of evidence and these should all be assessed. Much of the progress in evidence-based practice has revolved around means of collating and interpreting all the evidence relating to an issue. A range of approaches has been developed to collate evidence for evidence users to use to inform policy, decisions and practice. We summarise those approaches and provide an assessment of how the various outputs can be used and interpreted.

3.2 Systematic Maps

3.2.1 What is a systematic map?

Systematic maps, sometimes known as evidence maps, or evidence gap maps, are reproducible, transparent, analytical methods to collate and organise an evidence base using a decision-relevant framework. Unlike systematic reviews, systematic maps do not aim to answer questions of effectiveness or direction of impact, rather they are particularly useful to understand the extent and nature of the evidence base on a broad topic area (James et al., 2016). Systematic maps can help describe the distribution of existing evidence, highlighting areas of significant research effort and where key gaps exist. This can provide information to guide research, prioritise evaluation, and illustrate where there may be inadequate information to inform decision making (McKinnon et al., 2015). In recent years, systematic maps (and other types of evidence maps) have grown in popularity across different disciplines. In conservation and development, maps are often published in the journal *Environmental Evidence* and are conducted and published by organisations such as the International Initiative for Impact Evaluations (3ie).

3.2.2 How are systematic maps created?

Given the wide range of potential decisions and users that a systematic map may inform, the scope and framework of a systematic map are usually co-developed with a representative group of stakeholders. This is particularly critical for conservation as the challenges facing natural ecosystems require thinking across disciplines — for example linking climate change, nature conservation, and sustainable development. The evidence base for multi-disciplinary topics is likely to be heterogeneous, particularly in the types of terminology that different sectors use to describe interventions and outcomes. Previous work has shown that semantic variability is quite high even within a single discipline of conservation (Westgate and Lindenmayer, 2017), thus developing an agreed-upon framework of interventions/exposures and outcomes is critical to ensure that the finished product is interpretable and salient to the range of potential end users. For example, frameworks of interventions/exposures and outcomes are often grounded in some type of causal theory that describes how one expects these elements are linked (Cheng et al., 2020). This type of causal grounding is important to help the synthesis team interpret findings based on how plausible a causal relationship might be and thus whether an evidence gap is truly a gap or not.

Systematic maps share many characteristics, and therefore methodological processes, with systematic reviews. As the value of these approaches lies in their transparent and reproducible methods, which explicitly aim to account for potential biases, an *a priori* protocol should be developed that details systematic steps for designing and implementing a search strategy: screening, coding, and analysing included articles. This protocol ideally should be critiqued by those with methodological and topic expertise. While the methods for assembling an evidence base for a systematic map are similar to those for a systematic review, there are a few key differences. Systematic maps do not always include a critical appraisal of included studies and, instead of a formal synthesis of findings from included studies, the outputs may include summary tables, heatmaps and searchable databases or spreadsheets of the included evidence. Figure 3.1 shows the typical stages of a systematic map.

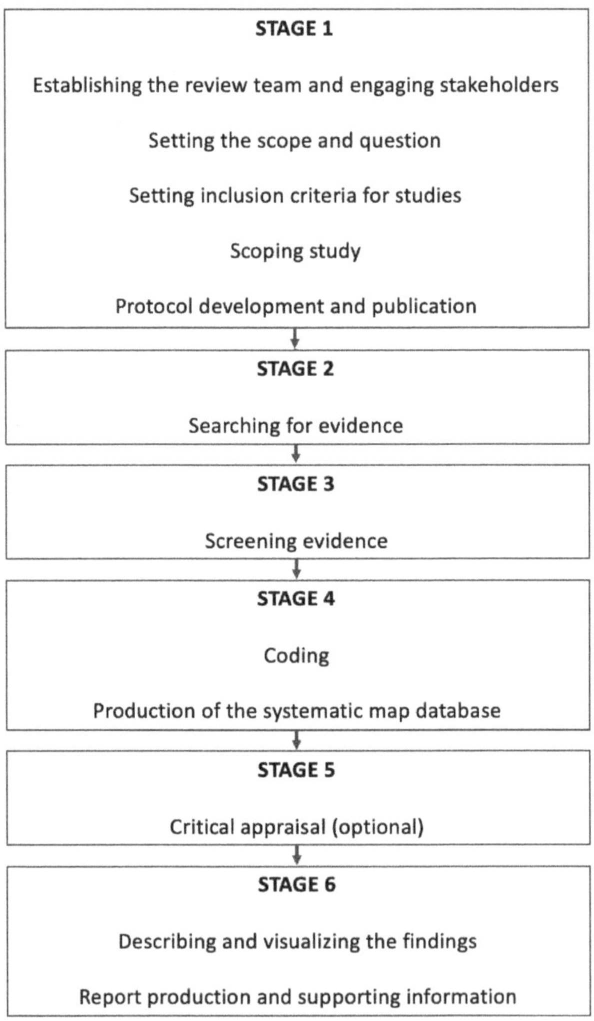

Figure 3.1 Typical stages of a systematic mapping process. (*Source:* James et al., 2016, CC-BY-4.0)

Searching for studies to be included in systematic maps may involve covering a broader range of sources and evidence types than a typical systematic review. For example, a systematic map may seek to characterise the range of existing knowledge on both interventions and outcomes in the framework, along with important contextual factors or conditions. Thus, this type of map may cover a range of various study types, not only studies focused on impacts, in order to capture all concepts within their framework. Some maps may also seek to characterise novel or emerging topic areas — so they may include evidence types such as datasets, citizen surveys, and practitioner knowledge, in addition to more traditional research studies.

3.2.3 How can systematic maps be used?

Systematic maps tend to have multiple dimensions given the wide range of potential queries that end users may have. Studies are typically categorised across the framework and represented as a 'heat map' that illustrates the distribution of studies. These maps are often disaggregated by a number of factors that can help better navigate where evidence exists and where it does not, for example by geography, publication date, demographic variables, and/or ecosystems. These factors can be used to 'slice' the data in the systematic map to look only at a subset of studies, such as those from a specific country or ecosystem. They can also be used to enable the reader of the map to drill down into a cell to find out more about the evidence contained within.

Given the potentially endless combination of factors that users may be interested to slice a map by, many have advocated interactive tools that can improve the accessibility and usability of maps for end users. For example, dedicated interactive websites have emerged featuring data from different systematic maps (e.g. Evidence for Nature and People Data Portal; Evidensia; 3ie). In addition, there is a range of different tools that aimed to help evidence synthesis teams produce interactive diagrams such as EPPI-Mapper (https://eppi.ioe.ac.uk/cms/Default.aspx?tabid=3790) and EviAtlas (Haddaway et al., 2019, https://estech.shinyapps.io/eviatlas/). As with question formulation and framework development, scoping and building an interactive platform should be undertaken in collaboration with a representative group of end users through user-centred design to ensure widespread usability.

Throughout this section, we have emphasised that systematic maps, given their broad nature, are useful for different purposes for different users. Thus, interpreting a systematic map can often be challenging for end users. Interpreting the distribution of the evidence base captured in a systematic map should be informed by individual decision needs and calibrated based on a set of key criteria. First, users should determine what volume of evidence is sufficient to inform their decision. For example, if a systematic map reveals that only a few studies exist for a specific linkage in the framework, an end user should consider how risky it might be to base a decision upon that. Second, users should determine how robust evidence needs to be to inform their decision. For example, if a systematic map reveals a high volume of studies but relatively few appear to meet the level of rigour required for a decision, more robust research may be needed. Third, users should consider how plausible that linkage is. Does it represent a causal relationship that is supported by theory? Ultimately, synthesis teams should collaborate with stakeholders to interpret maps and make clear what types of decisions the interpretations are relevant for and provide guidance for other users to calibrate their interpretations accordingly.

The Collaboration for Environmental Evidence Synthesis Appraisal Tool (CEESAT, https://environmentalevidence.org/ceeder/about-ceesat/) enables users to appraise the rigour, transparency and limitations of methods of existing reviews and includes a checklist designed specifically for evidence overviews such as systematic maps. The checklist (3.1) enables users to categorise each stage of the process reported in a systematic map or other evidence overview as either gold, green, amber or red to help inform users of the level of confidence that they may have in the reported findings.

3.1 Checklists for evidence reviews and systematic maps

From The Collaboration for Environmental Evidence Synthesis Appraisal Tool (CEESAT).

Type of evidence collation	**Systematic reviews/evidence reviews**	**Systematic maps/evidence overviews**
Version, date	Version 2.1, 29th July 2021	Version 2.1, 29th July 2021
The review question	☐ Are the elements of the review question clear?	☐ Are the elements of the review question clear?
The method/protocol	☐ Is there an *a-priori* method/protocol document?	☐ Is there an *a-priori* method/protocol document?
Searching for studies	☐ Is the approach to searching clearly defined, systematic and transparent? ☐ Is the search comprehensive?	☐ Is the approach to searching clearly defined, systematic and transparent? ☐ Is the search conducted in line with the defined search scope?

Type of evidence collation	Systematic reviews/evidence reviews	Systematic maps/evidence overviews
Including studies	☐ Are eligibility criteria clearly defined? ☐ Are eligibility criteria consistently applied to all potentially relevant articles and studies found during the search? ☐ Are eligibility decisions transparently reported?	☐ Are eligibility criteria clearly defined? ☐ Are eligibility criteria consistently applied to all potentially relevant articles and studies found during the search? ☐ Are eligibility decisions transparently reported?
Critical appraisal (not assessed for overviews)	☐ Does the review critically appraise each study? ☐ During critical appraisal was an effort made to minimise subjectivity?	
Data extraction/data coding	☐ Is the method of data extraction fully documented? ☐ Are the extracted data reported for each study? ☐ Were extracted data cross checked by more than one reviewer?	☐ Is the method of data coding fully documented? ☐ Are the coded data reported for each study? ☐ Were coded data cross checked by more than one reviewer?

Type of evidence collation	Systematic reviews/evidence reviews	Systematic maps/evidence overviews
Data synthesis (not assessed for overviews)	☐ Is the choice of synthesis approach appropriate? ☐ Is a statistical estimate of pooled effect (or similar) provided together with a measure of variance and heterogeneity among studies? ☐ Is variability in the study findings investigated and discussed?	
Review limitations	☐ Have the authors considered the limitations of the synthesis?	☐ Have the authors considered the limitations of the synthesis?

(*Source*: https://environmentalevidence.org/ceeder/about-ceesat)

These checklists can be downloaded from https://doi.org/10.11647/OBP.0321#resources. They can be modified and tailored for specific uses.

A database of systematic maps that have had CEESAT criteria independently applied can be viewed on the CEEDER website (https://environmentalevidence.org/ceeder-search). Eligible evidence reviews are rated by a pool of CEEDER Review College members who rate their reliability using CEESAT criteria. Several Review College members apply CEESAT for each evidence overview and disagreements in ratings are resolved by an editorial team.

3.3 Subject-Wide Evidence Syntheses

Subject-wide evidence synthesis was created as a methodology to collate evidence and make it easily accessible to decision makers (Sutherland et al., 2019). It enables the rapid synthesis of evidence across entire subject areas (comprising tens or hundreds of related review questions), whilst being transparent, objective and minimising bias. End users (practitioners, policy-makers and researchers) are involved in the process to ensure applicability and to encourage uptake.

Using the process of subject-wide evidence synthesis, Conservation Evidence (www.conservationevidence.com) provides a freely accessible, plain-English database, which contains evidence for the effects of conservation interventions (i.e. actions that have been or could be used to conserve biodiversity).

One of the first stages of subject-wide evidence synthesis is solution scanning (Section 7.6.1) to produce a comprehensive list of all actions that have been tried or suggested for the subject of the synthesis and that could realistically be implemented. In Conservation Evidence this list is developed in collaboration with an advisory board and is structured using IUCN Threat Categories and Action Categories (https://www.iucnredlist.org/resources/classification-schemes); systematic searches of the literature (e.g. so far 650 academic journals and 25 report series from organisational websites have been searched by Conservation Evidence) are undertaken and relevant studies that test the effectiveness of an action are summarised in a short standardised paragraph in plain English.

The broad diversity of output measures, such as species abundance, species diversity, behaviour, breeding performance, etc., however, represents a serious challenge hence formal analysis, such as meta-analysis, can often be impossible. To overcome this issue, Conservation Evidence uses a panel of experts who assess each action in terms of effectiveness, certainty (strength of the evidence), and harms to the subject under consideration (Sutherland et al., 2021). A modified Delphi Technique (see Section 5.5.1.) is used with at least two rounds of anonymous scoring. Figure 3.2 shows how the resulting median scores are converted into categories of effectiveness.

The main outputs of subject-wide evidence synthesis undertaken by Conservation Evidence are:

- Database of studies testing actions: By August 2022, over 8,400 studies had been summarised. Study summaries, although brief, aim to include sufficient detail of the study context and methods to allow users to begin assessing reliability and relevance to their own system (e.g. species, location, implementation method).
- Database of actions with summaries of studies testing them: By August 2022, over 3,650 actions for the conservation of habitats, species groups, and other conservation issues had been assessed. For each action, background information is provided where necessary to place in context or to refer to additional information other than tests of effectiveness. Key messages provide a brief overview or index to the studies summarised; a location map of studies is also provided. The overall 'category of effectiveness' for each action is generated through an assessment of the summarised evidence by an expert panel (academics, practitioners and policy-makers; Figure 3.2; Sutherland et al., 2021).

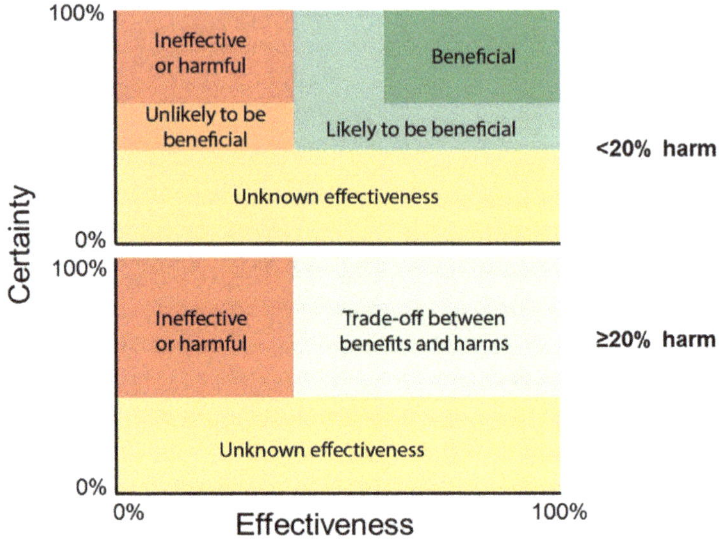

Figure 3.2 Categories of effectiveness based on a combination of effectiveness (the extent of the benefit and harm) and certainty (the strength of the evidence). The top graph refers to interventions with harms scored <20% and the bottom graph to interventions with harms ≥20%. (*Source:* Sutherland et al. 2021, CC-BY-4.0)

- Database of titles of non-English language studies testing actions: By August 2022, studies had been collated by scanning 419,679 paper titles from 330 journals in 16 different languages.
- Synopses: All of the actions relevant to a specific subject are grouped into a subject 'synopsis'. By August 2022, evidence for 24 different taxa or habitats had been collated including mammal, bird, forest, and peatland conservation.
- *What Works in Conservation*: An annual update of the information on the effectiveness of actions is produced as a book, *What Works in Conservation*. All the information within each update is also presented on the website.

The Conservation Evidence databases can be used for a range of issues, for example:

- *Determining whether a specific action is effective:* e.g. are bat bridges effective? The database can be used to provide an indication of the comparative effectiveness of different actions that could be taken (and where evidence is lacking). However, it is important to look at the evidence for each of the most relevant actions in more detail to determine how relevant it is to the user's system before making decisions (e.g. species studied, location, implementation methods tested, etc.).
- *Identifying actions to mitigate against a specific threat*: e.g. how could road deaths for large mammals be reduced? Refining a search on the Actions page (https://

www.conservationevidence.com/data/index) by 'Category: Terrestrial Mammal Conservation', and 'Threat: Transportation & service corridors' gives 39 actions with associated evidence. Relevant actions can also be found by using keywords — search terms should be kept simple and variants tried.

- *Identifying possible actions for a specific habitat or taxa:* e.g. peat bogs. There are 125 actions for managing peat bogs, with summaries of studies testing them.
- *Determining what actions have been tested for a specific species.*
- *As a source of literature in 17 languages for a literature review or systematic review.*
- *Identifying studies that relate to a specific habitat, taxa or country:* e.g. refining a search by 'Habitat: Deserts' indicates that there are currently 106 studies in deserts, similarly there are 13 studies on horseshoe bats and 14 in Zimbabwe. This might be useful if starting work on a given habitat, taxa or region.

3.4 Systematic Reviews

Systematic review entails using repeatable analytical methods to extract secondary data and analyse it (Pullin et al., 2020) and can also be carried out for qualitative studies (Flemming et al., 2019). In conservation, these are listed in the CCE library (https://environmentalevidence.org).

A systematic review typically involves the following eight stages (Collaboration for Environmental Evidence, 2018):

1. Planning a synthesis
2. Developing a protocol
3. Conducting a search
4. Eligibility screening
5. Data coding and data extraction
6. Critical appraisal of study validity
7. Data synthesis
8. Interpreting findings and reporting conduct

ROSES (RepOrting standards for Systematic Evidence Syntheses) provides forms that can be used during the preparation of systematic review and map protocols and final reports. The checklists can be downloaded from https://www.roses-reporting.com.

Although systematic reviews are designed to be more robust than traditional reviews, the methods used and the question addressed within a review may impact the confidence that can be placed in the findings of any review. The clarity of the question, the methods used for collating evidence, decisions around critical appraisal and synthesis of included studies, and transparency of reporting can all influence the interpretation of review findings.

The CEESAT checklist (3.1) has been developed to assess the rigour of each stage of any review (systematic and otherwise) in the field of environmental management. The tool can be used by users of reviews to assess the methods used, the transparency of reporting of a review, and the likely limitations that there may be. Criteria for assessing each element of a review are given in Woodcock et al. (2014); also see Section 3.2. A database of evidence reviews that have already had CEESAT criteria independently applied by a group of experts is available for reviews from 2018 onwards on the CEEDER website (https://environmentalevidence.org/ceeder-search) and can be easily searched by subtopic area.

Other questions should be asked when applying the results of a review. For example, ask when was the literature searched and does the gap in the literature matter, is there a geographical constraint to the review and how does that match the area of interest, how does the subject match the question being asked?

3.5 Rapid Evidence Assessments

Rapid evidence assessments (also sometimes described using similar terms such as rapid reviews, and rapid systematic reviews) are quick reviews of the evidence when resources are limited or the topic is urgent. Rapid evidence assessments may take a variety of forms, but typically follow the same processes as systematic reviews or systematic maps, with stages omitted or abbreviated (often the *level* of searching, and/or some of the quality checking stages). The abbreviated stages can vary between rapid reviews, since formal guidance for creation and quality appraisal of rapid evidence assessments is less established than for other evidence synthesis methods. Hamel et al. (2021) took definitions from 216 rapid reviews and 90 rapid review methods articles and found that no consensus existed in defining rapid reviews. There were also variations in the way in which each review was streamlined. Collins et al. (2015) outline guidance for creators of rapid reviews in the field of land and water management, but they separate this into processes that are aligned to systematic reviews (which they call rapid evidence assessments) and those that are more similar to systematic maps (which they call quick scoping reviews). The variability in rapid evidence assessment methods may make it difficult for users to assess the reliability and applicability of rapid reviews for their needs, but transparent reporting that outlines all stages of the methodology may help users evaluate individual rapid evidence assessments. The CEESAT checklist (3.1) may aid users in appraising rapid evidence assessments. The PROCEED open-access registry of titles and protocols for prospective evidence syntheses in the environmental sector https://www.proceedevidence.info/ allows for registration of rapid review protocols. The required structure for prospective authors to follow is based on standard systematic review structures supported by the Collaboration for Environmental Evidence, and although no guidance is provided for streamlining processes, the use of standardised formats such as this may help authors with reporting.

3.6 Meta-Analyses

Meta-analysis is a set of statistical methods for combining the magnitude of outcomes (effect sizes) across different studies addressing the same research question. The methods of meta-analysis were originally developed in medicine and social sciences (Glass et al., 1981) and then introduced in ecology in the early 1990s (Arnqvist and Wooster, 1995). Fernandez-Duque and Valeggia (1994) brought meta-analysis to the attention of conservation biologists and outlined several advantages of meta-analysis over narrative reviews. In particular, meta-analysis allows better control of type II errors (e.g. assuming that a particular human action has no effect when it in fact does), which can have more serious consequences for conservation than making a type I error (e.g. assuming that an action has an effect when it really does not). As a small sample size is a common limitation in conservation studies, many individual primary studies have low statistical power and might fail to demonstrate the effect even when it exists. Fernandez-Duque and Valeggia (1994) have demonstrated that meta-analysis enables demonstration of the overall effect even when this effect was not apparent from the individual studies included in the meta-analysis, thus increasing the power of primary studies.

Nowadays, meta-analyses are often conducted as part of systematic reviews in conservation biology (as in step 7 — data synthesis, see Section 3.4). One of the main applications of meta-analysis in conservation biology is to assess the effectiveness of management interventions. This is achieved by combining weighted effect sizes, with more weight given to some studies than others, from individual primary studies to calculate the overall effect and its confidence interval (CI). The sign (positive/negative effect), the magnitude, and the significance of the overall effect can then be assessed. The results of meta-analyses often challenge the conventional wisdom about management effectiveness (Stewart, 2010; Côté et al., 2013). Meta-analysis also allows an assessment of the degree of heterogeneity in effect sizes, revealing the factors causing variation in effect among studies. Such distinction is important for establishing the conditions under which the management interventions are effective.

As an example, to assess whether tree retention at harvest helps to mitigate negative impacts on biodiversity, Fedrowitz et al. (2014) conducted a meta-analysis of 78 primary studies comparing species richness and abundance between retention cuts and either clearcuts or unharvested forests. They found that overall species richness was higher in retention cuts than in clearcuts and unharvested forests. However, effects varied between different species groups (Figure 3.3). Retention cuts supported higher species richness and abundance of forest species compared to clearcuts, but lower species richness and abundance of open-habitat species. In contrast, species richness and abundance of forest specialists were lower in retention cuts than in unharvested forest while species richness and abundance of open-habitat species were higher than in unharvested forests. Species richness and abundance of generalists did not differ between retention cuts and clearcuts. The results support the use of retention forestry since it moderates negative harvesting impacts on biodiversity. However, retention forestry cannot substitute conservation actions targeting certain highly specialised species associated with forest-interior or open-habitat conditions.

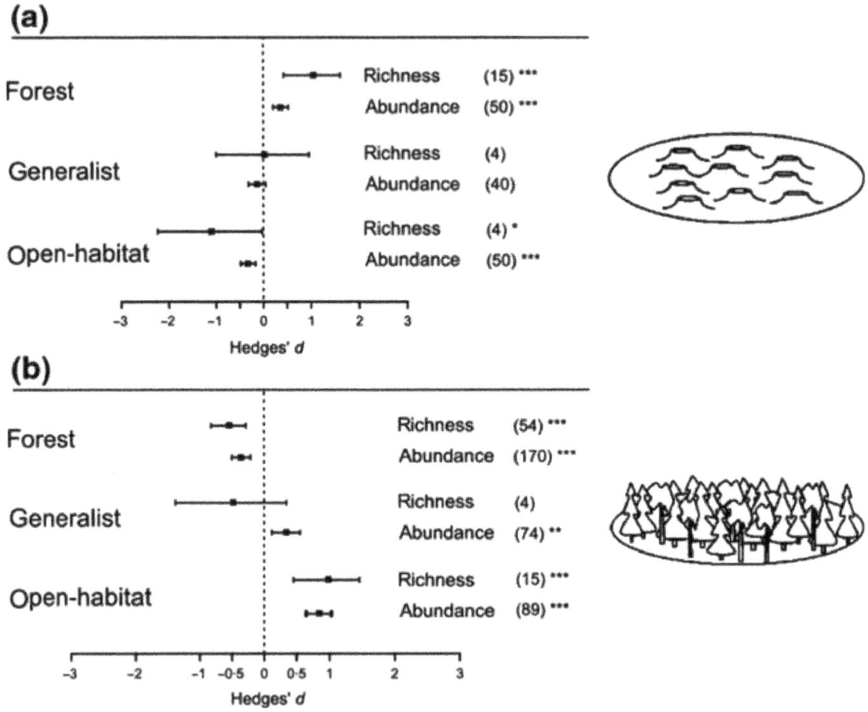

Figure 3.3 Effects of retention cuts (mean effect size 95% CI) on species richness and abundance of forest, generalist and open-habitat species when using (a) clearcut, or (b) unharvested forest as the control. Number of observations are stated in brackets. Effect size measure is a standardised mean difference (Hedges' d) between species richness and abundance in retention cuts and clearcut (a) or unharvested forest (b). Positive effects indicate higher species richness and abundance on retention cuts whereas negative effects indicate lower species richness and abundance on retention cuts. Effects are not significantly different from 0 when 95% CIs include 0. For significant effects, P-values are shown as *p < 0.05, **p < 0.01, ***p < 0.001. (*Source:* Fedrowitz et al., 2014, CC-BY-NC-3.0 https://creativecommons.org/licenses/by-nc/3.0/)

While the popularity of meta-analyses in ecology and conservation biology has grown considerably over the last two decades, repeated concerns have been raised about the quality of the published meta-analyses in these fields (Gates, 2002; Vetter et al., 2013; Koricheva and Gurevitch, 2014; O'Dea et al., 2021). Of particular relevance to policy and decision making in conservation, it has been argued that rapid temporal changes in magnitude, statistical significance and even the sign of the effect sizes reported in many ecological meta-analyses represent a real threat to policy making in conservation and environmental management (Koricheva and Kulinskaya, 2019). To improve the quality of meta-analyses in ecology and evolutionary biology, Koricheva and Gurevitch (2014) developed a checklist of quality criteria for meta-analysis (Checklist 3.1). A more extensive checklist covering both systematic review and meta-analysis criteria has been recently developed by O'Dea et al. (2021) as an extension of the Preferred Reporting Items for Systematic reviews and Meta-Analyses (PRISMA) to ecology and evolutionary biology (PRISMA-EcoEvo).

3.2 Checklist of quality criteria for meta-analyses

- ☐ Has formal meta-analysis been conducted (i.e. combination of effect sizes using standard meta-analysis methodology) or is it simply a vote count (comparison of % of statistically significant and nonsignificant findings)?
- ☐ Have standard metrics of effect size been used (e.g. standardised mean difference, correlation coefficient, response ratio, etc.) or, if non-standard metrics have been employed, is the distribution of these parameters known and have the authors explained how they calculated variances for such metrics?
- ☐ If more than one estimate of effect size per study was included in the analysis, has the potential non-independence of these estimates been taken into account?
- ☐ Have effect sizes been weighted by study precision or has the rationale for using an unweighted approach been provided?
- ☐ Has the statistical model for meta-analysis and the software used been described?
- ☐ Has heterogeneity of effect sizes between studies been quantified?
- ☐ Have the causes of existent heterogeneity in effect sizes been explored by meta-regression?
- ☐ If the effects of multiple moderators have been tested, have potential non-independence of and interactions between moderators been taken into account?
- ☐ If meta-analysis combines studies conducted on different species, has the phylogenetic relatedness of species been taken into account?
- ☐ Have tests of publication bias been conducted?
- ☐ If any meta-analysis combined studies published over a considerable time span then have possible temporal changes in effect size been tested?
- ☐ Has sensitivity analysis been performed to test the robustness of the results?
- ☐ Has the dataset used for meta-analysis, including effect sizes and variances/ sample sizes from individual primary studies and moderator variables, been provided in an electronic appendix?

(*Source:* Koricheva and Gurevitch, 2014)

This checklist can be downloaded from https://doi.org/10.11647/OBP.0321#resources. It can be modified and tailored for specific uses.

3.7 Open Access Effect Sizes

Metadataset (https://www.metadataset.com/), a database of effect sizes for actions, has been created, which, so far, contains over 15,000 effect sizes on invasive species control, cassava farming and aspects of Mediterranean agriculture, along with a suite of analytical tools (Shackleford et al., 2021). Metadataset enables interactive evidence synthesis, browsing publications by intervention, outcome, or country (using interactive evidence maps). It also allows filtering and weighing the evidence for specific options and then recalculating the results using only the relevant studies, a method known as 'dynamic meta-analysis' (Shackleford et al., 2021).

The database was created following a systematic review (Martin et al., 2020; see Section 3.4), systematic map (Shackleford et al., 2018; see Section 3.2) and subject-wide evidence synthesis (Shackleford et al., 2017; see Section 3.3) methodologies. Metadata was extracted from all studies that met each selection criteria. This included the mean values of treatments (e.g. plots with management interventions) and controls (e.g. plots without management interventions or which used alternative management interventions) and, if available, measures of variability around the mean (standard deviation, variance, standard error of the mean, or confidence intervals), number of replicates, and the P value of the comparison between treatments and controls.

Metadataset enables both narrative synthesis, in which the studies that met the inclusion criteria are described, and quantitative synthesis of the relevant results. The mean effects of each intervention on each outcome were calculated using standard meta-analytic methods (Borenstein et al., 2011; Martin et al., 2020). With the dynamic meta-analysis, users can filter the data to define a subset relevant to their situation and then the results for that subset are calculated using subgroup analysis and/or meta-regression (Shackleford et al., 2021; see Section 3.6 for further details). Figure 3.4 illustrates how Metadataset operates.

Individuals or organisations want to know whether an action will be effective for a specific option, for example for a particular species, but evidence summaries often provide general conclusions, for example for a group of species (Martin et al., 2022). Dynamic synthesis (Shackleford et al., 2021) was designed to help overcome this problem by enabling a process of creating an analysis tailored to the specific option, for example just considering studies within the user's region.

Dynamic meta-analysis enables analysis only using the subset selected by the user (subgroup analysis) or by analysing all of the data but calculating different results for different subsets while accounting for the effects of other variables (meta-regression). Dynamic meta-analysis also includes 'recalibration', a method of weighting studies based on their relevance, allowing users to consider a wider range of evidence, not just data that is completely relevant (as none may exist; Shackleford et al., 2021). Some 'critical appraisal' (i.e. deciding which studies should be included in the meta-analysis, based on study quality) and 'sensitivity analysis' (i.e. permuting the assumptions of a meta-analysis, to test the robustness of the results) is also possible. For example, where evidence was limited, a user could decide to include further relevant, but lower quality, studies.

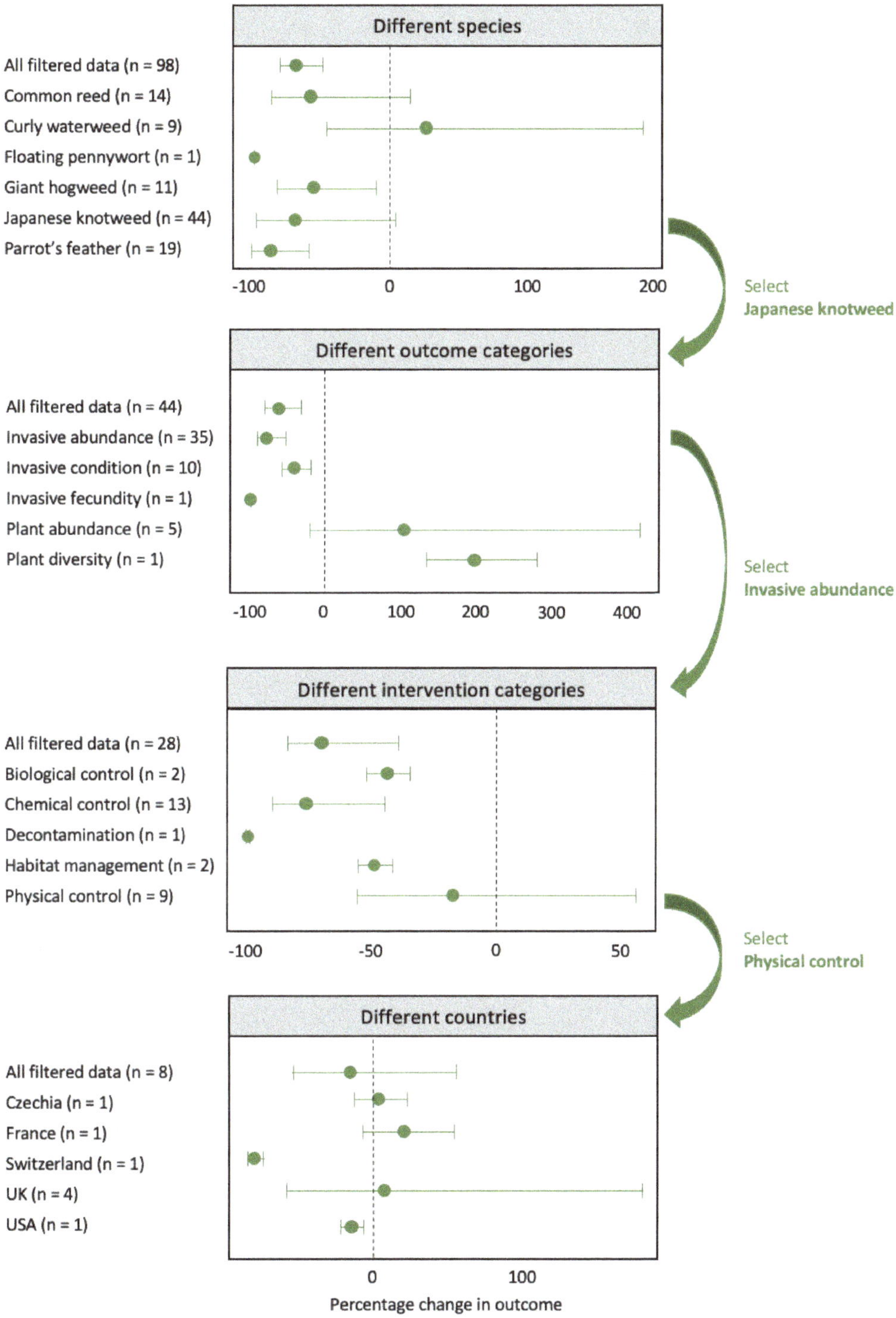

Figure 3.4 The process of adjusting analyses using Metadataset. From top shows the effect of all actions on all outcomes for all 17 invasive species and the six with the most data; Japanese knotweed is selected. Then shows the effect of all six most data rich outcome categories on Japanese knotweed; plant abundance is selected. Then shows the effect of all six actions on Japanese knotweed outcomes; physical control is selected. Finally shows the overall effect and the effect in five countries of physical control on Japanese knotweed. Number of effect sizes (n). (*Source:* authors)

3.8 Overviews of Reviews

Overviews of reviews (or reviews of reviews) use systematic methodologies to search for and identify systematic reviews within a topic area. The principles are similar to those used in traditional systematic reviews and maps, but instead of primary research, reviewers search for and identify existing systematic reviews, extracting, sometimes re-analysing, and reporting on their combined data.

In disciplines where multiple systematic reviews may be available within a topic area, overviews of reviews are recognised as useful methods of synthesis. For example, in the Cochrane handbook they are recommended for addressing research questions that are broader in scope than those examined in individual systematic reviews, for saving time and resources in areas where systematic reviews have already been conducted, and where it is important to understand any diversity present in the existing systematic review literature (Pollock et al., 2002). Overviews of reviews can also be used as part of multi-stage approaches to a question. For example, in the field of international development, Rebelo de Silva et al. (2017) carried out an overview of reviews to investigate the systematic review evidence base for the effects of interventions for smallholder farmers in Africa on various food security outcomes. This was used to inform a systematic map to assess the gaps and overlaps in a more focused sub-topic area, which was then used to inform a systematic review investigating the effects of specific interventions.

In the field of conservation, as the number of systematic reviews increases, overviews of reviews may become a useful tool for the future.

References

Arnqvist, G. and Wooster, D. 1995. Meta-analysis: Synthesizing research findings in ecology and evolution. *Trends in Ecology & Evolution* 10: 236–40, https://doi.org/10.1016/S0169-5347(00)89073-4.

Borenstein, M., Hedges, L.V., Higgins, J.P., et al. 2011. *Introduction to Meta-Analysis* (Chichester: John Wiley and Sons, Ltd), https://doi.org/10.1002/9780470743386.

Cheng, S.H., McKinnon, M.C., Masuda, Y.J., et al. 2020. Strengthen causal models for better conservation outcomes for human well-being. *PLoS ONE* 15: e0230495, https://doi.org/10.1371/journal.pone.0230495.

Collaboration for Environmental Evidence. 2018. *Guidelines and Standards for Evidence synthesis in Environmental Management. Version 5.0*, ed. by A.S. Pullin, et al., https://environmentalevidence.org/information-for-authors/.

Collins, A., Coughlin, D., Miller, J., et al. 2015. *The Production of Quick Scoping Reviews and Rapid Evidence Assessments: A How to Guide* (UK Defra; Joint Water Evidence Group), https://nora.nerc.ac.uk/id/eprint/512448/1/N512448CR.pdf.

Côté, I.M., Stewart, G.B. and Koricheva, J. 2013. Contributions of meta-analysis to conservation and management. In: *Handbook of Meta-Analysis in Ecology and Evolution*, ed. by J. Koricheva, et al. (Princeton: Princeton University Press), pp. 420–25, https://doi.org/10.23943/princeton/9780691137285.003.0026.

Fedrowitz, K., Koricheva, J., Baker, S.C., et al. 2014. Can retention forestry help conserve biodiversity? A meta-analysis. *Journal of Applied Ecology* 51: 1669–79, https://doi.org/10.1111/1365-2664.12289.

Fernandez-Duque, E. and Valeggia, C. 1994. Meta-analysis: A valuable tool in conservation. *Conservation Biology* 8: 555–61, https://doi.org/10.1046/j.1523-1739.1994.08020555.x.

Flemming, K., Booth, A. Garside, R., et al. 2019. Qualitative evidence synthesis for complex interventions and guideline development: Clarification of the purpose, designs and relevant methods. *BMJ Global Health* 4: e000882, https://doi.org/10.1136/bmjgh-2018-000882.

Gates, S. 2002. Review of methodology of quantitative reviews using meta-analysis in ecology. *Journal of Animal Ecology* 71: 547–57, https://doi.org/10.1046/j.1365–2656.2002.00634.x

Glass, G.V., McGaw, B. and Smith, M.L. 1981. *Meta-Analysis in Social Research* (Beverly Hills, CA: SAGE Publications).

Haddaway, N.R., Feierman, A., Grainger, M.J., et al. 2019. *EviAtlas*: A tool for visualising evidence synthesis databases. *Environmental Evidence* 8: Article 22, https://doi.org/10.1186/s13750-019-0167-1.

Hamel, C., Michaud, A., Thuku, M., et al. 2021. Defining rapid reviews: A systematic scoping review and thematic analysis of definitions and defining characteristics of rapid reviews. *Journal of Clinical Epidemiology* 129: 74–85, https://doi.org/10.1016/j.jclinepi.2020.09.041.

James, K.L., Randall, N.P. and Haddaway, N.R. 2016. A methodology for systematic mapping in environmental sciences. *Environmental Evidence* 5: Article 7, https://doi.org/10.1186/s13750-016-0059-6.

Koricheva, J. and Gurevitch, J. 2014. Uses and misuses of meta-analysis in plant ecology. *Journal of Ecology* 102: 828–44, https://doi.org/10.1111/1365-2745.12224.

Koricheva, J. and Kulinskaya, E. 2019. Temporal instability of evidence base: A threat to policy making? *Trends in Ecology & Evolution* 34: 895–902, https://doi.org/10.1016/j.tree.2019.05.006.

Martin, P.A., Shackelford, G.E., Bullock, J.M., et al. 2020. Management of UK priority invasive alien plants: A systematic review protocol. *Environmental Evidence* 9: Article 1, https://doi.org/10.1186/s13750-020-0186-y.

Martin, P.A., Christie, A.P., Shackelford, G.E., et al. 2022. Flexible synthesis to deliver relevant evidence for decision-making. Preprint, https://doi.org/10.31219/osf.io/3t9jy.

McKinnon, M., Cheng, S., Garside, R., et al. 2015. Sustainability: Map the evidence. *Nature* 528: 185–87, https://doi.org/10.1038/528185a.

O'Dea, R.E., Lagisz, M., Jennions, M.D., et al. 2021. Preferred reporting items for systematic reviews and meta-analyses in ecology and evolutionary biology: A PRISMA extension. *Biological Reviews* 96: 1695–1722, https://doi.org/10.1111/brv.12721.

Pollock, M., Fernandes, R.M., Becker, L.A., et al. 2022. Chapter V: Overviews of Reviews. In: *Cochrane Handbook for Systematic Reviews of Interventions, version 6.3* (updated February 2022), ed. by J.P.T. Higgins, et al. (Cochrane, 2022), www.training.cochrane.org/handbook.

Pullin, A., Cheng, S., Cooke, S., et al. 2020. Informing conservation decisions through evidence synthesis and communication. In: *Conservation Research, Policy and Practice*, Ecological Reviews, ed. by W.J. Sutherland, et al. (Cambridge: Cambridge University Press), pp. 114–28, https://doi.org/10.1017/9781108638210.007.

Rebelo Da Silva, N., Zaranyika, H., Langer, L., et al. 2017. Making the most of what we already know: A three-stage approach to systematic reviewing. *Evaluation Review* 41: 155–72, https://doi.org/10.1177/0193841X16666363.

Shackelford, G.E., Haddaway, M.R., Usieta, H.O., et al. 2018. Cassava farming practices and their agricultural and environmental outcomes: A systematic map protocol. *Environmental Evidence* 7: Article 30, https://doi.org/10.1186/s13750-018-0142-2.

Shackelford, G.E., Kelsey, R., Robertson, R.J., et al. 2017. *Sustainable Agriculture in California and Mediterranean Climates: Evidence for the Effects of Selected Interventions*, Synopses of Conservation Evidence Series (Cambridge: University of Cambridge), https://www.conservationevidence.com/synopsis/pdf/15.

Shackelford, G.E., Martin, P.A., Hood, A.S.C., et al. 2021. Dynamic meta-analysis: A method of using global evidence for local decision making. *BMC Biology* 19: Article 33, https://doi.org/10.1186/s12915-021-00974-w.

Stewart, G. 2010. Meta-analysis in applied ecology. *Biology Letters* 6: 78–81, https://doi.org/10.1098/rsbl.2009.0546.

Sutherland, W.J., Dicks, L.V., Petrovan, S.O., et al. 2021. *What Works in Conservation 2021* (Cambridge: Open Book Publishers), https://doi.org/10.11647/OBP.0267.

Sutherland, W.J., Taylor, N.G., MacFarlane, D., et al. 2019. Building a tool to overcome barriers in the research-implementation space: The Conservation Evidence database. *Biological Conservation* 238: 108199, https://doi.org/10.1016/j.biocon.2019.108199.

Vetter, D., Rücker, G. and Storch, I. 2013. Meta-analysis: A need for well-defined usage in ecology and conservation biology. *Ecosphere* 4: Article 74, https://doi.org/10.1890/ES13-00062.1.

Westgate, M.J. and Lindenmayer, D.B. 2017. The difficulties of systematic reviews. *Conservation Biology* 31: 1002–07, https://doi.org/10.1111/cobi.12890.

Woodcock, P., Pullin, A.S. and Kaiser, M.J. 2014. Evaluating and improving the reliability of evidence syntheses in conservation and environmental science: A methodology. *Biological Conservation* 176: 54–62, https://doi.org/10.1016/j.biocon.2014.04.020.

4. Presenting Conclusions from Assessed Evidence

William J. Sutherland[1,2], Tatsuya Amano[3,4], Nicolas Boenisch[5], Samantha H. Cheng[6,7], Alec P. Christie[1,2,8], H. Charles J. Godfray[9], Matthew J. Grainger[10], Angela R. McLean[9], William H. Morgan[1], Nancy Ockendon[11], Nicola P. Randall[12], Ullrika Sahlin[13], Thomas B. White[1]

Applying evidence builds on the conclusions of the assessment of the evidence. The aim of the chapter is to describe a range of ways of summarising and visualising different types of evidence so that it can be used in various decision-making processes. Evidence can also be presented as part of evidence capture sheets, argument maps, mind maps, theories of change, Bayesian networks or evidence restatements.

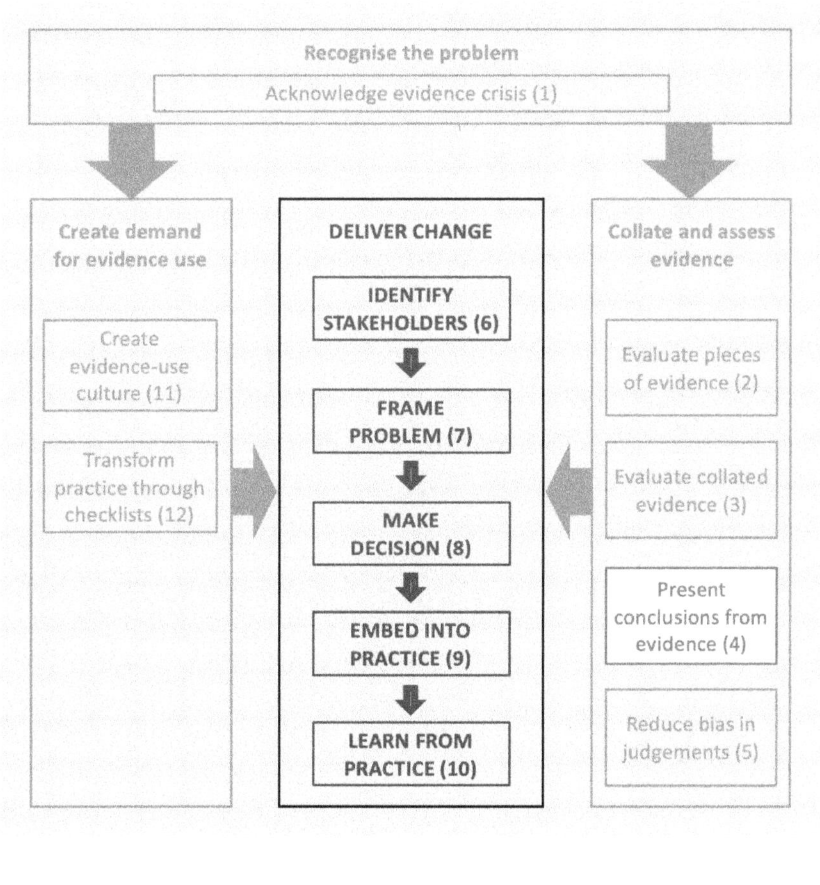

 https://doi.org/10.11647/OBP.0321.04

[1] Conservation Science Group, Department of Zoology, University of Cambridge, The David Attenborough Building, Pembroke Street, Cambridge, UK.
[2] Biosecurity Research Initiative at St Catharine's College (BioRISC), St Catharine's College, Cambridge, UK.
[3] School of Biological Sciences, University of Queensland, Brisbane, 4072 Queensland, Australia
[4] Centre for Biodiversity and Conservation Science, The University of Queensland, Brisbane, 4072 Queensland, Australia
[5] Foundations of Success Europe, 7211 AA Eefde, The Netherlands
[6] Center for Biodiversity and Conservation, American Museum of Natural History, New York, NY 10026, USA
[7] Global Science, World Wildlife Fund, Washington, D.C., USA
[8] Downing College, Regent Street, Cambridge, UK
[9] Oxford Martin School, University of Oxford, Oxford, UK
[10] Norwegian Institute for Nature Research (NINA), PO Box 5685 Torgarden, 7485 Trondheim, Norway
[11] Endangered Landscapes Programme, Cambridge Conservation Initiative, David Attenborough Building, Pembroke Street, Cambridge, UK
[12] Centre for Evidence-based Agriculture, Harper Adams University, Newport, Shropshire, UK
[13] Centre for Environmental and Climate Science, Lund University, Lund, Sweden

Contents

What does the evidence say? This is a common question and one that is critical to the process of decision-making (Chapter 8) and applying evidence, whether writing plans or deciding upon policies (Chapter 9). The aim of this chapter is to present a range of means of presenting evidence that gives a fair summary whilst making clear any limitations or biases.

This chapter considers the general principles, reviews the various elements that are involved in finding and presenting evidence, describes some of the main types of evidence and how these might be presented, and describes some approaches for presenting multiple pieces of evidence.

4.1 Principles for Presenting Evidence

In applying evidence, it is important to understand the conclusions drawn and the likelihood that those conclusions are accurate and relevant. The core general principles for presenting evidence (see Box 4.1) are designed to reduce the risk that the evidence is misapplied due to the conclusions being presented incorrectly; the limitations not being made clear; or because the conditions are very different.

In practice, evidence is presented in a range of formats that differ in the detail provided. In some cases, such as reports and assessments, the key evidence is likely to be well documented with details of the source and assessment of quality. However, in practice, much evidence is provided by experts. In this case, the responsibility is on the expert to ensure that if caveats are needed then they are provided (for example, 'although the evidence is weak it seems...', 'the evidence is conflicting but my interpretation is...', 'the only studies are on lizards but it seems likely that snakes would respond the same way'). Thus if a responsible evidence-based expert provides no caveat then it should be reasonable to assume the expert holds no such concerns. In practice, it is worth checking if the expert does this and also checking the details for key assumptions.

Box 4.1 General principles for presenting evidence

1. The means by which the evidence was obtained is clear.
2. The type and source of evidence is clear.
3. The manner in which it is communicated is tailored to the intended audience.
4. The presentation and interpretation of the evidence is, as far as possible, neutral and balanced.
5. Any biases or limitations are made clear.
6. Conflicting evidence is presented.
7. The strength of the evidence behind recommendations is transparent.
8. If an atypical study is presented then this is made clear ('communities vary in their enthusiasm for this project but some are very enthusiastic, for example...').

4.2 Describing Evidence Searches

Understanding how the evidence is presented is key for interpreting the results, as it indicates how complete the evidence is, and the likelihood of any bias.

4.2.1 Describing the types and results of searching

Here, we describe how different types of evidence could be obtained. Table 4.1 outlines possible content for describing a range of search methods.

Table 4.1 Suggested possible content, with examples, for presenting different means of searching for evidence.

Source	Possible content	Examples
Search of actions from Conservation Evidence	Describe search, give date of literature covered (given in the synopsis). Reflect on the actions and the quality, quantity, and relevance of evidence in relation to the topic of interest.	Using the search term *roads* and refining using *reptiles* gave 10 actions with at least one study. The action on crossing structures has 16 papers (the eight studies looking at survival include one randomised, controlled, before-and-after study and one review) and that on tunnels has 15 (the three studies examining survival comprise a site comparison studies, a before-and-after study and a replicated test). The reptile synopsis covers literature up to 2018.

Source	Possible content	Examples
Search for reviews from CEEDER	Describe the search giving the date of retrieved reviews. Reflect on the actions and the quality, quantity and relevance of evidence.	Searching for *glyphosate* on 27.05.2022 gave two reviews. A 2020 systematic review of the literature in English on the effect of applying glyphosate on crop yield in Europe found 67 experimental comparisons. A 2018 narrative review discusses the impact of glyphosate on microorganisms, plants and animals.
Search for studies from Conservation Evidence	Describe the search and types of paper and relevance to the issue of interest.	A 27.05.2022 search for *meles* gave the following studies relating to the European Badger *Meles meles*: 9 roads and infrastructure, 2 reintroductions, 3 reducing impact as predators, 2 reducing the impact of TB, 5 reducing the impact on crops.
Systematic literature searches	Describe what was searched, how and when.	A systematic search carried out as part of a paper, broad review, or plan but not published as a systematic review.
Informal search	Describe search and outputs.	A quick search of Google for *status of smooth snake in Finland* provided a 1981 provisional atlas and a map compiled by IUCN in 2017.

4.2.2 Constraints on generality statements

Evidence collations often provide general conclusions, for example, to cover wide geographic areas or groups of species. However, these are based on the evidence available, which often has taxonomic, habitat and geographic biases (Christie et al., 2020a, 2020b). For example, an action may have been assessed as being beneficial for birds, but all of the studies testing the action may have been carried out in one habitat type for a small number of species. Therefore, before making any decisions it is vital to consider the evidence in more detail to assess its relevance for your study species or system and to state the relevance and any constraints in any outcome, for example, *there are numerous good studies showing this resulted in higher breeding success, but all the studies are on Salmonids.*

4.3 Presenting Different Types of Evidence

The aim of this section is to consider how different types of evidence, including describing costs, may be summarised.

4.3.1 Conservation Evidence summaries

Conservation Evidence collates and summarises individual studies on a topic in a consistent manner to ease comparisons and ensure all the key components are included. For example:

A search of Conservation Evidence for 'tree harvesting' found seven individual studies and seven relevant actions.

Conservation Evidence uses a standard format for describing studies that test interventions, which may be of wider use. The following colour-coded format is here illustrated with an example.

A [TYPE OF STUDY] in [YEARS X-Y] in [HOW MANY SITES] in/of [HABITAT] in [REGION and COUNTRY] [REFERENCE] found that [INTERVENTION] [SUMMARY OF ALL KEY RESULTS] for [SPECIES/HABITAT TYPE]. [DETAILS OF KEY RESULTS, INCLUDING DATA]. In addition, [EXTRA RESULTS, IMPLEMENTATION OPTIONS, CONFLICTING RESULTS]. The [DETAILS OF EXPERIMENTAL DESIGN, INTERVENTION METHODS and KEY DETAILS OF SITE CONTEXT]. Data was collected in [DETAILS OF SAMPLING METHODS].

For the type of study there is a specific set of terms that are used (Table 4.2). For the results, for the sake of brevity, only key results relevant to the effects of the intervention are included. For Site context only those nuances that are essential to the interpretation of the results are included.

As an example:

> A replicated, randomized, controlled, before-and-after study in 1993–1999 of five harvested hardwood forests in Virginia, USA (1) found that harvesting trees in groups did not result in higher salamander abundances than clearcutting. Abundance was similar between treatments (group cut: 3; clearcut: 1/30 m^2). Abundance was significantly lower compared to unharvested plots (6/30 m^2). Species composition differed before and three years after harvest. There were five sites with 2 ha plots with each treatment: group harvesting (2–3 small area group harvests with selective harvesting between), clearcutting and an unharvested control. Salamanders were monitored on 9–15 transects (2 x 15 m)/plot at night in April–October. One or two years of pre-harvest and 1–4 years of post-harvest data were collected.

(1) Knapp S.M., Haas C.A., Harpole D.N. and Kirkpatrick, R.L. 2003. Initial effects of clearcutting and alternative silvicultural practices on terrestrial salamander abundance. *Conservation Biology* 17: 752–62.

Table 4.2 The terms used to describe study designs in Conservation Evidence summaries.

Term	Meaning
Replicated	The intervention was repeated on more than one individual or site. In conservation and ecology, the number of replicates is much smaller than it would be for medical trials (when thousands of individuals are often tested). If the replicates are sites, pragmatism dictates that between five and ten replicates is a reasonable amount of replication, although more would be preferable. The number of replicates is stated when given. Replicates should reflect the number of times an intervention has been independently carried out, from the perspective of the study subject. For example, 10 plots within a mown field might be independent replicates from the perspective of plants with limited dispersal, but not independent replicates for larger motile animals such as birds. In the case of translocations/release of captive bred animals, replicates should be sites, not individuals.
Randomized	The intervention was allocated randomly to individuals or sites. This means that the initial condition of those given the intervention is less likely to bias the outcome.
Paired sites	Sites are considered in pairs, within which one was treated with the intervention and the other was not. Pairs, or blocks, of sites are selected with similar environmental conditions, such as soil type or surrounding landscape. This approach aims to reduce environmental variation and make it easier to detect a true effect of the intervention.
Controlled*	Individuals or sites treated with the intervention are compared with control individuals or sites not treated with the intervention. The treatment is usually allocated by the investigators (randomly or not), such that the treatment or control groups/sites could have received the treatment.
Before-and-after	Monitoring of effects was carried out before and after the intervention was imposed.
Site comparison*	A study that considers the effects of interventions by comparing sites that have historically had different interventions (e.g. intervention vs no intervention) or levels of intervention. Unlike controlled studies, it is not clear how the interventions were allocated to sites (i.e. the investigators did not allocate the treatment to some of the sites).
Review	A conventional review of literature. Generally, these have not used an agreed search protocol or quantitative assessments of the evidence.
Systematic review	A systematic review follows structured, predefined methods to comprehensively collate and synthesise existing evidence. It must weigh or evaluate studies, in some way, according to the strength of evidence they offer (e.g. sample size and rigour of design).
Study	If none of the above applies, for example, a study measuring change over time in only one site and only after an intervention or a study measuring the use of nest boxes at one site.

* Note that 'controlled' is mutually exclusive from 'site comparison'. A comparison cannot be both controlled and a site comparison. However, one study might contain both controlled and site comparison aspects e.g. study of fertilised grassland, compared to unfertilised plots (controlled) and natural, target grassland (site comparison).

4.3.2 Systematic maps

Systematic maps (Section 3.2.1) are overviews of the distribution and nature of the evidence found in literature searches. Standardised descriptive data from individual studies are usually extracted and presented within searchable databases or spreadsheets, which accompany written systematic reports and the descriptive statistics. As critical appraisal of included studies does not usually take place in systematic maps, to avoid vote-counting, systematic maps authors do not usually summarise the outcomes of included studies.

Their description includes statistics about the studies included, such as the number of papers on different sub-topics or from different regions, a summary of the quantity and the quality of the evidence identifying areas of strong evidence and knowledge gaps. It also includes a description of the process used (including whether Collaboration for Environmental Evidence guidelines and ROSES reporting standards are adopted), date carried out, and sources and languages searched. For example:

> *Adams et al. (2021) mapped out the literature on the effects of artificial light on bird movement and distribution and found 490 relevant papers. The most frequent subjects were transportation (126 studies) and urban/suburban/rural developments (123) but few were from mineral mining or waste management sectors. Many studies are concerned with reducing mortality (169 studies) or deterring birds (88). The review followed a published protocol and adopted the Collaboration for Environmental Evidence guidelines and ROSES reporting standards. The search, on August 21 2019, was for literature in English in the Web of Science Core Collection, the Web of Science Zoological Record, Google Scholar and nine databases.*

Or for a brief version:

> *One good quality, recent systematic map of the English literature (Adams et al., 2021) found 88 papers on using light to deter birds.*

4.3.3 Systematic reviews

Systematic reviews use repeatable analytical methods to extract secondary data and analyse it and can also be carried out for qualitative studies. Their description includes the objective, number of studies, statistics and the main results. Description of any meta-analysis should describe the summary or averaged effect size (e.g. the mean difference between having the focal intervention and not having it), the uncertainty around this (e.g. 95% confidence limits) and the heterogeneity (variability in reported effect sizes) between studies included in the analysis and the main results of any meta-regression used to explore the drivers of the between-study heterogeneity. A description of the process used (whether Collaboration for Environmental Evidence guidelines and ROSES reporting standards are adopted), date carried out and sources and languages searched, should also be described.

For example, for the systematic review given in Chapter 3:

> *An English language systematic review, following Collaboration for Environmental Evidence guidelines, produced 78 primary studies that assess whether tree retention at harvest helps*

to mitigate negative impacts on biodiversity, (Fedrowitz et al., 2014). Meta-analysis showed that overall species richness (15 studies) and abundance (50 studies) were significantly (hedges d p<0.001) higher with tree retention at harvest compared to clear cutting. However, there were significant differences among taxonomic groups in richness-response to retention cuts compared with un-harvested forests.

Or more briefly:

A good quality meta-analysis of 78 studies showed tree harvest resulted in more biodiversity than clear cutting, but the responses varied between groups.

4.3.4 Dynamic syntheses

Metadataset (see Section 3.7) provides a means for running meta-analysis for the subject of concern. For example for invasive species the user can select the species, countries included action and outcome. It generates a Forest plot with effect sizes and confidence intervals, a paragraph summarising the main conclusions and a funnel plot showing the precision of studies against their effect size results as one check for publication bias. Their description includes the comparison selected and geography included, the response ratio, statistical significance, the number of studies and papers and any significant heterogeneity.

For example, the example shown in Figure 3.4, of analysing the effectiveness of different action on invasive Japanese knotweed could be summarised as follows.

*The effect of chemical and mechanical control on the abundance of invasive Japanese knotweed (*Fallopia japonica*) was selected for analysis. Chemical control reduced abundance by 63% lower compared to the untreated control (response ratio = 0.37; significance P = 0.0084; 193 data points from 13 studies in 12 publications). For physical disturbance, the abundance was 16% lower than the control but this was not statistically significant (response ratio = 0.84; P = 0.579; 27 data points from 8 studies in 8 publications). For both analyses, there was significant heterogeneity between data points (P < 0.0001).*

Or for a brief version

Analysis using metadataset showed a significant 63% reduction in Japanese knotweed abundance when treated with herbicide but a non-significant 16% reduction with physical disturbance.

4.3.5 Conservation Evidence actions

State date, name of action, assessment category and describe the nature of the evidence in relation to the topic of interest. One key principle is to be careful of 'vote counting' — presenting the number of studies showing different results when these might differ in quality.

The action 'Add mixed vegetation to peatland surface' (2018 synopsis) has 18 studies and is classified as Beneficial. Seventeen replicated studies looked at the impact on Sphagnum *moss cover (five were also randomised, paired, controlled, before-and-after): in all studies* Sphagnum *moss was present (cover ranged from <1 to 73%), after 1–6 growing seasons, in at least some plots. Six of the studies were controlled and found that* Sphagnum *cover was higher in plots sown with vegetation including* Sphagnum *than in unsown plots. Five studies reported that* Sphagnum *cover was very low (<1%) unless plots were mulched after spreading fragments.*

Or for a brief version:

The 17 replicated studies on Conservation Evidence looking at adding mixed vegetation showed in all cases there was some Sphagnum *in at least some plots but in five cases at low density.*

4.3.6 Costs of actions

Most conclusions will wish to give some indication of cost, yet, as described in Chapter 2, the reporting of costs is usually unsatisfactory. Either costs are not given or are given without stating what is included. One solution is the standardised collation and reporting of costs to help think through and report the direct costs of conservation actions. Iacona et al. (2018) developed a framework to help record the direct accounting costs of interventions. These steps are outlined below. Major accounting costs include the costs of consumables, capital expenditure, labour costs and overheads.

- Step 1: State the objective and outcome of the intervention.
- Step 2: Provide context, and the method used for the intervention.
- Step 3: When, where and at what scale was the intervention implemented?
- Step 4: What categories of cost are included? (see Table 2.4) What are these costs and how might they vary with context?
- Step 5: State the currency and date for the reported costs.

White et al. (2022b) created a further 6-step framework for thinking through the wider economic costs and benefits of different conservation actions that can be used to assess cost-effectiveness:

- Step 1: Define the intervention/programme.
- Step 2: Outline the costing perspective and reporting level.
- Step 3: Define the alternative scenario.
- Step 3: Define the types of cost and benefits included or excluded (see Table 2.4).
- Step 5: Identify values for the included cost categories.
- Step 6: Record cost metadata (e.g. date, currency).

In a similar style to the layout of the CE effectiveness summary paragraphs (Section 4.3.2) the following suggestions give suggested standard frameworks for reporting different elements of costs in line with cost reporting frameworks (e.g. Iacona et al. 2018). This is a flexible format covering a range of levels of detail.

For marginal intervention costs for the conservation organisation (probably the majority of the reported costs):

First sentence on total costs and what is included:

> Total cost was reported as XX in [currency;] in [year] ([XX USD]) at [information on the scale e.g. XXkm, per hectare; breakdown not provided]. Costs are given over a [XX] time horizon and a discount rate of [XXX] applied. They included [types of cost] and excluded [types of cost] /

Possible additional sentence on the financial benefits of action.

> Financial benefit was reported as: XX in [currency; [XX equivalent USD]; breakdown not provided), resulting in a net cost of XX. The benefits only included [types of benefit].

Additional sentence on the breakdown of the specific costs if provided.

> Cost breakdown: XX [XX USD] for XX materials, XX [XX USD} for XX hours/days of labour, XX [XX USD] for overheads, XX [XX USD] for capital expenditure... [continue for all types of cost reported] (For further details see original paper).

For project level or organization costs, where the costs of interventions are given as part of a broader package:

> Total cost of the action was reported as part of a [describe type and scale of conservation project]. The cost of this project was reported as XX in [currency; no breakdown provided] including [specific activities]. Costs are given over a [XX] time horizon and a discount rate of [XXX] applied. They included [types of cost] and excluded [types of cost]

Note: *Where costs or benefits are incurred by actors other than the conservationist/researcher, this should be noted. E.g., 'the total included .... Opportunity costs to local communities, payments made by volunteers to travel to the nature reserve.'*

As examples:

> Total costs reported as £1,000 GBP in 2022 ($1,200 USD) for 2km of fencing. Costs are given over a 5 year time horizon with no discount rate applied. Cost breakdown: £100/2km for wire fencing and £900 for 5 day of labour for 4 staff. Financial benefit was reported as £200 ($240 USD), resulting in a net cost of £800 ($960 USD). Benefits included the reduced loss of livestock to local stakeholders.

> Total cost reported as €10,000 EUR in 2008 ($12,000 USD) for the translocation of 10 animals (£1000 / animal). Cost breakdown: €2000 [$2400 USD] for tracking devices, supplementary food, and veterinary supplies, €6000 [$7200 USD} for one staff employed by the project for two months and €2000 [$2400 USD] for overheads (project management). Financial benefits reported as €10,000 ($12,000 USD), resulting in a net cost of €0 ($0 USD). Benefits included the estimated tourism value of each reintroduced eagle, although benefits are accrued by the wider sector.

As an example of a brief version:

> *In 2008, this project cost an extra US$550 comprising $250 for fuel and $300 for chainsaw and safety equipment.*

4.4 Presenting Evidence Quality

4.4.1 Information, source, relevance (ISR) scores

Figure 4.1 shows a range of information, source, reliability (ISR) cuboids as introduced in Chapter 2. For a particular assumption in a particular context, the ISR score is an assessment of whether a particular piece of evidence, whether a systematic review or observation, is relevant, and was generated in a reliable manner and from a source considered reliable. Such assessments will often not be public, especially if it would be embarrassing for the source to discover how their reliability is assessed. In other cases, it may be possible to compare different examples of the same type of evidence using predefined criteria or tools available for specific evidence types. One such tool is the CEESAT checklist, https://environmentalevidence.org/ceeder/about-ceesat/, which enables users to appraise and compare the reliability of the methods and/or reporting of evidence reviews and evidence overviews (such as systematic maps) by assessing the rigour of the methods used in the review, the transparency of the reporting, and the limitations of the review or overview.

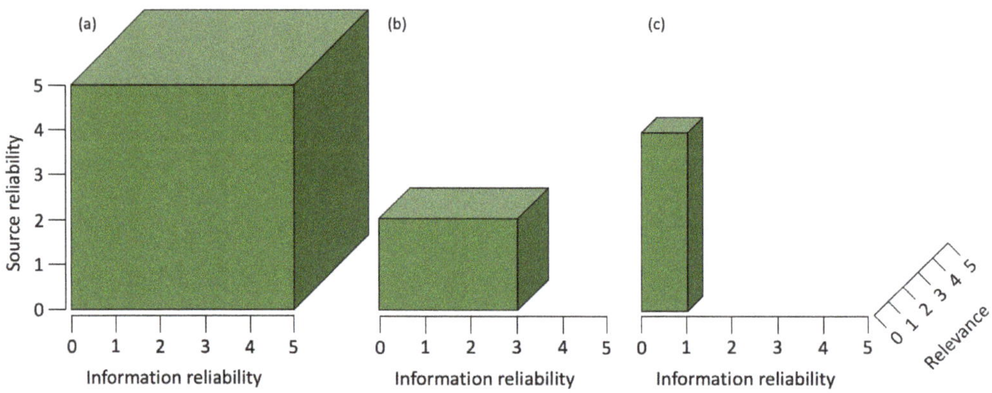

Figure 4.1 Cuboids of different strengths of evidence can range in weight from 0 (no weight) to 125 (5^3). In practice they will usually be cuboids of different dimensions. (a) strongly supported on all three axes; ISR = 5.5.5, (b) limited information and source reliability but highly relevant; ISR = 3,2,2 (c) high source but little information reliability or relevance; ISR = 1,1,5. (*Source:* authors)

Chapter 2 outlines some factors that influence the scores given for each of the three ISR elements. Our suggestion is that ISR becomes a standard way of assessing the weight of different pieces of evidence including different types of knowledge with the ISR score at the end of the claim: *During lockdown, when the reserve was closed, cranes were said to have nested within 3m of the trackway in part of a reserve where they have never nested before (I = 3, S = 4, R = 5).*

The weight of a piece of evidence can then be presented by multiplying the three ISR axes up to a maximum weight of 125 (5^3). Of course, if the value of one axis is zero, meaning that an element of the evidence either cannot be trusted or is judged to be irrelevant, then the weight is zero. Table 4.3 suggests how these weights can be converted into descriptions of the strength of evidence for a single piece of evidence.

Table 4.3 Conversion of weights of single pieces of evidence (from multiplying three axes) into descriptions of evidence strengths.

Weight	Description of evidence strength
0–1	Unconvincing piece of evidence
2–8	Weak piece of evidence
9–27	Fair piece of evidence
28–64	Reasonable piece of evidence
65–125	Strong piece of evidence

4.5 Balancing Evidence of Varying Strength

Whilst in some cases, evidence will either clearly refute or support an assumption (as in the above examples), in many situations there is a range of relevant evidence that varies in its strength and direction in relation to a particular assumption. This is often the case when assessing the effectiveness, costs, acceptability, or feasibility of actions, where evidence can, for example, show an action to be beneficial to different extents, have no effect, or have negative impacts. Where this is the case, one approach, modified from Salafsky et al. (2019), is to imagine placing the various pieces of evidence along a balance according to the direction and magnitude of the effect they report, from large negative effects, through to large positive effects (in this case, effectiveness; Christie et al., 2022) (Figure 4.2).

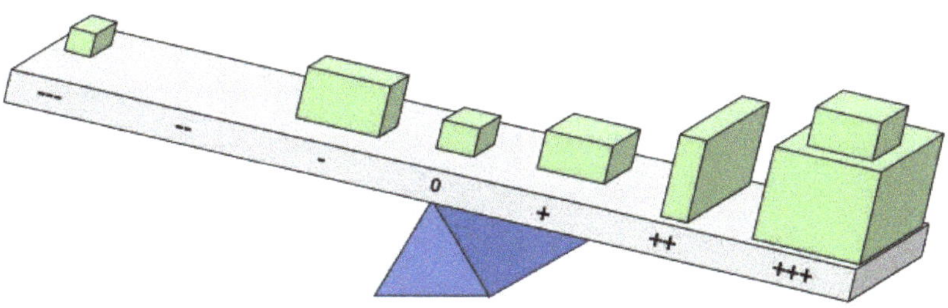

Figure 4.2 A means of visualising the balance of evidence behind an assumption. The pieces of evidence are placed on the balance according to the extent to which it supports or rejects the assumption. The greater the tilt the greater the confidence in accepting (or rejecting) an assumption. (*Source:* authors)

How can this visualisation be converted into a conclusion? Table 4.4 provides a way of assessing the evidence for an assumption about an action. The cumulative evidence score is the sum of the weights of each block of evidence. The weight of each block can be calculated by multiplying the ISR (information, source, relevance) scores (Section 4.2.2). Adding the weights of all evidence blocks together gives the cumulative weight of evidence. The mean effectiveness category comes from assessing the effectiveness associated with each piece of evidence and then taken as the average (the weighted median — place the evidence in order of effectiveness, add up the ISR values until you reach the figure that is half the value for the cumulative evidence, take the effectiveness linked to piece of evidence).

Table 4.4 Converting the combined evidence into statements of the strength of evidence. The maximum score for a piece of evidence is 125 (5 x 5 x 5) so strong evidence requires, say, one really strong study conducted in a manner that minimises bias and overwhelming evidence requires over two high-quality studies.

Cumulative evidence score	Evidence category
>250	Overwhelming evidence
101–250	Strong evidence
51–100	Moderate evidence
11–50	Weak evidence
1–10	Negligible evidence

4.5.1 Statements of effectiveness

The cumulative evidence and mean effectiveness can be combined to give statements of the effectiveness for actions — the Strategic Evidence Assessment (SEA) model (Table 4.5). These statements can then be used in other processes, such as writing management plans or funding applications.

4.5.2 Words of estimative probability

When President Kennedy was deciding whether to proceed with the proposal for the disastrous Bay of Pigs invasion of Cuba, the US Joint Chiefs of Staff stated that the proposal had a 'fair chance' of success, which Kennedy interpreted as likely to work. The Chiefs actually meant that they judged the chances of success as '3 to 1 against' (Wyden 1980). As a result of such lessons, the accepted least ambiguous approach is to attempt estimating probabilities. Studies have shown, at least in medicine, that giving natural frequencies 'Three out of every 10 patients have a side effect from this drug' leads to fewer problems of interpretation than probabilities, such as 'You have a 30% chance of a side effect from this drug' (Gigerenzer and Edwards, 2003).

Some dislike judging probabilities (how certain we are about something) or estimating frequencies (how often something happens), partly because it seems to give unjustified accuracy. An alternative approach is to express likelihoods in the form of standardised terms. Table 4.6 gives the language for communicating probabilities recommended by the Intergovernmental Panel on Climate Change (IPCC, 2005). Table 4.7 gives the opposite: examples of some 'weasel' words that are likely to be ambiguous and so should be avoided.

Table 4.5 The Strategic Evidence Assessment model. Converting the combined evidence into statements of effectiveness combining the scores from Table 4.4 with effectiveness.

Cumulative evidence score	Median evidence category	Considerable harms	Moderate harms	Minor harms	No effect	Little benefit	Moderate benefit	Considerable benefit
>250	Overwhelming evidence	Overwhelming evidence of considerable harms	Overwhelming evidence of moderate harms	Overwhelming evidence of minor harms	Overwhelming evidence of no effect	Overwhelming evidence of little benefit	Overwhelming evidence of moderate benefit	Overwhelming evidence of considerable benefit
101–250	Strong evidence	Strong evidence of considerable harms	Strong evidence of moderate harms	Strong evidence of minor harms	Strong evidence of no effect	Strong evidence of little benefit	Strong evidence of moderate benefit	Strong evidence of considerable benefit
51–100	Moderate evidence	Moderate evidence of considerable harms	Moderate evidence of moderate harms	Moderate evidence of minor harms	Moderate evidence of no effect	Moderate evidence of little benefit	Moderate evidence of moderate benefit	Moderate evidence of considerable benefit
11–50	Weak evidence	Weak evidence of considerable harms	Weak evidence of moderate harms	Weak evidence of minor harms	Weak evidence of no effect	Weak evidence of little benefit	Weak evidence of moderate benefit	Weak evidence of considerable benefit
1–10	Negligible evidence	Negligible evidence of considerable harms	Negligible evidence of moderate harms	Negligible evidence of minor harms	Negligible evidence of no effect	Negligible evidence of little benefit	Negligible evidence of moderate benefit	Negligible evidence of considerable benefit

Table 4.6 Terms suggested by the IPCC (2005) for referring to probabilities.

Probability	Recommended term
>99%	Virtually certain
90–99%	Very likely
66–90%	Likely
33–66%	About as likely as not
10–33%	Unlikely
1–10%	Very unlikely
<1%	Exceptionally unlikely

Table 4.7 Examples of 'weasel' terms likely to be ambiguous.

A chance	Cannot dismiss	Could	Maybe	Perhaps	Somewhat
Believe	Cannot rule out	Estimate that (or not)	Might	Possibly	Surely
Cannot discount	Conceivable	May	Minor	Scant	Suggest

4.6 Visualising the Balance of Evidence

In some cases it may be particularly informative to visualise the collated evidence base, ideally communicating both the weight of the evidence pieces and the results they found. Here we present two different options for visualising a collated evidence base that are best suited to slightly different types of results.

4.6.1 Ziggurat plots

Ziggurat plots (Figure 4.3) are best suited to results where the outcomes can be expressed categorically (e.g. varying discrete levels of effectiveness). The width of each bar represents its ISR score (out of 125) with studies piled in order of ISR score to give a ziggurat shape. This has the merit of showing the distribution of the evidence.

In some cases, the median may be a poor representation of the distribution, for example, because the results are strongly bimodal. In this case add a comment e.g. 'median effect size mixed, but results strongly bimodal'.

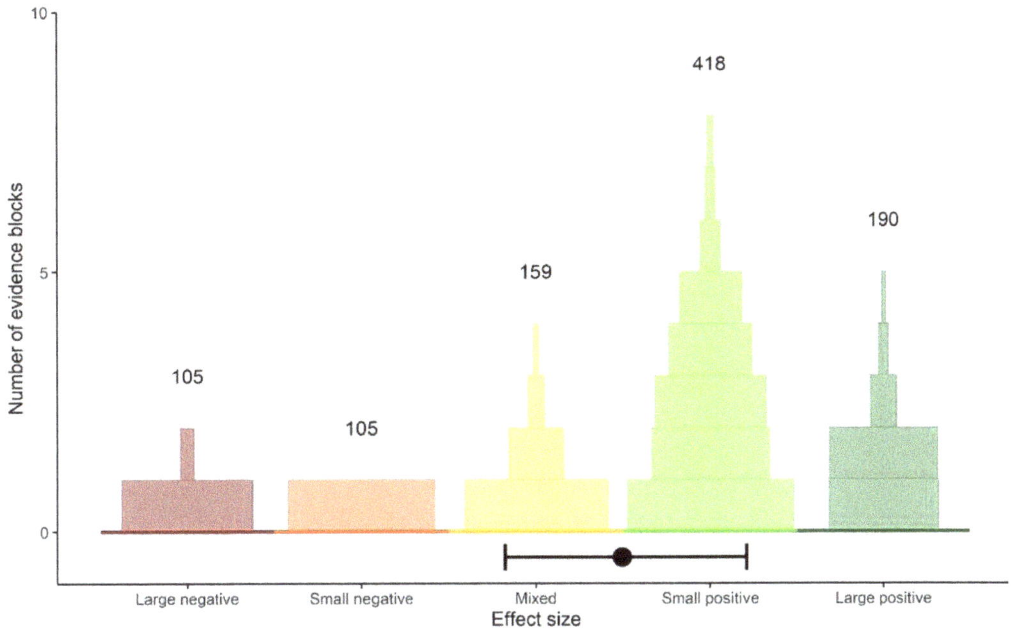

Figure 4.3 A ziggurat plot in which each study is a horizontal bar whose width is the information reliability, source reliability, and relevance (ISR) score (up to 125). The studies are collated for different categories of effectiveness. The number above each pile of evidence blocks shows the total evidence score for that pile. In order to derive an average effect size, we defined effect size categories as consecutive integers (large negative = -2; small negative = -1; mixed = 0; small positive = 1; large positive = 2) and calculated the weighted mean (filled black point) and 95% confidence intervals. The confidence interval was calculated by bootstrapping the weighted mean. (*Source:* authors). The following R code can be used to carry out this process, which we have adapted from code produced by Stackoverflow users Tony D and Ben https://stackoverflow.com/questions/46231261/bootstrap-weighted-mean-in-r. Text in quote marks must be altered by the user:

```
library(boot)
df <- data.frame(x= “degree of support”, w= “ISR score”)
wm <- function(d,i){
  return(weighted.mean(d[i, 1], d[i, 2]))
}
bootwm <- boot(df, wm, R=10000)
boot.ci(boot.out = bootwm)
```

Figure 4.4 shows a range of possible outcomes. In some cases, the evidence may be clear and convincingly show either a positive effect or no effect. In others the results may be bimodal, for example, if combining studies of reptiles but lizards and snakes show very different responses. In other cases, the responses may be unclear — perhaps as a range of different unknown variables are important.

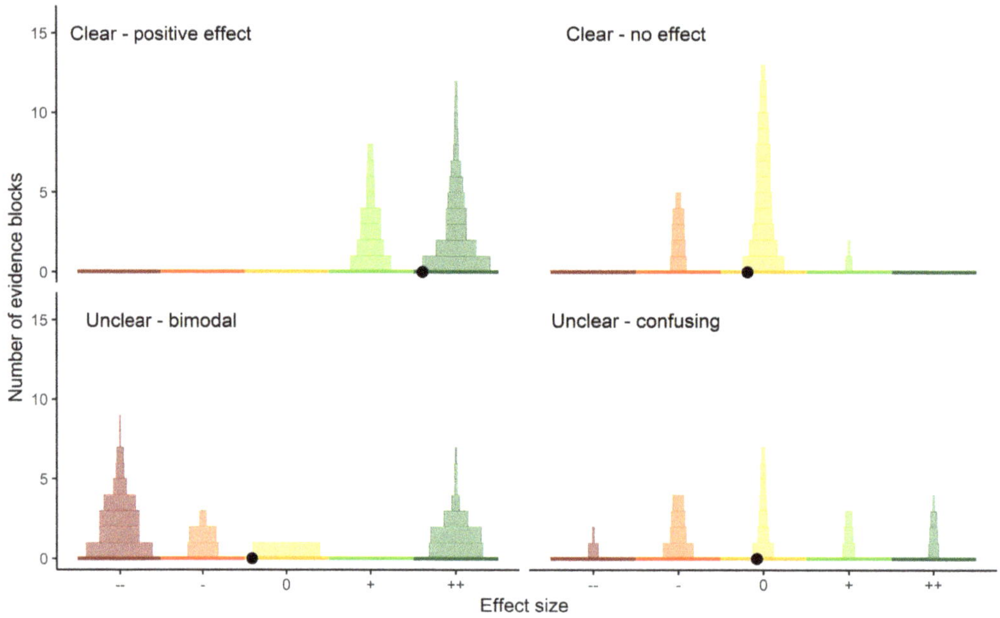

Figure 4.4 A range of possible outcomes of ziggurat plots. Those on the top show clear results, with either a moderately large effect size (left) or evidence for negligible effect (right). Those underneath show unclear effects as either bimodal (left) or scattered results (right). The filled black point shows the average effect size, calculated as the weighted mean. (*Source:* authors)

4.6.2 Weighted histogram plots

For continuous results (an effect size, or counts of a particular species, for example) however, a ziggurat plot may hide useful information, and the collated evidence base can be better visualised with a weighted histogram plot (Figure 4.5). Sources of evidence are ordered by the magnitude of the result they report, and the height of the bar is the IRS score (out of 125). Evidence pieces that report the same effect sizes are stacked on top of each other.

4.7 Synthesising Multiple Evidence Sources

Combining knowledge sources is key. The evidence that decision makers draw on will usually be a combination of scientific knowledge, local knowledge and experience. For example, when managing a reserve, much of the decision making will be based on the experience and local knowledge of the reserve manager and team ('I think this area is too close to the forest for the species to breed', for example), informed by knowledge from the local community, local naturalists, fishers, etc and scientific knowledge that has generality.

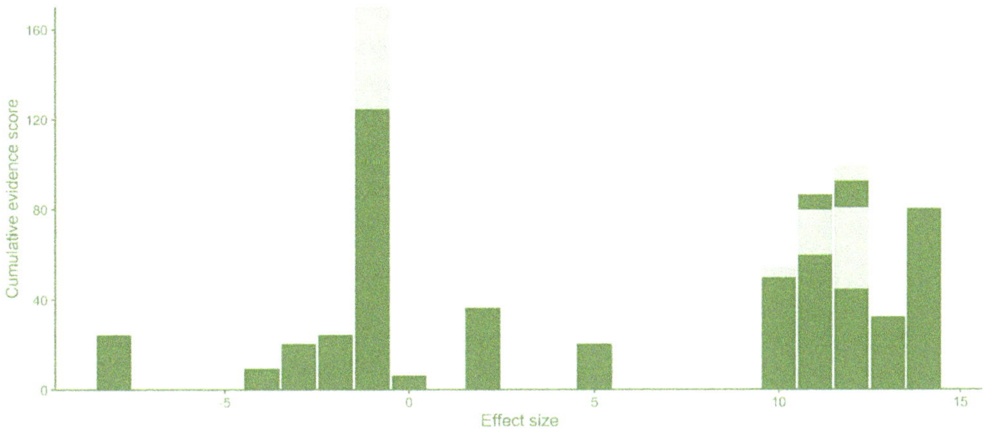

Figure 4.5 A weighted histogram plot in which each piece of evidence is represented by a vertical bar, whose height is the information reliability, source reliability, and relevance (ISR) score. Evidence pieces are arranged in order of the magnitude of the effect size, and where they report the same effect size, they are stacked on top of each other, with alternating colours for each evidence piece in the stack. (*Source:* authors)

4.7.1 Evidence review summary statements

The result of the collation or analysis can also be presented in a summary. The aim is to describe the results so that they provide a summary of the amount and nature of the evidence (e.g. quality/reliability, coverage: species, habitats, locations), what it tells us (e.g. effects on key outcome metrics), and any limitations. The different elements of this chapter can be combined, as shown in Table 4.8, to provide a summary of the evidence.

Table 4.8 The different main elements of summarising evidence described in this chapter with an illustrative sentence.

Element	Example
Type of search	Searching for 'earthworm' on CEEDER to look for the impact of invasive earthworms
Type of evidence	produced one good quality meta-analysis from 2018 based on 430 observations from 30 independent studies,
Constraints on generality statement	with many studies looking at impacts on Collembola and oribatid mites but few on mycorrhizal fungi,
Study conclusions	showing compelling evidence for negative effects on below-ground biodiversity of recipient ecosystems,
Information on costs	but provided no figures on costs.

Such summaries can highlight knowledge clusters and thus gaps in the evidence, which are important to take into consideration. However, it is important that summaries are not used to vote count, i.e. draw conclusions based on the number of studies showing positive vs negative results, which is usually a misleading method of synthesis (Stewart and Ward, 2019). Studies are generally not directly comparable or of equal value, and factors such as study size, study design, reported metrics, and relevance of the study to your situation need to be taken into consideration, rather than simply counting the number of studies that support a particular interpretation. Chapter 2 provides suggestions for describing single studies.

> *In the UK three replicated randomised controlled experiments showed that adding a collar and bell to domesticated cats reduced the predation rate of small mammals by about a half; one of these also showed that effects were similar for bells and a collar-mounted sonic device. Experiments outside the UK showed that a pounce protector (a neoprene flap that hangs from the collar) stopped 45% of cats from catching mammals altogether. A study in the USA showed cats provided with brightly patterned collars brought home fewer mammals than did cats with no collars in autumn, but not in spring. In all cases the cats had a history of bringing back whole prey and the catches were recorded by their owners.*

Or for a brief version:

> *Domesticated cats took roughly half as many small mammals if fitted with collar and bell, collar and sonar device or a neoprene flap hanging from the collar. Brightly coloured collars were less effective.*

4.7.2 Evidence capture sheets

Perhaps the most straightforward method for bringing together multiple sources of evidence is to use an evidence capture sheet. Table 4.9 presents an example of such a sheet, with each row constituting a different source of evidence, and each column providing vital information about the validity of the source and the result that it presents. While evidence capture sheets comprehensively communicate the key information of each source of evidence, they stop short of providing an overall conclusion of what the evidence base says.

4.7.3 Argument maps

Argument maps, also called argument diagrams, have three main roles: easing the creation of a logical argument, presenting the basis of an argument so it is easy to follow and reframing an existing argument so making the assumptions, logic and any gaps transparent. The heart of the argument map is to present the case for and against a position in a logical manner. Their merit is that, if done well, they organise and clarify the evidence and reasoning for a contention. These can be done collectively as a way of understanding the basis of any disagreement. They have the considerable advantage that they can easily be converted into text for a decision maker to make the final decision. If using a software package this conversion can be done automatically.

Table 4.9 Example of tabular presentation of evidence for a proposed project that plans to introduce natural grazing with ponies to the montado habitat in Iberia to increase biodiversity. I = information reliability, S = source reliability, R = relevance all on 0–5 scales. Note the evidence strength is for the particular problem rather than an overall assessment.

Evidence source[a]	Type of evidence[b]	Direction and strength of results[c]	I[d]	S[e]	R[f]	Description of evidence strength*
Experience of colleague	Anecdotal	Colleague has been told that there used to be higher densities of livestock and also more flowers in the area 50 years ago	1	2	5	Weak
Scientific study (Leal et al. 2019)	Experimental	A replicated controlled study found that moderate levels of grazing benefited most species of bird.	3	5	5	Strong
Grazelife report www.grazelife.com	Observational? (not specified)	Herbivory leads to more diverse vegetation and benefits many species	3	3	4	Moderate
https://www.conservationevidence.com/actions/1628	Synthesis	Three of five studies across Europe found that increasing grazing intensity increased plant diversity in shrublands	2	5	3	Moderate

a. E.g. peer-reviewed article, expert opinion, grey literature report, personal experience.
b. E.g. synthesis, experimental, observational, anecdotal, theoretical/modelling.
c. Was the result strongly positive, weakly positive, mixed, or no effect?
d. How convincing is the evidence? Depends on the type of evidence, for example, may depend on experimental design and sample size. (0–5)
e. Is the source biased or independent? (0–5)
f. Does it apply to your problem? This will depend on habitat, geography or socio-economic similarity. (0–5)

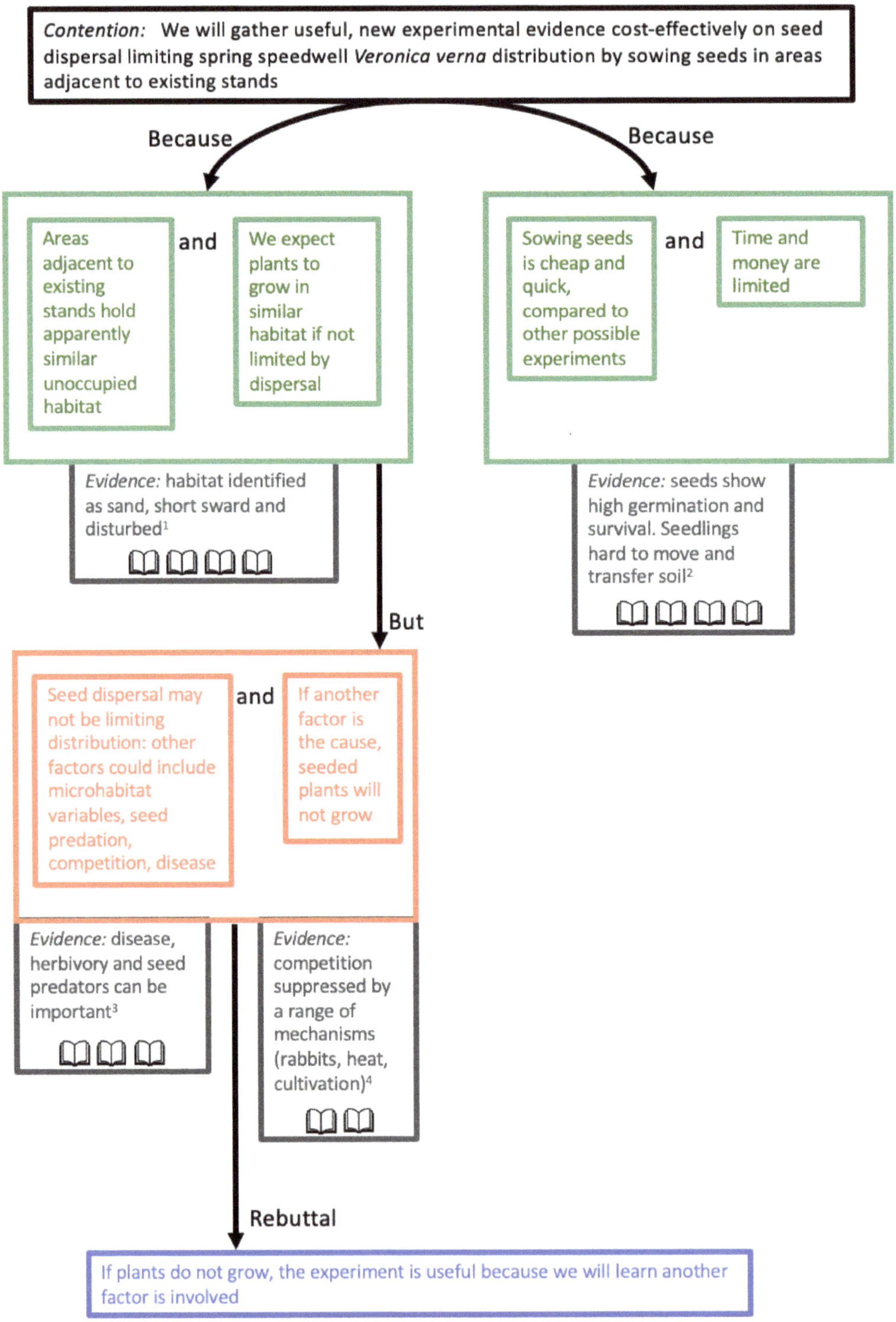

Figure 4.6 An example of an argument map to decide whether or not to introduce a plant to new locations. The main elements are: main contention, premise, counterargument, rebuttal and evidence. 📖📖📖📖📖 = overwhelming evidence; 📖📖📖📖 = strong evidence; 📖📖📖 = moderate evidence; 📖📖 = weak evidence; 📖 = negligible evidence. The evidence is elaborated in Box 4.2. (*Source:* authors)

Box 4.2 Evidence sources for plant reintroduction argument map

1. Heathland sites with *Veronica verna* all have low skeletal vegetation and open patches (Watt 1971). Experiments show it benefits from disturbance (Nabe-Nielsen et al. 2021).

2. In an experiment 22% of 700 *V. verna* seeds germinated in the autumn after sowing (Norman Sills pers. comm.). Practitioner experience is that seedlings are harder to transplant and result in transferring soil (Tim Parkhurst pers. comm.).

3. Extensive ecological literature shows there can be a serious impact of disease, herbivory and seed predators on annual plants. Herbivory by unknown invertebrates and slugs occurred on closely related species (Boutin and Harper, 1991). Five of 11 seedlings in an outdoor box were grazed (by slugs?), but four regrew (Norman Sills pers. comm.). Potentially could limit distribution but entirely speculative.

4. All existing sites with little competition. Rabbit grazing is traditionally important (Back from the Brink, 2021). At Weeting Heath arable weed reserve is cultivated. Hot dry conditions seem likely to limit competitors (*V. verna* grows over winter and flowers in spring) but no evidence.

They start with a statement, the main contention, that could be correct or not. The arguments for it being right, the premises, are then listed followed by the arguments for it being wrong, the counterarguments. Co-premises are two statements that both have to be correct. Lines are only drawn between boxes to specify that something is a reason to believe or a reason not to believe something else. The process forces the logic as to whether you should believe the claim and why or why not.

There are three basic rules (https://www.reasoninglab.com/wp-content/uploads/2013/10/Argument-Maps-the-Rules.pdf):

1. The **Rabbit Rule** (you cannot pull a rabbit out of the hat by magic): every significant word, phrase or concept appearing in the contention must also appear in one of the premises; i.e. every term appearing above the line has to appear below the line. In the example here the terms are sow, spring speedwell, seeds, adjacent areas, distribution, and dispersal.
2. The **Holding Hands Rule**: every term appearing below the line has to appear above; if something appears in a premise but not in the contention, it must appear in another premise.
3. **The Golden Rule:** every simple argument has at least two co-premises.

Figure 4.6 shows an argument map created by members of the Breckland Flora Group (Jo Jones, David Dives, Julia Masson, Tim Pankhurst, Norman Sills, William Sutherland, and James Symonds) guided by Mark Burgman. They are interested in planning a reintroduction programme for spring speedwell *Veronica verna*. Creating this map helped think through the logic for carrying out an experiment to assess whether the species is restricted in range due to lack of dispersal or whether there it is constrained by very specific, but not fully understood, habitat requirements.

Argument maps can easily be sketched out, but there are numerous argument-mapping tools available. Argument mapping has been adopted by the New Zealand government for biosecurity analyses.

4.7.4 Mind maps

Mind maps, as popularised by Buzan (1974), express the relationships between different aspects of a problem. They are also sometimes called spider diagrams or brainstorms. Although often considered as a means of note taking, they can also be used for generating new ideas. Of relevance here they can be used to summarise information around an issue. They are a less structured process than a theory of change or argument map and so are useful for pulling together information to provide a clear account of the components of the issue. One advantage of summarising with such a visual model is that it makes the logic explicit and thus invites further comments. They can be drawn by hand or created using various packages available. These then can be created by a person or team putting together the strands of evidence.

These are created by starting with a central idea or topic, placed in the middle of the map. Branches are then added to this central topic. Each branch is then explored. If the evidence is extracted it can be assessed and added. Another option is to draw a preliminary map based on expert opinion and then adjust as evidence is added.

Figure 4.7 shows an example of a mind map created after a group discussion with experts (Malcolm Ausden, Jennifer Gill, Rhys Green, Jennifer Smart, William Sutherland, and Des Thompson). Box 4.3 lists some supporting evidence. This is intended to be illustrative of the process rather than a comprehensive review of the literature and concepts.

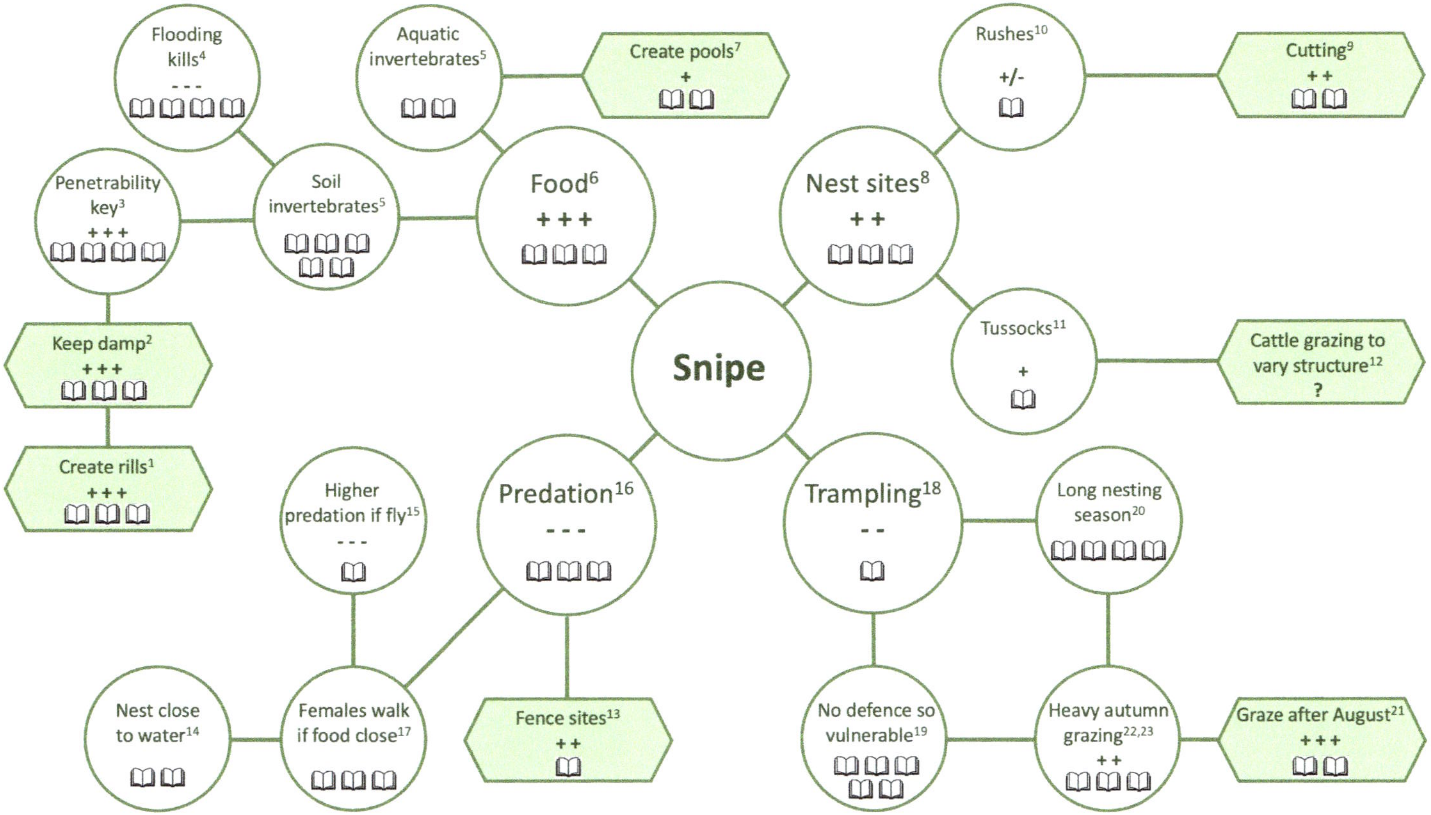

Figure 4.7 A mind map of possible means of managing lowland grassland habitat to improve the common snipe *Gallinago gallinago* population. 📖📖📖📖📖 = overwhelming evidence; 📖📖📖📖 = strong evidence; 📖📖📖 = moderate evidence; 📖📖 = weak evidence; 📖 = negligible evidence; ? = no evidence; - - - = considerable harm; - - = moderate harm; - = minor harm; 0 = no effect; +/- = mixed effect; + = little benefit; ++ = moderate benefit; +++ = considerable benefit. See Box 4.3 for the evidence sources. (*Source:* authors)

Box 4.3 Evidence sources for mind map on snipe management

1. Creating rills (shallow linear features) initially beneficial in Otmoor enabling snipe to breed even where soil type doesn't allow them to probe but numbers subsequently crashed (Malcolm Ausden, Rhys Green, pers. comm.)
2. Penetration resistance of the soil related to depth of water table (Green, 1988)
3. Breeding season shorter if soil becomes impenetrable (Green, 1988)
4. Earthworms killed if flooded (Ausden et al., 2001)
5. Mainly feed on earthworms and Diptera larvae from soil but some aquatic invertebrates especially Chironomid larva (Green, 1988; Cramp and Simmons, 1983)
6. Largely restricted to areas where food is available; snipe were more likely to persist in fields where the soil conditions were wet and soft (Smart et al., 2008)
7. Pools used when cannot probe ground or there is low soil invertebrate biomass nearer to the nest (Green et al., 1990) Pools used when cannot probe ground so last resort (Rhys Green, pers. comm.)
8. Restricted to sites with some long grass (widespread experience; Cramp and Simmons, 1983)
9. Rush cutting benefits both lapwing and redshank (Robson and Allcorn, 2006); in Cumbria, England, number nesting increased from 1 to 11 following several interventions: rush cutting, higher water levels, more intensive grazing regime, and scrape creation (Holton and Allcorn, 2006). Unknown if it also benefits snipe.
10. Rushes provide some cover but deleterious if too abundant; upland snipe prefer rushes (Hoodless et al., 2007)
11. Some say prefer long, tussocky grass for nesting (Mason and MacDonald, 1976; Cramp and Simmons, 1983) but Rhys Green (pers comm) has found about 300 snipe nests but never one in a tussock and no relation between vegetation height and snipe density in wet grassland studied (Green 1986)
12. Widespread experience shows suitable habitat with muddy poached areas and suitable height for nesting best created by cattle grazing
13. Predator-exclusion fencing led to increased productivity of lapwing (Malpas et al., 2013)
14. In Otmoor 90% nest within 1 m of rill (Rhys Green, pers. comm) but no relation in upland site (Hoodless et al., 2007)

15. One of 10 radio-tracked snipe taken by kestrel; at least another eight snipe remains found around that kestrel's nest; likely to be more vulnerable if fly (Green et al., 1990)
16. 60% loss of eggs to predators (Green, 1988); some adults predated in the breeding season (Green et al. 1990)
17. Females walk to feed unless sites are within 70 m, are too dry, or have insufficient food (Green et al., 1990)
18. Nest loss is correlated with cattle density with 19% lost to trampling in one study (Green, 1988); in USA experiment, trampling is greater where cattle are present but overall success is unaffected (Popotnik and Giuliano, 2000)
19. Do not defend nest (widespread field experience; Cramp and Simmons, 1983) so particularly vulnerable to trampling
20. Long nesting season (Mason and MacDonald, 1976; Cramp and Simmons, 1983)
21. As can still be nesting in August if earlier clutches fail and fields remain suitably wet, it makes sense to delay grazing until late summer, but this can mean the cut hay is of low quality and therefore difficult to dispose of, and might affect sward quality for Snipe and other species in the longer-term (Malcolm Ausden, pers. comm.).
22. Grazing needed to maintain a suitable sward but can trample nests, so follows that heavier grazing in autumn can be used to help create a suitable sward for the following spring. Although with recent warmer winters autumn grazing may be insufficient to maintain short sward for lapwing creating, which may create demand for early spring grazing (Green, pers. comm.)

In practice most decisions appear to comprise an exploration of components of the issue, rather like an informal mental version of a mind map, followed by a pronouncement of the conclusion. Our experience is that it is often unclear what is justifying the decision. A sketched version of the discussion provides a reminder of the issues and helps force consideration of the components.

4.7.5 Evidence underlying theories of change

A theory of change (CMP, 2020) is a series of causally linked assumptions about how a team thinks its actions lead to intermediate results and to target outcomes (see Chapter 7 for further details). Displaying a theory of change in a strategy pathway diagram (see an example in Figure 4.8 below, repeated from 7.3) facilitates the identification of analytical questions and assumptions that can be tested with evidence.

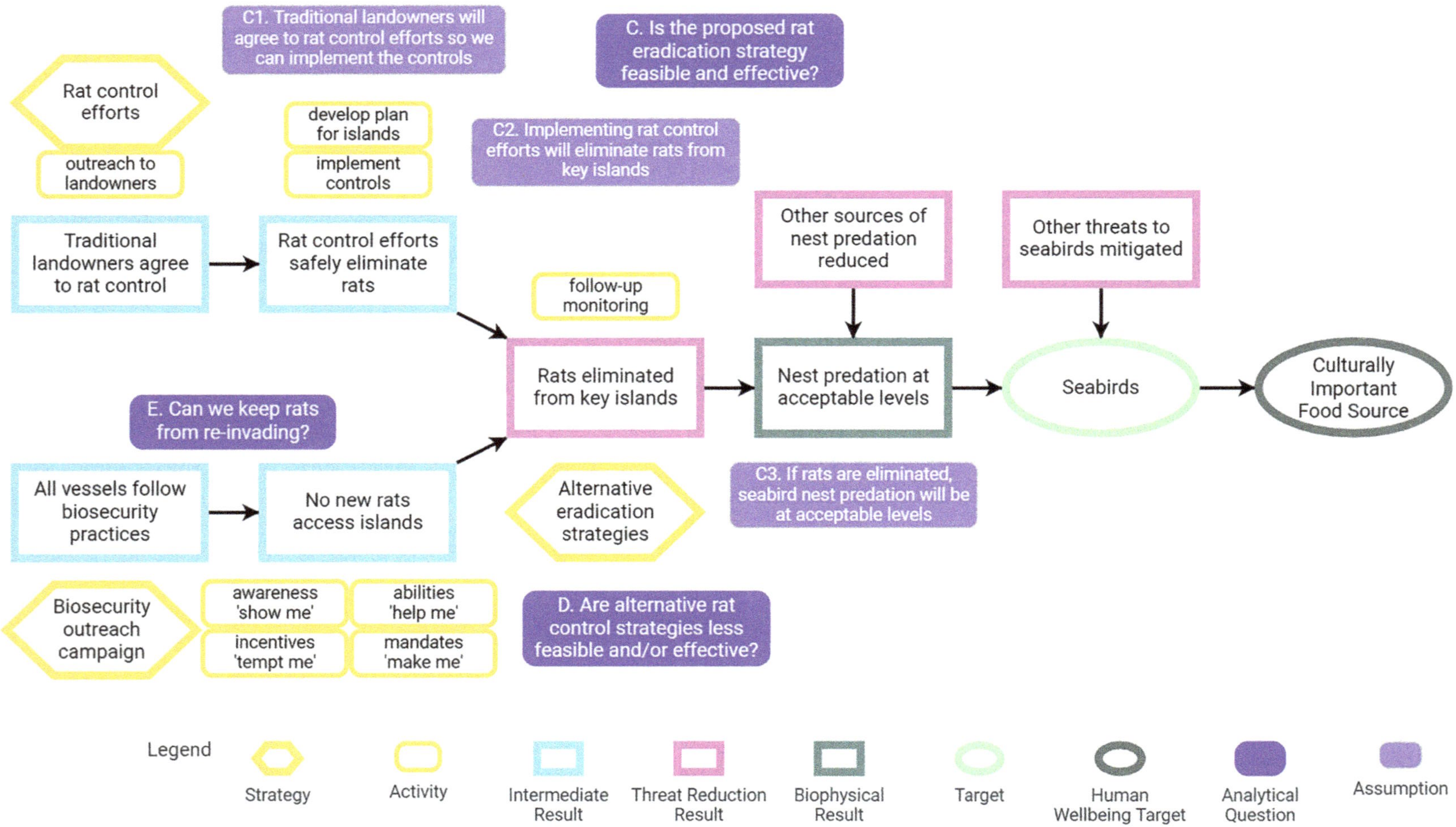

Figure 4.8 A theory of change pathway showing analytical questions and assumptions. This example is derived from several different real-world rat eradication projects. Note, assumptions for questions D and E are not shown. (*Source*: Salafsky et al., 2022, CC-BY-4.0)

Table 4.10 Summary of current evidence for analytical questions relating to the theory of change shown in Figure 4.7. (*Source*: Salafsky et al., 2022)

Analytical question	Who, date	Summary of current evidence for question	Current confidence in evidence	Implications	Additional information needs
C. Is the proposed rat eradication strategy feasible and effective?	Team, Jun 2021	While it is clear from other sites that rat eradication would be technically feasible in the short-term, it is not yet clear whether this will be socially acceptable given the potential lack of durability of the solution.	Need more info	Explore with the community whether eradication is acceptable.	Compare rat population levels in treated and untreated areas. Conduct key informant interviews with tribal elders.
D. Are alternative rat control strategies less feasible and/or effective?	Team Nov, 2021	Our experience is that other means of controlling rats such as trapping or introducing natural predators would not be technically feasible in our islands.	Confident, but...	Continue to monitor the emergence of new strategies that may be more effective.	Stay abreast of new literature/examples of rat eradication strategies.
E. Can we keep rats from re-invading?	Team, Dec 2021	It is not clear whether it is feasible to keep rats from re-invading; experience from other sites is that this requires a major investment of resources and that re-invasion may actually be facilitated by the previous invasive species.	Not confident	Unless we can figure out how to solve this problem, eradication will only be a temporary fix at best.	Stay abreast of new literature/examples of biosecurity strategies.

Once evidence is collected, theories of change can provide the backbone to assess the evidence and contextualise it for a respective conservation action. Often, evidence capture sheets (Table 4.10) are used to relate available evidence to the theory of change and form the basis for drawing overall conclusions about key questions and assumptions, and the effectiveness of the strategy pathway.

4.7.6 Bayesian networks

Bayesian networks (BNs), as outlined in Box 4.4, are powerful statistical models that represent suspected, theorised or known causal relationships between variables supported by data and knowledge (i.e. evidence) (Jensen, 2001; Stewart et al., 2014). They comprise a network of nodes (i.e. variables or factors) connected by directed arcs (i.e. edges, links or arrows) contextualising the underlying probabilistic relationships. BNs can be built by combining a range of data sources including empirical evidence and expert elicitation. These models are particularly useful because they explicitly and mathematically incorporate uncertainty in a transparent way (Landuyt et al., 2013). Due to the graphical nature of BNs, complex information can be

Box 4.4. A simple example of a Bayesian network

Imagine a scenario where we arrive home from work and notice the grass in our front garden is wet. There are only two possible causes of the wet grass; it has rained, or our sprinkler (automatic watering system) has been on. How can we decide what is the cause of the wet lawn?

First, we build a causal network representing the relationships between the three nodes (Figure 4.9).

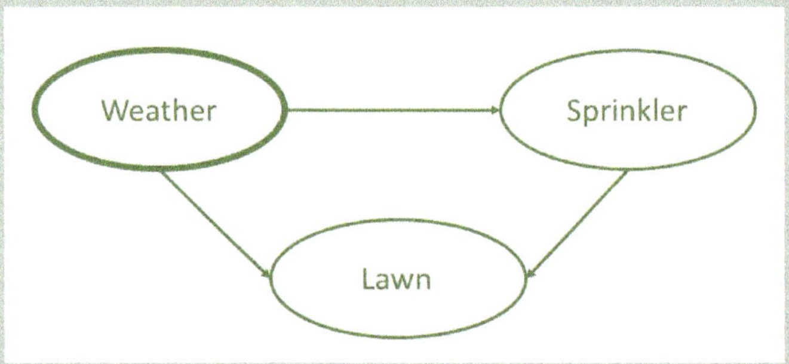

Figure 4.9 A causal diagram of the relationship between the three nodes in the system. The state of the Sprinkler node and the Lawn node are conditionally dependent on the state of the Weather node. (*Source:* authors)

Each node has a discrete number of states it can be in. The weather node can be in one of three states: Sunny, Cloudy or Rainy. The Sprinkler node can be in one of two states: On or Off and the Lawn node can also be in one of two states: Wet or Dry.

We are not sure what the weather was but we can use the weather forecast which tells us the probability that it was sunny is 40%, that it was cloudy is 30% and that it was raining 30%. We can use expert opinion or data to build the underlying conditional probabilities tables which encode the relationships between the nodes (Table 4.11). The probability that the Sprinkler is On when the state of the Weather node is Sunny is 80%. The probability that the Sprinkler is On when the state of the Weather node is Rainy is 10%. Note that each column in the conditional probability table sums to 100%.

Table 4.11 *A conditional probability table for the Sprinkler node given the three states of the Weather node.*

Weather	Sunny	Cloudy	Rainy
Sprinkler On	80	30	10
Sprinkler Off	20	70	90

If we have no evidence about the state of any of the nodes, we can see that the model displays the probability of each state for each node (Figure 4.10). When we have no evidence about the system, we can still gain useful information from the conditional probabilities. In this case, the state of the Lawn node is more likely to be Wet.

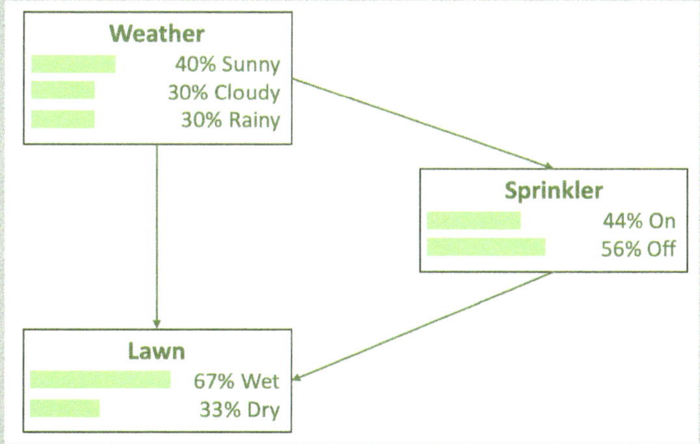

Figure 4.10 A Bayesian network of the lawn and sprinkler system shown in Figure 4.9 including the expected weather conditions, with the resulting consequences for sprinkler use. The probability of rain and the probability of the sprinkler being on gives the overall probability that the grass will be wet. (*Source:* authors)

The true power of Bayesian networks is that they can not only calculate complex combinations of conditional probabilities to deduce the probable state of a node (something that humans struggle to do) but that we can use evidence from any node in the network to determine the probable state of any other node (called abductive learning).

As we know the state of the Lawn node (it was wet when we arrived home) we can set the node (known as 'setting evidence') to 100% and that will give us the probability of each state in the Weather and Sprinkler nodes (Figure 4.11). If the Lawn is 100% Wet then with no knowledge of the state of the Sprinkler we know that the probability that it was sunny is higher than the probability that it was rainy or cloudy. Therefore the probability that the Sprinkler is the cause of our wet lawn is 59%.

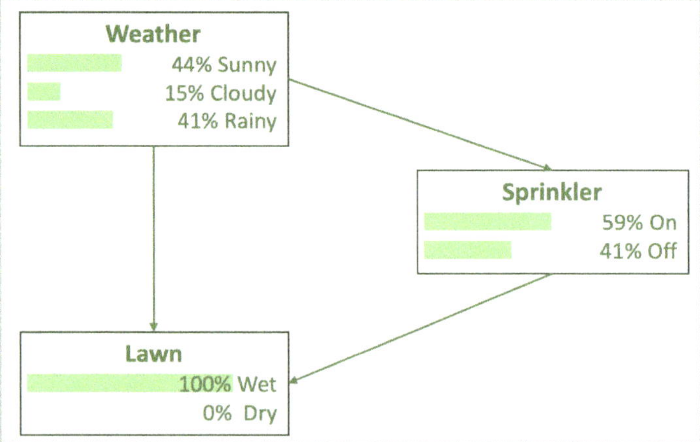

Figure 4.11 The Bayesian network shown in Figure 4.10 but with the lawn set to wet (as observed). From setting the lawn node can then determine the state of the weather and sprinkler nodes. (*Source:* authors)

represented in an intuitive manner that is easily interpreted by non-technical stakeholders (Spiegelhalter et al. 2004).

Developments in computing technology has allowed the rapid development of BNs over the last 10 to 15 years. BNs can now include both categorical and continuous variables and can be dynamic (across space and/or time), which has widened the scope for the application in applied conservation. For example, Stephenson et al. (2018) developed a dynamic BN to investigate the drivers of changes in pot-fishing effort distribution along the Northumbrian coastline of Northeast England across seasons (Figure 4.12). Empirical data on fishing vessel characteristics, habitat type and catch statistics were combined with qualitative data from interviews with fishers. This allowed for the identification of the key drivers of the change in fishing activity between 2004 and 2014; technological changes have increased the ability of fishers to fish in poor weather conditions. Stephenson et al. (2018) could then use their BN to develop management scenarios and test the possible outcomes of changing, for example, regulations on the number of fishing days or the exclusion of fishing from certain areas along the coast.

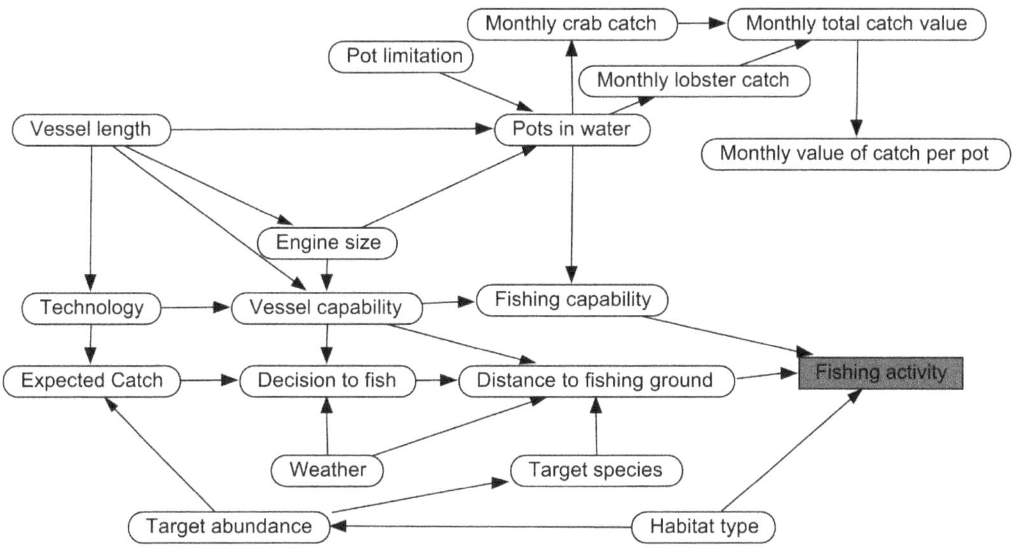

Figure 4.12 Inference diagram of the factors influencing pot-fishing activity along the Northumbrian coast, UK. (*Source:* Reprinted from Stephenson et al., 2018, with permission from Elsevier)

BNs can be used in evidence synthesis to contextualise evidence into a decision. A well conducted systematic review and meta-analysis can often only provide evidence for a small part of the full scope of a decision problem. For example, Carrick et al. (2019) asked the question 'is planting trees the solution to reducing flood risks?'. Using a systematic review and meta-analysis they could only answer a lesser question about the effect of trees on stream flow. Flood risk is a wider issue and consists of many other factors that can influence the outcome. Using a BN to place the results of the systematic review into the full decision context, Carrick et al. (2019) were able to show the sensitivity of Flood Risk node to uncertainty in the other nodes in the network (Figure 4.13). This, in effect, allows one to quantify which uncertainties should be reduced first to get a better understanding of flood risk. These included, for example, infrastructure density (where roads and houses are placed on the landscape) and the capacity of reservoirs to slow down water flow into river catchments.

Bayesian networks can encompass decision nodes and utility nodes to make them explicit decision support tools (sometimes called Bayesian Decision Networks — see Nyberg et al. 2006). A decision node represents two or more choices that influence the values of outcome nodes. For example, the decision to mow or not to mow a grassland or to irrigate or not irrigate a crop could be included in the network. Utility nodes which show the value of an action in units of importance to the decision maker (this could be financial cost or expected change in the number of species) are often linked to these decision nodes to allow trade-offs between costs and benefits to be explored.

Several commercial packages (with limited free options) exist such as HUGIN, NETICA and GENIE and open source R packages such as bnlearn (Scutari, 2010) and caret (Kuhn, 2022) that can be used to develop BNs. Hybrid networks (combining continuous and categorical data) can

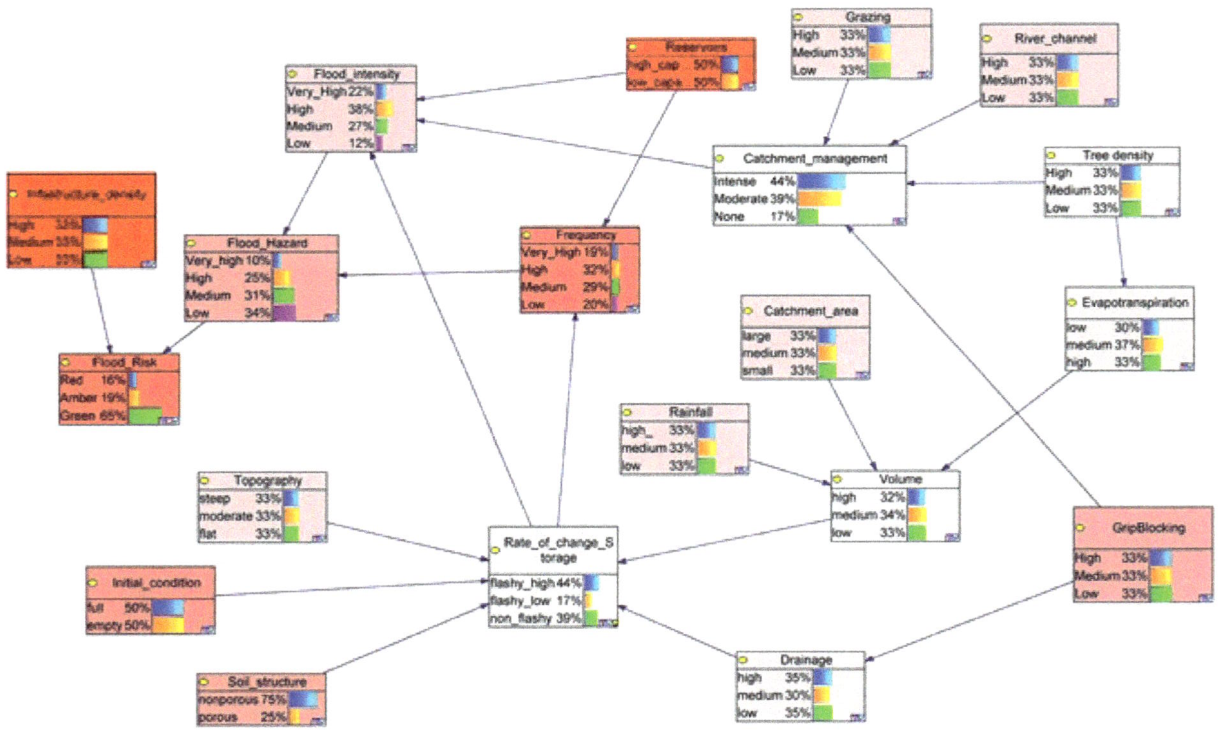

Figure 4.13 Bayesian networks are used to contextualise evidence for decision making. Here, 'Flood Risk' is dependent upon physical factors (e.g. soil structure, rainfall etc.) combined with human "interventions" such as storage dams ('Reservoirs') and changing upland drainage regimes ('Grip Blocking'). An added benefit of using Bayesian networks is that the sensitivity of the focal node ('Flood Risk') to uncertainty in the other nodes in the network (darker shades of red indicate greater importance for reducing uncertainty). Reducing uncertainty in, for example, the 'Reservoirs' node will have a greater impact on our understanding of 'Flood Risk'. Primary research could be targeted at these (darker red) nodes to reduce uncertainty in decisions about 'Flood Risk'. The systematic review and meta-analysis of Carrick et al. (2019) represent a single node in the network ('Tree density') whilst all other nodes come from the literature. (*Source:* Carrick et al., 2019, © 2018 The Chartered Institution of Water and Environment Management (CIWEM) and John Wiley & Sons Ltd, reproduced with permission from John Wiley & Sons Ltd).

now be developed using JAGs software (Plummer, 2003) through the R package Hydenet (Dalton and Nutter, 2019).

4.7.7 Evidence restatements

Policy decisions often need to be made in areas where the evidence base is both complex and heterogeneous (precluding formal meta-analysis) and highly contested. Evidence 'Restatements' seek to summarise the evidence base in a manner as policy neutral as possible while being clear about uncertainties and evidence gaps. If successful they can clarify the role of economic considerations, value judgements and other considerations in addition to the evidence base in the process of policy-making.

There are different ways to construct restatements and here we describe a model developed by the Oxford Martin School at Oxford University.

1. A topic is chosen after discussion with policy makers.
2. A panel of experts is then convened, typically from universities and research organisations.
3. A draft evidence summary is then prepared which takes the form of a series of numbered paragraphs written to be intelligible to a policy maker who is familiar with the subject but is not a technical expert. The summary is based on a systematic review of the literature and each statement is accompanied by an opinion on the strength of the evidence using a reserved vocabulary that varies between projects.
4. An extensive annotated bibliography is produced, linked to each paragraph, allowing policy makers access to the primary literature should they require further information.
5. The draft restatement is discussed paragraph by paragraph by the expert group and revised over several iterations.
6. This version is then sent out for review to a large group of people including other subject experts but also policy makers and interested parties (for example from the private sector and NGOs). They are asked to comment on the evidence itself, the assessment of the strength of the evidence, and also the aspiration that the restatement is policy neutral.
7. The restatement is revised in the light of these comments and the final version is drawn up after further iterations with the expert panel.
8. The restatement is submitted for publication to a journal (appearing as an appendix to a short 'stub' paper with the annotated bibliography as supplementary material online) where it receives a final review. Restatements are independently funded (by philanthropy).

Topics that have been the subject of restatements include the control of bovine TB in cattle and wildlife (Godfray et al., 2013) and the biological effects of low-dose ionising radiation (McLean et al., 2017). Other topics covered have included the effects of neonicotinoid insecticides on pollinators, landscape flood management, endocrine-disrupting chemicals, and control of *Campylobacter* infections.

Restatements have been popular with policy makers because (i) they strive to be policy neutral; (ii) they are conducted completely independently of the policy maker (important in areas of great contestation); (iii) they carry the authority of the expert panel and a scientific publication; and (iv) they are relatively brief and written to be understandable by policy makers. Their disadvantages include reliance on expert judgement (the heterogeneous evidence base precluding a more algorithmic approach) and the time and effort required of typically senior experts.

References

Adams, C.A., Fernández-Juricic, E., Bayne, E.M. et al. 2021. Effects of artificial light on bird movement and distribution: a systematic map. *Environmental Evidence* 10: Article 37, https://doi.org/10.1186/s13750-021-00246-8.

Ausden M., Sutherland W.J. and James R. 2001. The effects of flooding lowland wet grassland on soil macroinvertebrate prey of breeding wading birds. *Journal of Applied Ecology* 38: 320–38, https://doi.org/10.1046/j.1365-2664.2001.00600.x.

Back from the Brink. 2021. *Looking After Spring Speedwell* Veronica verna. Ecology and Conservation Portfolio (Salisbury: Plantlife), https://naturebftb.co.uk/wp-content/uploads/2021/06/Spring_speedwell_FINAL_LORES.pdf.

Boutin, C. and Harper, J.L. 1991. A comparative study of the population dynamics of five species of Veronica in natural habitats. *Journal of Ecology* 79: 199–221, https://doi.org/10.2307/2260793.

Buzan, T. 1974. *Use Your Head* (London: BBC Books).

Carrick, J., Abdul Rahim, M.S.A.B., Adjei, C., et al. 2019. Is planting trees the solution to reducing flood risks? *Journal of Flood Risk Management* 12(S2): e12484, https://doi.org/10.1111/jfr3.12484.

Christie, A.P., Amano, T., Martin, P.A., et al. 2020a. The challenge of biased evidence in conservation. *Conservation Biology* 35: 249–62, https://doi.org/10.1111/cobi.13577.

Christie, A.P., Amano, T., Martin, P.A., et al. 2020b. Poor availability of context-specific evidence hampers decision-making in conservation. *Biological Conservation* 248: 108666, https://doi.org/10.1016/j.biocon.2020.108666.

Christie, A.P., Chiaravalloti, R.M., Irvine, R., et al. 2022. Combining and assessing global and local evidence to improve decision-making. OSF, https://osf.io/wpa4t/.

CMP. 2020. Conservation Standards, https://conservationstandards.org/about/.

Cramp, S. and Simmons, K.E.L. 1983. *Handbook of the Birds of Europe, the Middle East and North Africa: The Birds of the Western Palearctic. Vol. III: Waders to Gulls* (Oxford: Oxford University Press).

Dalton, J.E. and Nutter, B. 2020. HydeNet: Hybrid Bayesian Networks Using R and JAGS. R package version 0.10.11, https://cran.rproject.org/web/packages/HydeNet/index.html

Fedrowitz, K., Koricheva, J., Baker, S.C. 2014. Can retention forestry help conserve biodiversity? A meta-analysis. *Journal of Applied Ecology* 51: 1669–79, https://doi.org/10.1111/1365-2664.12289.

Gigerenzer, G. and Edwards, A. 2003. Simple tools for understanding risks: From innumeracy to insight. *BMJ (Clinical research ed.)* 327: 741–44, https://doi.org/10.1136/bmj.327.7417.741.

Green, R.E. 1986. The management of lowland wet grassland for breeding waders. Unpublished RSPB Report.

Green, R.E. 1988. Effects of environmental factors on the timing and success of breeding of common snipe *Gallinago gallinago* (Aves: Scolopacidae). *Journal of Applied Ecology* 25: 79–93, https://doi.org/10.2307/2403611.

Green, R.E., Hirons, G.J.M., and Cresswell, B.H. 1990. Foraging habitats of female common snipe *Gallinago gallinago* during the incubation period. *Journal of Applied Ecology* 27: 325–35, https://doi.org/10.2307/2403589.

Godfray, H.C.J., Donnelly, C.A., Kao, R.R., et al. 2013. A restatement of the natural science evidence base relevant to the control of bovine tuberculosis in Great Britain. *Proceedings of the Royal Society B, Biological Sciences* 280: 1634–43, https://doi.org/10.1098/rspb.2013.1634.

Hoodless, A.N., Ewald, J.A., and Baines, D. 2007. Habitat use and diet of common snipe *Gallinago gallinago* breeding on moorland in northern England. *Bird Study* 54: 182–91, https://doi.org/10.1080/00063650709461474.

Holton, N. and Allcorn, R.I. 2006. The effectiveness of opening up rush patches on encouraging breeding common snipe *Gallinago gallinago* at Rogersceugh Farm, Campfield Marsh RSPB reserve, Cumbria, England. *Conservation Evidence* 3: 79–80, https://conservationevidencejournal.com/reference/pdf/2230.

Iacona, G.D., Sutherland, W.J., Mappin, B., et al. 2018. Standardized reporting of the costs of management interventions for biodiversity conservation. *Conservation Biology* 32: 979–88, https://doi.org/10.1111/cobi.13195.

IPCC. 2005. *Guidance Notes for Lead Authors of the IPCC Fourth Assessment Report on Addressing Uncertainties*, July 2005 (Intergovernmental Panel on Climate Change), https://www.ipcc.ch/site/assets/uploads/2018/02/ar4-uncertaintyguidancenote-1.pdf.

Jensen, F.V. 2001. Causal and Bayesian networks. In: *Bayesian Networks and Decision Graphs* (New York: Springer), pp. 3–34.

Kuhn, M. 2022. Caret: classification and regression training. R package version 6.0-93, https://CRAN.R-project.org/package=caret.

Landuyt, D., Broekx, S., D'hondt, R., et al. 2013. A review of Bayesian belief networks in ecosystem service modelling. *Environmental Modelling & Software* 46: 1–11, https://doi.org/10.1016/j.envsoft.2013.03.011.

Malpas, L.R., Kennerley, R.J., and Hirons, G.J.M., et al. 2013. The use of predator-exclusion fencing as a management tool improves the breeding success of waders on lowland wet grassland. *Journal for Nature Conservation* 21: 37–47, https://doi.org/10.1016/j.jnc.2012.09.002.

Mason, C.F. and MacDonald, S.M. 1976. Aspects of the breeding biology of Snipe. *Bird Study* 23: 33–38, https://doi.org/10.1080/00063657609476482.

McLean A.R., Adlen, E.K., Cardis E., et al. 2017. A restatement of the natural science evidence base concerning the health effects of low-level ionizing radiation. *Proceedings of the Royal Society B* 284, 20171070, https://doi.org/10.1098/rspb.2017.1070.

Nabe-Nielsen, L.I., Reddersen, J., and Nabe-Nielsen, J. 2021. Impacts of soil disturbance on plant diversity in a dry grassland. *Plant Ecology* 222: 1051–63, https://doi.org/10.1007/s11258-021-01160-2.

Nyberg, J. B., Marcot, B.G., and Sulyma, R. 2006. Using Bayesian belief networks in adaptive management. *Canadian Journal of Forest Research* 36: 3104–16, http://wildlifeinfometrics.com/wp-content/uploads/2018/02/Nyberg_et_al_2006_Bayesian_Models_in_AM.pdf.

Plummer, M. 2003. JAGS: A program for analysis of Bayesian graphical models using Gibbs sampling. In: *DSC 2003. Proceedings of the 3rd International Workshop on Distributed Statistical Computing*, ed. by K. Hornik, et al., 124: 1–10, https://www.r-project.org/nosvn/conferences/DSC-2003/Drafts/Plummer.pdf.

Popotnik, G.J. and Giuliano, W.M. 2000. Response of birds to grazing of riparian zones. *Journal of Wildlife Management* 64: 976–82, https://doi.org/10.2307/3803207.

Robson, B. and Allcorn, R.I. 2006. Rush cutting to create nesting patches for lapwings *Vanellus vanellus* and other waders, Lower Lough Erne RSPB reserve, County Fermanagh, Northern Ireland. *Conservation Evidence* 3: 81–83, https://conservationevidencejournal.com/reference/pdf/2231.

Salafsky, N., Boshoven, J., Burivalova, Z., et al. 2019. Defining and using evidence in conservation practice. *Conservation Science and Practice* 1: e27, https://doi.org/10.1111/csp2.27.

Salafsky, N., Irvine, R., Boshoven, J., et al. 2022. A practical approach to assessing existing evidence for specific conservation strategies. *Conservation Science and Practice* 4: e12654, https://doi.org/10.1111/csp2.12654.

Scutari, M. 2010. Learning Bayesian networks with the bnlearn R package. *arXiv Preprint*, https://doi.org/10.48550/arXiv.0908.3817.

Smart, J., Amar, A., O'Brien, M., et al. 2008. Changing land management of lowland wet grasslands of the UK: Impacts on snipe abundance and habitat quality. *Animal Conservation* 11: 339–51, https://doi.org/10.1111/j.1469-1795.2008.00189.x.

Spiegelhalter, D.J., Abrams, K.R., and Myles, J.P. 2004. *Bayesian Approaches to Clinical Trials and Health-Care Evaluation*, Vol. XIII (Chichester: John Wiley & Sons), https://doi.org/10.1002/0470092602.

Stewart, G. and Ward, J. 2019. Meta-science urgently needed across the environmental nexus: a comment on Berger-Tal et al. *Behavioural Ecology* 30: 9–10, https://doi.org/10.1093/beheco/ary155.

Stephenson, F., Mill, A.C., Scott, C.L., et al. 2018. Socio-economic, technological and environmental drivers of spatio-temporal changes in fishing pressure. *Marine Policy* 88: 189–203, https://doi.org/10.1016/j.marpol.2017.11.029.

Stewart, G.B., Mengersen, K., and Meader, N. 2014. Potential uses of Bayesian networks as tools for synthesis of systematic reviews of complex interventions. *Research Synthesis Methods* 5: 1–12, https://doi.org/10.1002/jrsm.1087.

Watt, A.S. 1971. Rare species in Breckland; their management for survival. *Journal of Applied Ecology* 8: 593–609, https://doi.org/10.2307/2402895.

White, T., Petrovan, S., Booth, H., et al. 2022. Determining the economic costs and benefits of conservation actions: A framework to support decision making. *Conservation Science & Practice*, https://doi.org/10.31219/osf.io/kd83v.

Wyden, P. 1980. *The Bay of Pigs: The Untold Story* (New York: Simon & Schuster).

5. Improving the Reliability of Judgements

Bonnie C. Wintle[1], Nibedita Mukherjee[2], Victoria Hemming[3], Stefano Canessa[4], Marissa McBride[5]

Judgements underpin all aspects of decision making, whether assessing the nature of the problem, interpreting evidence, or deciding upon the risks and benefits of different actions. Serious problems arise with conventional means of deriving judgements to inform decisions, such as confusing values and facts, listening to a single expert, or deploying processes prone to individual and group biases such as forced consensus. Thankfully there are methods that reduce the impact of such biases including processes such as the Delphi Technique and the IDEA protocol. This chapter outlines these methods to aid in eliciting better judgements in decision making.

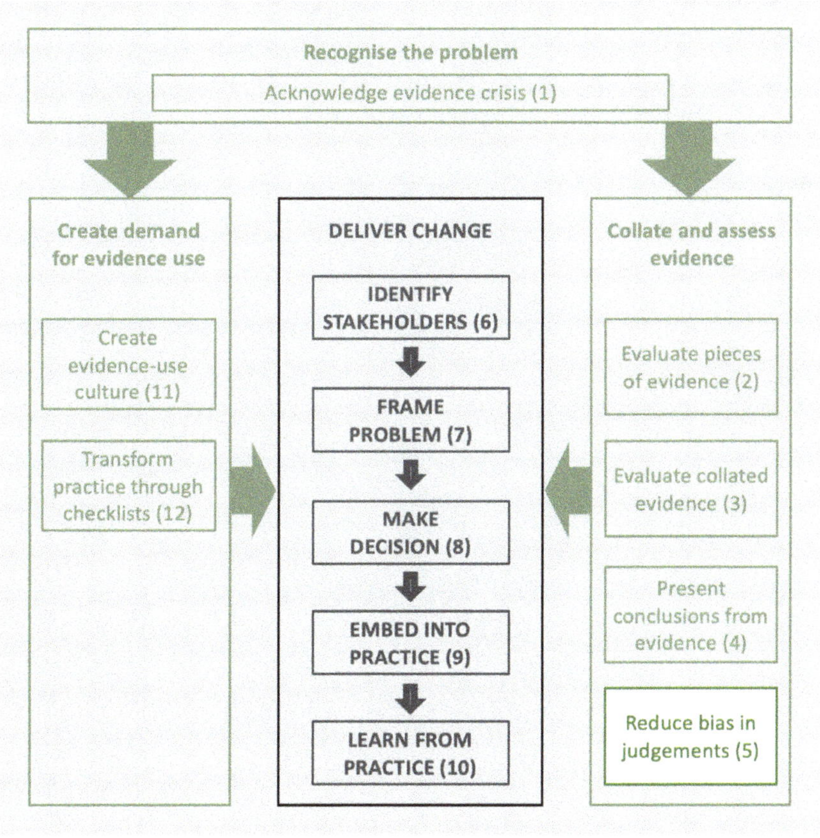

https://doi.org/10.11647/OBP.0321.05

[1] School of Ecosystem and Forest Sciences, University of Melbourne, Melbourne, VIC, Australia
[2] College of Business, Arts and Social Sciences, Brunel University, London, UK
[3] Department of Forest and Conservation Sciences, The University of British Columbia, Forest Sciences Centre, 3041–2424 Main Mall, Vancouver BC, V6T 1Z4 Canada
[4] Division of Conservation Biology, Institute for Ecology and Evolution, University of Bern, Switzerland
[5] Centre for Environmental Policy, Imperial College, Princes Gardens, South Kensington, London, UK

Contents

5.1 The Role of Judgements in Decision-Making

In an ideal world, many of us wish decision makers would 'just follow the evidence'. Yet, sometimes the evidence is less than clear. As described in the previous three chapters, the evidence is often mixed in format, reliability and relevance, and often insufficient to give a confident, straightforward, definitive answer. In order to proceed with less than perfect data, decision makers routinely rely on the judgements of people, namely experts (Burgman, 2015; Martin et al., 2012; Morgan, 2014).

We take a broad definition of expertise to include anyone who has sufficient familiarity with the subject matter to be able to grasp and knowledgeably answer the question at hand. In this chapter, we take a look at the performance of experts and why some judgements and decisions go horribly wrong. We describe the main sources of cognitive biases that can hinder judgements and then describe methods that can be employed to obtain more reliable judgements. These lessons and techniques can be applied widely, whether the judgement relates to agreeing on the status of a species, predicting the likely benefits from a conservation programme, or deciding which candidate will perform best for the task.

Despite our desire for decision makers to just follow the evidence, there is no such thing as a purely 'evidence-based' decision. As described in Chapter 8, all decisions involve judgements about the evidence combined with judgements about values (what are we trying to achieve? what trade-offs are we willing to make?) (Gregory et al., 2012; Hemming et al., 2022). We contend that the use of experts lies in helping to estimate figures and assess facts and evidence; this is our focus. For example, the invasive nature of the cane toad has been declared a failure of experts, but it was in fact predicted and clearly articulated by Froggatt (1936). Unfortunately, their advice was ignored as political and public pressure pushed for the widespread release of the cane toad (Keogh, 2011).

5.2 When Experts Are Good (and Not so Good)

There are competing narratives about the nature of expertise. On the one hand, there is literature describing experts as demonstrating 'elite, peak, or exceptionally high levels of performance on a particular task or within a given domain' (Bourne Jr et al., 2014), often due to superior cognitive functioning. For example, chess experts can recognise patterns faster than novices (Bilalic et al., 2010). They have developed expertise through 'deliberate practice': that is, a regimen of effortful activities, usually over ten years or more, designed to optimise improvement (Ericsson et al., 1993; Charness et al., 2005). After this, experts are equipped with skills and experience that allow them to solve complex problems with greater accuracy and in a fraction of the time than novices. Not only do they store a great deal of content, but also the best pathways to access that information, which can be likened to a book's index. For well-defined questions (e.g. physics problems), experts can select and apply appropriate mental algorithms that lead them to a solution faster (Larkin et al., 1980). Rather than working backwards from the unknowns as novices tend to do, experts identify what is known and work forwards until they have solved the desired unknown. However, though experts can show impressive knowledge within their domain, once outside their narrow area of expertise expert performance is surprisingly limited (Ericsson and Lehmann, 1996), and the delineation between expert and lay knowledge is not a sharp one (Jasanoff, 2006).

On the other hand, there are studies that paint a bleak picture of experts who are 'not immune to the cognitive illusions that affect other people' (Kahneman, 1991, p.144) (see also Box 5.1). In fact, Tversky and Kahneman (1971) first evidenced the 'law of small numbers' bias (i.e. assuming that a small sample is highly representative of the population) with survey responses from a group of mathematical psychologists.

Box 5.1 The serious challenge of relying on experts: three examples

Tetlock (2005) identified 284 people who made their living from making judgements and asked them to make a total of over 80,000 predictions in their area of expertise. They were asked whether something, say the oil price, would be higher, lower or the same on a given date. The experts' predictions performed barely better than random chance and those who knew more were less reliable — even in their domain — because of their overconfidence.

Burgman et al. (2011) looked at how well groups of specialists, from epidemiologists to frog biologists, answered questions of fact within their domain of expertise. Each participant introduced themselves and described their own experience and credentials, judged each other's ability to get answers right, and then was given a series of factual questions to answer. Those perceived as possessing the highest levels of expertise were no more accurate than those perceived as less expert.

One of the reasons it is difficult to translate substantive expertise into accurate judgements is that experts are human, and humans are biased, and highly influenced by context. In a famous example of anchoring, Tversky and Kahneman (1974) showed people a roulette wheel rigged to stop on either 10 or 65, and then asked them to state the percentage of UN countries that are in Africa. Those whose wheel stopped at 10 guessed 25% on average, while those whose wheel stopped at 65 guessed an average of 45%.

Study after study has found expert judgements to be overconfident (McKenzie et al., 2008), and lacking in validity (Oskamp, 1965) and reliability (Trumbo et al., 1962), to name a few problems. Disastrous decisions have been linked back to flawed judgements, resulting in nuclear reactor disasters (Hollnagel and Fujita, 2013), poor military strategy (Kent, 1964), and diagnostic errors in medicine and clinical psychology (Graber, 2005). Experts are prone to the cognitive biases that befall us all (see Section 5.3). On top of this, they can also be influenced by social and political pressures.

Shanteau (1992) explained the apparent contradiction between the optimistic and pessimistic views of expertise by pointing out that the domains of expertise in each camp vary.

The pessimistic story is told by researchers who have focused on judgements in domains that are more dynamic and unpredictable (e.g. behavioural economics, clinical psychology) (e.g. Dawes, 1994). The more optimistic story comes out of studies on judgement in more static, rule-based domains (e.g. physics, chess) (Anderson, 1981).

Studies in naturalistic (non-experimental) settings also provide some evidence that the pessimistic view is indeed too bleak. For example, using a sample of 19,396 observations, Johnson and Bruce (2001) found an almost perfect correlation between the subjective probability judgments of horses' success (implicit in the bettors' wagering activities), and the objective probability of success as determined by race outcomes. Murphy and Winkler (1977) found that weather forecasters were strikingly accurate, despite their poor reputation. Similarly, Stewart et al. (1997) found experts were better than an operational model at precipitation and temperature forecasts. One reason why these examples show such good expert performance is because the task is reasonably predictable and the experts in these environments receive constant feedback, giving them an opportunity to learn the regularities of the task (Kahneman and Klein, 2009). If a naturalistic setting is irregular (e.g. a 1-in-100-year flood occurs that year), an expert's knowledge will not translate to the same high levels of performance.

It follows, then, that the difference between *when* experts appear to be good and bad is task dependent. In predictable tasks (like chess), experts perform well and can learn from deliberate practice and feedback. Unfortunately, the sorts of judgements that are made in environmental science, such as probability judgements for risk assessments, *do not* generally develop from deliberate practice. Forecasts, in particular, are difficult because projections — such as future disease rates — are made over relatively long time frames and are not easily validated. The *best* forecast can never be truly determined because the truth is unknowable at the current time. This introduces yet another challenge: it is often in these dynamic fields (e.g. emerging technologies) where conflict arises, because (a) there is much at stake, and (b) it is near impossible to prove the *best* judgement or even the best *judge*.

Problematically, attempts to select better experts by vetting them based on their professional and demographic credentials, such as speciality, years of experience, self or peer recommendation, and age, have often been in vain since these credentials are unreliable predictors of the quality of judgements provided by individuals. That is to say, studies have repeatedly shown there is typically no correlation between these attributes and the performance of experts under uncertainty (Burgman et al., 2011; Hemming et al., 2018b, 2020a; Tetlock, 2005; Tetlock and Gardner, 2016).

The less-than-perfect performance of experts when making judgements and forecasts in unpredictable environments naturally raises questions about their role in decision making, and what, if anything, can be done to improve judgements when experts are required. Despite the pessimistic outlook we've just painted, there is hope. A deeper dive into the track record of experts indicates many of the problems may not be with using experts per se, but rather with us using experts inappropriately (Morgan, 2014). For example, asking questions which relate to values rather than facts, giving questions with vague language (Kent, 1964; Morgan, 2014), relying on a single expert (Keogh, 2011), assuming credentials will align with more reliable expertise (Burgman et al., 2011; Shanteau et al., 2003), failing to elicit or account for uncertainty in decision making (Gregory and Keeney, 2017), failing to subject the elicitation of expert judgement to the same level of transparency and reproducibility as is required for other forms of empirical data (Drescher and Edwards, 2018; French, 2012), and ignoring or overlooking techniques that have been shown to help experts provide their best judgements under uncertainty (for example structuring group interactions).

The good news is that the judgements of experts, and the decisions that rely on them, can be improved by understanding why experts may make mistakes (biases and heuristics, Section 5.3), and by employing structured approaches that help experts to provide their best judgements under uncertainty (Section 5.5).

5.3 Blind Spots of the Human Mind

When faced with decisions, the human brain typically seeks to conserve energy by simplifying problems and making them more manageable, applying a range of mental short-cuts known as 'heuristics' (Simon, 1977). These short-cuts help us handle thousands of daily decisions effectively and effortlessly, but in some cases they can lead to a range of cognitive biases and poor decisions. Research on such phenomena, initiated by Amos Tversky and Daniel Kahneman in the 1970s, has highlighted a wide range of biases and situations where they can affect our judgments. This is particularly true when we face severe uncertainty, time pressure, strong emotions, and high stakes. When turning to experts to assist decisions, the risk of bias is higher when judgments are elicited in an unstructured, ad hoc way, and when reasoning and decision processes are opaque. Studies in other fields have shown experts are also prone to biases (Berthet, 2022), and there is no reason to assume conservation experts are any different.

Considerable research has investigated the impact of biases in medicine (Saposnik et al., 2016), negotiation (Caputo, 2013), computer engineering (Mohanani et al., 2018), tourist decisions (Wattanacharoensil and La-ornual, 2019), and forensic science (Cooper and Meterko, 2019). Below, we discuss a few examples of common biases that are likely to influence conservation decisions through biased expert judgments, individually or interacting with one another (additional biases can be found in Table 5.1).

Experts are often consulted for their (true or perceived) superior access to information and interpretation of it. Heuristics are typically invoked when interpreting evidence, which can lead to irrational judgements that sometimes violate probability axioms and logic. For example, we tend to give too much importance to events that are easier to recall, possibly because of personal experience or because they have occurred recently ('availability heuristic'; Tversky and Kahneman, 1973). For example, we may overestimate the danger of flying if an airline crash has been in the news recently. We also tend to classify events or people based on preconceived classes or stereotypes ('representativeness heuristic'; Kahneman and Tversky, 1972). As a result, we may discard general or baseline information in favour of specific details ('base rate neglect'; Bar-Hillel, 1980), and even conclude that a specific case is more likely than a general one, which contradicts logic ('conjunction fallacy'; Tversky and Kahneman, 1983). We also tend to place too much faith in small samples, inferring causality from extreme outcomes that are more likely to be caused by random fluctuations ('belief in the law of small numbers'; Tversky and Kahneman, 1971). If those random fluctuations are then followed by more normal events, we might be tempted to conclude — incorrectly — that this return to normality was caused by interventions we adopted ('regression to the mean'; Barnett et al., 2005). All these biases may be especially relevant in conservation, where samples are often small, environmental variability high, and causality difficult to infer, leading to flawed judgements in general (e.g. when interpreting time series; Fournier et al., 2019) and specific cases (e.g. when evaluating management of endangered species; Margalida et al., 2017).

Presenting and asking for information in different ways can shift people's preferences and change the way they interpret probabilities ('framing bias'; Tversky and Kahneman, 1985). For example, when asked about the effectiveness of a conservation action, experts give different estimates when the question is framed in terms of mortality than survival, although they *should* be complementary (Perneger and Agoritsas, 2011). The framing bias is compounded in judgements reflecting loss aversion, that is, our tendency to respond to losses more strongly than to gains, underpinning 'prospect theory' (Kahneman and Tversky, 1979), for which Kahneman won a Nobel Prize in Economics in 2002. As with the almost farcical example of estimates being influenced by watching a roulette wheel (Box 5.1), a range of other studies have shown how responses can be affected by irrelevant numbers given in background information, or unrelated information in the previous question with a tendency to anchor on such values and simply adjust estimates up or down from that point, in the direction that seems intuitive given the question ('anchoring heuristic'; Tversky and Kahneman, 1974).

It is no surprise that we tend to interpret information in a way that is consistent with our pre-existing view of the world ('motivated reasoning'; Kunda, 1990). For example, a geneticist and a demographer might focus on different evidence to explain a species decline. Further, we tend to seek information that confirms our beliefs, and discard information that does not ('confirmation bias'; Nickerson, 1998). For example, reviewers who disagree with the conclusions of a scientific paper seek flaws in the methods to justify rejecting it (Mahoney, 1977). Discarding newly acquired cognitions (or pieces of knowledge) that are inconsistent with our existing cognitions helps reduce cognitive dissonance. This might happen, for example, when the results of a conservation management trial contradict prior expectations, but we are reluctant to change those expectations or do so inconsistently (Canessa et al., 2020).

It is tempting to think that we, as rational individuals, are more immune to these well-known biases than are others; studies show that people see others as more susceptible to cognitive and motivational biases than themselves (Ehrlinger et al., 2005). But awareness of these issues is only the first step in properly addressing them, and the best antidote for individual biases is groups.

Although, as discussed below, it is good practice to consult multiple experts (Sutherland and Burgman, 2015), additional social psychological biases can arise if individuals interact in an unstructured way. For example, groups tend to make more extreme and risky judgments than would individuals ('group think'; Janis, 1982). Groups also tend to become more optimistic about the accuracy of their collective judgement ('overconfidence'; Sniezek, 1992), particularly when faced with difficult tasks (Sniezek et al., 1990) that result in low accuracy (Puncochar and

Fox, 2004). Group decisions can be led astray by charismatic, convincing or extroverted people ('halo effects'; Thorndike, 1920) who may override the influence of those with better expertise. Kuran and Sunstein (1998) describe the problem of 'informational cascades' and 'reputational cascades' in which the group consensus is distorted. In informational cascades, early adopters of an idea or perspective can overly influence the acceptance or rejection of it as 'fact', with others questioning their own perspective if it is in disagreement, and sometimes failing to disclose potentially important contradictory information. In reputational cascades, people retain private beliefs that they consider correct, although against the emerging consensus, but do not reveal them in case it damages their reputation or causes hostility. Cascades are especially likely when the topic has an emotional resonance. When there is momentum behind a falsehood, it becomes easier to buy into.

Table 5.1 Some of the most common sources of bias. (*Source:* adapted from Mukherjee et al., 2018)

Name of bias	Description
Anchoring	Individuals take an initial piece of information as a benchmark, and then give it disproportionate weight in judging subsequent information and making decisions.
Availability heuristic	Individuals more easily recall more recent information, and therefore give it disproportionate weight in judging subsequent information and making decisions.
Base rate neglect	Individuals ignore base rate information (how likely an event is in general) in favour of specific information (details about a specific case), possibly because of 'representativeness'.
Belief in the law of small numbers	Individuals overestimate how much small samples represent the general population.
Confirmation bias	Individuals tend to selectively search for, interpret or recall information that confirms their pre-existing beliefs.
Conjunction fallacy	Individuals incorrectly assume that two conditions (events A and B) are more likely than one of them (event A), possibly because of representativeness.
Dominance effect	Individuals who are perceived to be dominant (even though they might not have better decision-making abilities) have a disproportionate influence in group decision making.
Egocentrism	Individuals tend to preferentially rate their own opinion higher than that of others.
Evaluation apprehension	In a group, individuals are concerned about how they are being judged by others and this affects their decision outcomes.
Framing effect	Individuals make a different decision depending on how the same information is presented.

Name of bias	Description
Free riding or social loafing	People reduce their effort when working in a group, as opposed to working alone, expecting other group members to make the assessment.
Group think	At the expense of independent critical thinking, individuals in a group seek concurrence, avoid creating disunity, and support the decisions taken by the majority or the perceived leader of the group. The desire or pressure to be accepted as a good group member leads to acceptance of a majority decision although a better decision was possible with better group dynamics.
Halo effect	An individual's decisions or perceptions are coloured by perceptions of attributes (e.g. charisma or attractiveness) that are totally unrelated to the topic being evaluated.
Hidden profile	In a group discussion some information is shared by all members but other relevant information is not.
Hindsight bias	Individuals believe that they 'knew it all along' i.e. an event is more predictable after it has already occurred than before.
Information cascade	An individual modifies actions or decisions based on observations of others in the group at the cost of their own information or judgement.
Myopic loss aversion	Individuals temporarily lose sight of the big picture and concentrate on the immediate problem at hand. This may lead to erratic decisions that are not beneficial in the long term.
Naïve realism	An individual thinks that their reality is more objective and unbiased compared to those who hold a different opinion.
Overconfidence effect	Tendency of an individual to have higher subjective confidence in their judgement than objective accuracy would allow.
Prospect theory	Individuals are risk-averse when facing gains (they prefer lower wins with more certainty), but risk-seeking when facing losses (they prefer higher losses with more uncertainty, that is, even a small chance of minimising losses).
Representativeness	Individuals can incorrectly classify people or events, giving disproportionate weight to how representative of that class they believe it to be, rather than to the actual probability that it belongs to it.
Semmelweis reflex	Individuals reject new evidence that contradicts a paradigm.
Shared information bias	The tendency of individuals in a group to discuss preferentially the information that is familiar to all compared to information that only a few know.

5.4 Strategies for Improving Judgements

Recognising the serious problems identified in the previous section, researchers have studied how to make the decision-making process better (more accurate, less biased, more factual, more empathetic to diversity, more rational, and fundamentally evidence-driven) (Berthet, 2022; Saposnik et al., 2016). Some simple strategies can considerably overcome biases during elicitation. The first is to restrict expert judgements to the estimation of facts, for example, estimating quantities or probabilities, rather than asking experts what to do. The second is to seek the advice of more than one expert. The third is to apply protocols to help experts provide their best judgements under uncertainty, acknowledging that experts' beliefs are not in the form of clear probability distributions waiting to be elicited, rather, they are partially shaped by the elicitation procedure (Winkler, 1967).

Such insights can be used to explore methods for improving the quality of judgments. Most of these techniques are adapted from the human judgement literature, some of which is based on experimental results and some based on theory or known good practice generally.

5.4.1 Improving individual judgements

Although the most effective way to mitigate biases is to consult a group of people, it is not always possible to recruit multiple experts. With or without groups, there are still simple strategies that can be used to get the best out of individual experts. These suggestions are described below and summarised in Table 5.2.

Consider the opposite

Biases, such as overconfidence, can be mitigated by explicitly considering counter evidence, or reasons for the alternative view (Hoch, 1985). For example, when asking participants for judgments and their level of confidence in their answer, Koriat et al. (1980) found that calibration improved when participants were also asked specifically for contradicting reasons against their answer.

Reduce linguistic uncertainty

Language-based ambiguity, vagueness and under-specificity in elicitation problems create confusion (Regan et al., 2002) and can be reduced by carefully defining terms, specifying context and thresholds, and asking for quantitative likelihoods rather than qualitative ones (e.g. does *very unlikely* mean 0.1 or 0.3 probability?) (Wintle et al., 2019). See also the interpretation of terms and probabilities in Table 4.3.

Linguistic uncertainty can be problematic because it also allows for the expression of motivational biases, especially in judgements about risk. Experts — if given the space to do so through ambiguous, vague or underspecified language — may give inflated or conservative judgements of likelihood and consequence, either to be on the safe side of environmental protection, or to avoid obstacles to development. By clarifying language and context, experts will be less inclined to conflate value judgements with factual ones when answering elicitation questions.

Present evidence and questions in frequency formats

Evidence is sometimes presented as a probability, such as the likelihood that a threatened species is present although not detected in surveys. Research shows that most people find these statements easier to comprehend using frequency formats (e.g. 3 in 100) rather than probabilities (e.g. 0.03). When dealing with frequency formats, people make fewer errors of probabilistic reasoning, such as base-rate neglect (Gigerenzer et al., 2008).

Use neutral problem frames

Debiasing may also be achieved by using balanced or neutral problem frames. Williams and Mandel (2007) reframed questions to include the probability complement — for example, consider the chance of having a virus as well as the chance of not having a virus. They found that adding this extra element improved the accuracy and coherence of probability judgements. Where possible, avoid framing the decision problem in purely positive or negative terms (e.g. in terms of loss or gain), or at least, attempt a balanced presentation of information within the question or problem context.

Elicit uncertainty with free choice interval judgements

Judgments about facts and parameters are often elicited in the form of an interval (Lin and Bier, 2008). This expression of uncertainty is essential for decision makers who wish to exercise their risk attitude and base their decision on the best or worst case scenario (the uncertainty bounds), or the nominal case (best estimate) (Burgman, 2005). Interval judgements generally have an attached confidence, for example, they provide an interval that the person is 80% sure contains the true GDP of Canada. Unfortunately, overconfidence is particularly prevalent in interval judgments (Moore and Healy, 2008), with 90% intervals typically containing the answer only 50% or less of the time (Soll and Klayman, 2004). People tend to use a constant interval width over a wide range of confidence levels, leading to a high degree of overconfidence for 90% intervals, but much less for 50% intervals or for intervals without a pre-assigned degree of confidence (free choice intervals) (Teigen and Jørgensen, 2005). For this reason, we suggest allowing experts to assign their own confidence to their interval. Low confidence intervals (e.g. 50%) can later be converted (stretched) to a higher level of confidence (e.g. 80%) as required (Speirs-Bridge et al., 2010).

Elicit estimates over multiple steps

Dividing the question into multiple steps improves the chances that people will think about different kinds of evidence, and be less biased (Soll and Klayman, 2004). Instead of asking for a range of values in which the answer is thought to lie, the question can be broken into three: what is the highest plausible number of individuals, what is the lowest plausible number of individuals, and what is your best estimate of the number of individuals? This helps avoid answers that are too precise, or intervals that are too narrow and overconfident. Question order is also thought to affect such judgments: eliciting a best estimate before the bounds can lead to anchoring on the best estimate and producing overly narrow (overconfident) bounds around it, compared to when the interval is elicited first (Soll and Klayman, 2004). This approach is known as the 3-point interval elicitation method (for quantities) (Speirs-Bridge et al., 2010).

Ask the same person repeatedly for the same judgement

Herzog and Hertwig (2009) demonstrated that simply averaging two estimates of the same quantity *from the same person* leads to more accurate judgments than either judgement alone, by harnessing the 'wisdom of the crowd within'. Using techniques to increase the independence of the second estimate, such as encouraging people to consider why their first one might be wrong (as above), they found an accuracy improvement of 4.1 percentage points (a medium effect size (Cohen's d) of 0.53), which is a substantial gain when compared with 7 percentage points achieved through total independence (from averaging the estimates of two individual people).

Moreover, leaving a longer time between estimates from one individual also increases their independence, further enhancing the accuracy of the internal average (from 6% to 16% error reduction after three weeks) — similar to 'sleeping on it' (Vul and Pashler, 2008). Of course, improved accuracy could also be due to more evidence having been acquired in that time, but participants in Vul and Pashler's experiment were unaware that they would be tested a second time, and the questions were such that an incidental acquisition of evidence was unlikely.

Give individual feedback

Feedback is important for learning and can improve estimation (Kopelman, 1986). A meta-analysis by Kluger and DeNisi (1996) (607 effect sizes; 23,663 observations) showed that while feedback interventions improved performance on average (d = 0.41), over one-third of the feedback interventions decreased performance. This suggests that not all types of feedback are equally effective.

Outcome feedback refers to learning the results or true value (e.g. *actual* length of the Nile River) and, although commonly provided, tests show it to be ineffective in improving probability forecasts (Fischer, 1982) if the information does not contribute to a pattern that can inform further answers (Benson and Önkal, 1992).

Cognitive feedback focuses on the problem solving process and is more successful (Newell et al., 2009). One effective approach is calibration feedback, which compares a person's overall correct answers — known as percentage 'hits' — with their confidence levels (e.g. Lichtenstein and Fischhoff, 1980). Giving calibration feedback about people's hit rates (i.e. asking them to compare the percentage of their intervals that captured the 'true value' with their levels of confidence) improved participants' subsequent estimates of species abundance (Wintle et al., 2013).

Table 5.2 Summary of strategies for improving individual experts' judgements.

Strategy	Description
Consider the opposite	Thinking of reasons why you might be wrong reduces overconfidence and improves accuracy.
Resolve linguistic uncertainty: clarify and quantify	Resolving vague, ambiguous, underspecified or context-dependent terms helps disentangle value judgements and minimises linguistic uncertainty.
Present questions in frequency formats	Presenting the question in frequencies rather than probabilities overcomes problems like base-rate neglect.
Use neutral problem frames	Framing the question or problem in neutral or balanced terms improves probability judgements.
Elicit uncertainty with free choice interval judgements	Eliciting uncertainty using interval judgements, and allowing experts to assign their own confidence.
Elicit estimates over multiple steps	Asking the question over multiple steps reduces overconfidence.
Ask for the judgement twice	Averaging two estimates of the same quantity from the same person improves accuracy.
Give feedback	Systematic feedback improves judgement performance, particularly cognitive feedback
Stay within the expert's domain	An expert who performs well in one domain will not necessarily perform well in another, even if it is closely related.
Be transparent about values and motivational biases	Disclosing agendas allows experts to cross-examine the arguments of others. Broader discussions — even amongst experts — are more likely to encompass the values of the public.

5.4.2 Improving group judgements

Numerous studies have shown that the combined judgements of a group of individuals are usually more accurate than any given individual, even those that might be considered to be the 'best expert' (e.g. Burgman et al., 2011; Wintle et al., 2013; Mellers et al., 2014; Hemming et al., 2020b). Individual biases tend to cancel each other out when aggregated, leading to a wisdom of crowds effect (Surowiecki, 2004). This is particularly powerful in the group average when different group members' judgements 'bracket' the true value they are attempting to estimate (Larrick and Soll, 2006). If not, groups may give precise but inaccurate judgements.

Eliciting judgements from groups introduces a whole new suite of considerations, including the problems and benefits of interaction, and group composition. The section below touches upon some important considerations and commonly used strategies to reduce bias in groups (some of these are summarised in Table 5.3).

Ask for the judgement from multiple experts

Taking the average of quantitative estimates provided by two people typically achieves better percentage accuracy improvement than choosing the estimate of the best performing judge (e.g. Soll and Larrick, 2009 [experiment 1: 17% improvement from averaging, 12.9% improvement from choosing]).

Benefits of discussion

There is good evidence that group interaction and discussion usually improves the accuracy of judgements, particularly for quantitative tasks, with group members becoming more accurate individually during group interaction (Schultze et al., 2012; Mercier and Claidière, 2022). Deliberation and social influence can also improve the crowd's collective accuracy when structured in small, independent groups (Navajas et al., 2017). Discussion offers the potential to improve group performance by resolving misunderstanding of the question, providing opportunities for people to introduce new information and learn from each other (Mojzisch and Schulz-Hardt, 2010), encouraging critical thinking (Postmes et al., 2001), and encouraging counterfactual reasoning (Galinsky and Kray, 2003). A study with 211 teams engaged in 969 face-to-face discussions (Silver et al., 2021) found that group interaction improved group accuracy, but only when groups were already collectively well calibrated (i.e. more accurate people were more confident, and less accurate people were less confident going into discussion). Under these conditions, the most knowledgeable (and confident) people were more likely to influence the answers of the less knowledgeable people in the group.

Successful group interaction — that is, interaction that improves the performance of the group — requires ensuring that individuals do learn from the group. This means encouraging group members to share information, and avoiding 'hidden profiles' (where pieces of relevant information are not shared by everyone in the group) (Stasser and Titus, 2003). Several factors moderate this, including the type of task (Stasser and Stewart, 1992).

Foster dissent and argumentation

Disagreeing participants are poised to produce an often beneficial discussion, but such a discussion may not necessarily follow unless all group members are encouraged to contribute their perspectives. There are benefits of dissent within groups. Mercier and Sperber (2011) contend that reasoning in defence of an argument, to persuade another, is more rigorous than reasoning in pursuit of truth or to solve abstract problems. Benefits also come from the division of cognitive labour — it is more efficient for one person to adopt a position and gather evidence supporting that position. Schulz-Hardt et al. (2006) found that groups with dissenting members (pre-discussion) made better decisions than homogenous groups.

When pre-existing dissent is lacking, it may need to be artificially fostered using structured methods. Two specific approaches have been explored for stimulating disagreement within a group, which could also be useful for groups of experts: Devil's Advocacy (DA) and Dialectical Inquiry (DI) (reviewed by Katzenstein, 1996). Dialectical Inquiry involves the development of two plans that rest on a conflicting set of assumptions. Each side then engages in a debate about the relative merits of each. In Devil's Advocacy, the prevailing plan is criticised by the devil's advocate, but no alternative plan is proposed. In a meta-analysis of five studies that investigated the effects of DA and DI, Schwenk (1990) found that DA leads to superior decision making, over an expert-based approach (n = 432, effect sizes ranging from d = 0.64 to 0.13), but the effect of DI over the expert-based approach was smaller and not statistically significant (n = 393, average d = 0.16, confidence interval included zero: -0.10 to 0.42).

Critical and counterfactual thinking, not consensus

Postmes et al. (2001) primed 'consensus' groups and 'critical' groups by participating in different lead up tasks (collaborating in producing a poster and reaching a consensus about the design, or discussing differences of opinion about a new tax law). The experimental task itself involved selecting a job candidate, where there was an optimal solution to be uncovered via information sharing, to reduce hidden profiles. In both consensus and critical thinking groups, 11% of group members selected the correct candidate for the job before discussion. After discussion, the critical thinking primed groups substantially outperformed the consensus primed groups — 67% of the former made the correct decision, compared to only 22% of the latter. In a similar study, Galinsky and Kray, (2003) found that counterfactually primed groups made the correct decision 40–67% of the time, compared to 15–23% for non-counterfactual primed groups (the range here reflects lowest to highest best estimates from multiple experiments).

Anonymity

Senior, high-status team members often dominate the lower-status members, even though the junior members are often more accurate (Kerr and Tindale, 2011). In workshop conditions, junior members are often reluctant to speak up if their view contradicts more senior members, to the detriment of group accuracy (Thomas and Fink, 1961; Galinsky et al., 2008), and halo effects mean that junior people defer to the wisdom of senior people (Thorndike, 1920), even though seniority does not tend to correlate with performance (Burgman et al., 2011). Anonymity, such as adding material online, can buffer against this, reducing social barriers to participation in group deliberations by insulating group members from reputational pressures (Valacich et al., 1992) and reducing the pressure to conform with scoring.

Give group feedback

Providing feedback, such as seeing the distribution of anonymously provided judgements, may improve estimation performance (Mukherjee et al., 2015; Wintle et al., 2013). Encouraging discussion on the distribution, and getting feedback on the judgements and reasoning of other group members before revising judgements, is an essential component of the Delphi Technique and IDEA protocols (described in more detail in Section 5.5).

Hold group members accountable (for the process)

Group members who are held accountable for their decisions show increased motivation, self-criticism, and information processing (Scholten et al., 2007). For example, professional auditors who are held accountable make more accurate classifications of industrial bond issues into financial rating categories (Ashton, 1992). But not all kinds of accountability are effective.

'Process accountability' (being accountable for the way judgements were reached) is typically thought to improve information processing because people need to demonstrate a comprehensive understanding of the decision problem (e.g. Lerner and Tetlock, 1999; De Dreu et al., 2000). It has been found to improve accuracy by encouraging people to take more of the available information into account; without it, people are more likely to formulate judgements based on insufficient evidence. 'Outcome accountability' involves evaluating groups based on the outcomes of their decisions. It affects information processing in a different way because people focus on pleasing an audience, introducing a question of whether accountability alters how people think or merely what people say they think (Lerner and Tetlock, 1999). It can erode accuracy by increasing the amount of noise (or 'scatter') in participants' judgments (Siegel-Jacobs and Yates, 1996) and might cause people to focus on irrelevant details (Stewart et al., 1998).

Evaluate each other's judgements

Studies have shown that people are better at 'evaluation' (assigning a probability that their selection is true or contains the truth, in the case of intervals) than 'production' (also known as 'generation'). Winman et al. (2004) give an example: imagine you are about to buy a house and you want to know what the interest rates will be over the coming year; you could either ask your financial advisor for an interval of probable interest rates, or you could create your own interval and ask the financial advisor to evaluate the probability that your interval will contain the true value. These are formally equivalent ways of expressing uncertainty about the same event, yet Winman et al. show that the latter format almost abolishes overconfidence. This prompted Aidan Lyon, Fiona Fidler, Bonnie C. Wintle and Mark A. Burgman (unpublished) to conduct an experiment where participants produced and evaluated their own interval judgments for a series of quantitative questions before also evaluating someone else's. Four replications of this experiment (total n = 111) showed that judgement swapping resulted in a 6.88% percentage point reduction in overconfidence compared to evaluating their own interval.

Consider the judgements of others

Group members could practically benefit from simple elicitation techniques that encourage them to think about what other people would do, or what sort of judgements they might make. Yaniv and Choshen-Hillel (2012a) asked participants to estimate the calories contained in 20 target foods — what is the calorie value of an orange? — then view five additional randomly drawn answers suggested by group members displayed below their own. After viewing the five advisory estimates, participants made a second, final judgement. Participants had a mean absolute error of 91.2 for their initial judgements and a more accurate subsequent final judgement (76.2), indicating that they integrated — to a good extent — the judgments of the five anonymous advisors.

This is similar logic to another study (Yaniv and Choshen-Hillel, 2012b) showing that people improve their accuracy when putting themselves in someone else's shoes, that is, when asked to predict the judgement that another (matched) participant would give if they were shown the same set of estimates. Related research shows that voting polls are more accurate when people are asked for their expectations (who do you think will win the upcoming election?) rather than their intentions (if the election were held today, who would you vote for?) (Rothschild and Wolfers, 2011). Similarly, 'social circle' polls that ask participants about the voting intentions of their social contacts are also more accurate (Galesic et al., 2018).

In expert elicitation, there is a benefit to seeking an initial, independent judgement from each expert that genuinely reflects that expert's knowledge and perspective, before integrating other perspectives.

Diversity

The benefit of groups depends on the diversity of their members (Hong and Page, 2004, 2020; Page, 2007), partly because biases in different directions cancel each other out in the group average, but also because individual members of a diverse group bring different perspectives and ignite interesting debates. When planning for a hydroelectric facility in Canada, Failing et al. (2007) show that combining technical expertise with other knowledge sources drawn from local residents and Indigenous communities promoted a broader understanding of causal pathways and the consequences of flow changes to the river system, and made it more acceptable to a broader range of stakeholders. Theoretically, it has been shown that the benefit of groups is greatest where the overlap between the knowledge bases of individual members is least (i.e. members possess independent knowledge) (Clemen and Winkler, 1985). Homogenous groups induce spurious consensus and interdependence, producing unduly high confidence without the accuracy to match (Yaniv et al., 2009). Selecting for member diversity using information on demographics, experience, worldview, and cognitive reasoning style may be one way to reduce dependency between members, and studies have shown diverse groups to outperform homogenous groups in terms of quality of problem solutions (Worchel et al., 1992).

Where group members are not already diverse, then it may be possible to stimulate cognitive diversity (different ways of thinking within a group), for example, different mindsets can be fostered or primed via framing a problem positively for some group members (e.g. money saved), and negatively for others (e.g. money spent) (Yaniv, 2011).

Role playing within a group has been found to improve forecasting accuracy (Armstrong, 2001). This might involve giving members different cost functions for the question at hand or getting them to act out an interaction that they might expect to see from representatives of different parties.

Managing communication and conflict in diverse groups

Much of the literature on group decision making assumes that members will cooperate, share information objectively and work towards more informed decisions. However, diverse groups can be prone to conflict and communication breakdowns (e.g. O'Reilly and Chatman, 1986). Process losses such as these will affect group performance. In highly value-laden cases, there is likely to be some extent of polarisation. This may be exacerbated where experts are involved, because they have a long history of gathering evidence to support their position (Mercier and Sperber, 2011). Taber and Lodge (2006) studied how people evaluate arguments about affirmative action and gun control, finding attitude polarisation to be more pronounced in subjects who are more knowledgeable.

But conflict is not all bad. It can be a source of cognitive growth, where individuals have different responses to the same problem and are motivated to achieve a joint solution. For example, certain forms of interpersonal disagreement can facilitate intellectual development in children (Azmitia and Perlmutter, 1989). Jehn et al. (1999) hypothesised that informational diversity would lead to task conflict in groups. That is to say, group members with different expertise, education and training may take different approaches to the task, both in content (what to do) and process (how to do it). Their results suggested that informational diversity did indeed lead to task conflict when it came to content, but not process. But nonetheless, informational diversity improved group performance and efficiency, especially for complex tasks.

When groups are not communicating well, the potential benefits of informational diversity often go unrecognised (Steiner, 1972). For example, Dougherty (1992) found that product teams of group members with different functional training struggled to get their products to market. This is the same reason why group members often fail to uncover hidden profiles (Stasser and Titus, 1985), because they fixate too much on common ground, which — in the case of diverse groups — is rather limited. Importantly, it illustrates that structured approaches to elicitation and discussion are critical for diverse groups, by explicitly eliciting unique information and reasons for/against from each group member, individuals are prompted to share information.

Providing feedback on positive performance and achievement to diverse groups is likely to increase motivation and relieve frustration, even if some level of conflict remains. Indeed, Jehn et al. (1999) found that diverse groups that were performing well reported high morale, yet still reported relationship conflict. Therefore, the positive effect of group performance overwhelmed the negative effect of relationship conflict on morale.

Socratic questioning

'Dialectical bootstrapping' as outlined by (Herzog and Hertwig, 2009) is akin to 'Socratic questioning'. Socratic questioning is a way to get experts to consider alternative models, explore hidden assumptions and sources of information, consider alternative ways to express their opinions or provide meaning, and consider the relevance of questions themselves (Elder and Paul, 1998). This type of questioning is commonly used in the IDEA protocol (see Section 5.5.3) by the facilitator, however, it could be taught to participants as a way of cross-examining one another's reasoning and assumptions. While more evidence is required, Yang (2008) demonstrated a statistically significant improvement in how groups of students equipped to question reasoning and assumptions through Socratic questioning improved their cognitive thinking skills more than groups that were enabled to simply undertake unstructured deliberations without the training.

Consider the aggregation method

When combining the judgements of multiple experts, different methods can yield better-aggregated judgements in different contexts. In his review of methods for combining forecasts, Clemen (1989) reports that simple aggregation methods are more effective in most cases, and the median and the mean often perform similarly (e.g. Palan et al., 2020). An important consideration is that the median is more robust to outliers, whereas the mean is sensitive to outliers. This is also an issue if individuals are trying to manipulate the outcome by giving extreme values. If you have reason to believe that outliers, or more extreme judgements, contain important information, then the mean may be more appropriate than the median, perhaps even an extremised aggregate or the geometric mean of odds, which have been found to perform well in probability aggregation (Satopää et al., 2014; Baron et al., 2014; see also Hanea et al., 2021b). For a useful summary that includes empirical data, see https://forum.effectivealtruism.org/posts/acREnv2Z5h4Fr5NWz/my-current-best-guess-on-how-to-aggregate-forecasts. Means are usually recommended if the judgements are normally distributed, and medians better reflect skewed data distributions. Taking an equally weighted average of the group has been found to be a robust method for achieving accurate and well calibrated expert judgements (Hemming et al., 2020b).

Scoring and weighting individuals

The alternative to equally weighting judgements from each expert is to give greater weight to the judgements of those who you have reason to believe might perform better. One of the few methods for weighting experts that has support is to use 'seed' questions (also known as 'calibration' or 'test' questions) (Aspinall and Cooke, 2013; French, 2011). Seed questions are questions which relate to main elicitation questions, but for which a resolution to the question can be obtained by the time the judgements of experts need to be aggregated. Experts are scored on the seed questions. The performance of experts on seed questions is used to develop weights for aggregating expert judgements (Hemming et al., 2020b). This method underpins the Classical Model for Structured Expert Judgement (Cooke, 1991). The Classical Model is most often used for assessing continuous random variables (quantities). However, it can be used to assess probabilities assigned to discrete events (Hanea et al., 2016). A recent review by Colson and Cooke (2017), demonstrated that across the 73 case studies to which the Classical Model has been applied, performance-weights out-performed equal weights in 74% of those cases.

In order to develop weights for experts based on test questions, scoring rules are required. An important property of a scoring rule is that it should not influence the forecaster in an undesirable way (Brier, 1950). When applying scoring rules it is important that the assessor and the experts understand the scoring rules, as different scoring rules have different underlying assumptions and properties for which they aim to maximise (e.g. different theoretical distributions (Cooke, 1991), and can lead to a different ranking of experts (Winkler and Murphy, 1968).

Table 5.3 Summary of strategies for improving group judgements.

Strategy	Description
Benefits of discussion	Allowing groups to interact improves judgement performance under certain conditions (e.g. quantitative tasks, when large groups are divided into smaller ones, and group members are collectively well calibrated). Benefits are seen through resolving misunderstandings and providing more immediate opportunities for people to introduce new information and learn from each other.
Anonymity	Anonymity reduces social barriers to participation in group deliberations by insulating group members from reputational pressures.
Foster dissent and argumentation	Dissenting groups introduce and share new information; Devil's Advocacy and Dialectical Inquiry can foster dissent and lead to better decisions in dynamic, poorly understood environments.
Group feedback	Performance feedback is more effective than outcome feedback for improving group members' judgments; group averages can be used as feedback to improve estimates.
Critical and counterfactual thinking	Groups that are primed to think critically and counterfactually make the correct decision more often.
Accountability	Process accountability improves information processing and motivation in groups (people need to demonstrate understanding of the decision problem), whereas outcome accountability can have perverse consequences.
Evaluating each other's judgements	People are better calibrated when evaluating someone else's judgements than their own.
Consider the judgements of others	Accuracy can be improved by predicting the judgement that someone else would give.
Diversity (cognitive)	Diversity in background, training, age, gender, political ideology, personality traits may be proxies for cognitive diversity (ensure a range of perspectives, generate counter-arguments, avoid confirmation bias). Role playing and utilising different problem frames can create cognitive diversity and improve decision quality.
Manage conflict	Providing performance feedback to members of a diverse group can improve morale, despite relationship conflict.
Anonymity	Anonymity reduces social barriers to participation in group deliberations by insulating group members from reputational pressures.
Socratic questioning	Training people to question assumptions and reasoning can improve the quality of their judgements.

Strategy	Description
Consider the aggregation method	Simple methods of aggregation are sufficient in most cases (e.g. mean or median), but extremising can improve aggregated probabilities. The median may be preferred if you wish to reduce the influence of extreme values, e.g. from individuals attempting to manipulate the system.
Scoring and weighting individuals	Rather than equally weighting judgements from each expert, give greater weight to the judgements of those who you have reason to believe might perform better (e.g. based on their performance on seed questions).

Consider group size

Studies on the number of people required to see the wisdom of crowds effect draw different conclusions, with diminishing returns found beyond anywhere from 5–6 (Hora, 2004), 8–12 (Hogarth, 1978), and 50 participants (Satopää et al., 2014). Hemming et al. (2018b) found that groups containing 5–9 participants considerably outperformed the median individual, with six of the eight groups being more accurate than 75% of individuals in round two. On combining the judgements of all 58 participants into a 'supergroup', it performed as well as the average group, but no better.

The wisdom of the crowd effect can be largely explained as a statistical phenomenon whereby judgements of individuals are random independent samples. If the samples are diverse then the information pool related to the questions will increase (Clemen and Winkler, 1999), and the errors of individuals are likely to cancel each other out (Larrick and Soll, 2006). Fewer than five individuals can lead to groups which are heavily influenced by outliers, while beyond nine individuals the contribution of each new member often bears little weight on the final aggregation.

The ideal group size depends on a great many factors (e.g. whether interactions are allowed, whether the discussion is facilitated, whether the elicitation is online or face-to-face, synchronous or asynchronous, etc.). Group size effects are particularly moderated by the type of task (see Kerr and Tindale, 2011, for a discussion). For example, it can depend on the extent to which the correct answer can be demonstrated (Bonner and Baumann, 2008), or the creativity of the task; when brainstorming, larger groups can generate more creative ideas (Coskun, 2011). Group size can also moderate the extent to which participants share information (Stasser et al., 1989). Bonner and Baumann (2008) suggest that when group size is small (e.g. 3 people), all members may participate in the discussion and consider the input of others while as group sizes become large (6+), evaluating each individual contribution becomes burdensome and so expert status becomes more influential (underscoring the importance of anonymity). Larger groups may also be more prone to 'social loafing' (relying on others to contribute).

Taking these considerations into account, we suggest 5–12 people is an appropriate group size for expert elicitation. Advice for IDEA groups (see Section 5.5.3) is to aim for about 8–12 participants to account for the loss of a few from attrition, and no more than 20 (Hemming et al., 2018a). This strikes a useful balance between the sensitivity of the aggregated estimate to the assessments of single individuals (seen with smaller groups), with the washing out of individual contributions (seen with larger groups).

5.5 Structured Frameworks for Making Group Judgements

The approaches to reduce bias described above can be easily integrated into a broader framework for eliciting expert judgements. It may be tempting to just use unstructured face-to-face groups: a minimally-facilitated discussion where participants sit at a single table and discuss a question until consensus is considered reached. While unstructured groups are ubiquitous (Graefe and Armstrong, 2011; Kerr and Tindale, 2011), and an improvement on relying on a single expert judgement, the lack of use of any of the strategies outlined above for mitigating biases greatly reduces the effectiveness.

This section describes five *structured* methods for incorporating a set of principles, including anonymity, considering the opposite/counterfactual thinking, diversity, feedback, information sharing, mathematical (rather than behavioural) aggregation and weighting. A comprehensive controlled experiment (Graefe and Armstrong, 2011, n = 227) comparing four group formats, Face-to-Face (FTF), Nominal Group Technique (NGT), Delphi technique (Delphi), and Prediction Market (PM), found that judgments from the most structured elicitation formats (NGT, Delphi) were more accurate (lower mean absolute error, MAE) than PMs or unstructured formats (FTF). This supports similar findings from earlier reviews (e.g. Rowe and Wright, 1999). Graefe and Armstrong also compared the accuracy of structured group judgements to their prior individual estimates. For all three structured techniques (NGT, Delphi, PMs), the mean group result was statistically significantly ($p<.01$) more accurate than the average of the initial estimates. The MAE reductions were: NGT =3.1; Delphi = 3.6 and PM = 3.05. Differences *between* these techniques were small, and not statistically significant. FTF was not included, as this method provided no benchmark to assess error reduction against.

Each of these techniques is briefly outlined in the following sections together with a variant of the Delphi technique (the IDEA protocol) and 'superforecasters'.

5.5.1 Delphi technique

The Delphi technique (Linstone and Turoff, 1975) is an elicitation procedure developed in the mid-1940s to improve forecasting technology during the Cold War. The purpose of the Delphi design is to mitigate dominating individuals, and problems such as the halo effect and groupthink (Mukherjee et al., 2015). In traditional Delphi groups, participants make estimates remotely and anonymously. That is, they do not meet face-to-face or learn each other's identities. In practice, this may vary, for example, not all Delphi groups will be entirely anonymous. Participants are usually invited to make written comments justifying or explaining their estimates, so there may be minimal interaction between group members, but it is substantially less than in FTF and NGT. The Delphi process is often described as iterative, meaning that estimates are made in 'rounds', with feedback about each other's estimates, and the accompanying comments, provided to participants between rounds by a facilitator. The number of rounds can vary, but is typically limited to two or three, although some stop once there is little change in the judgements by experts (e.g. behaviour consensus is achieved). The group result is usually the mean or median of the final round of estimates.

Delphi technique is also used for structuring group discussions for qualitative judgements (e.g. horizon scanning, problem formulation, objective setting, and identifying options for action) so is not exclusively used for numerical estimates (Mukherjee et al., 2015). See Box 5.2 for a step-by-step guide to the Delphi technique.

Box 5.2 Process for running a Delphi technique

1. Recruit a diverse group of individuals (n < 20). Diversity should be represented in domain expertise, experience, and demographic diversity. Experts need to know enough to understand the questions being asked.
2. Ask experts to answer the first round of questions (provide initial, private, individual judgements, together with rationales for their judgements).
3. Collate the responses into an anonymous feedback document or platform and present the results (usually circulated by a facilitator), including plots of judgements where possible, for the experts to review.
4. Ask experts to provide a second round of judgements and rationales, with the aim of converging on consensus.
5. Circulate again to participants.
6. Conduct more elicitation rounds as necessary, until the desired level of consensus is reached.
7. Aggregate estimates, most often using a mean or median.

5.5.2 Nominal Group technique

The Nominal Group Technique (NGT) was developed in the early 1970s (Van de Ven and Delbecq, 1971). In an NGT, participants are asked by a facilitator to individually reflect and generate ideas, typically based on a structured questionnaire. Subsequently, participants are asked to collectively prioritise the ideas and suggestions issued by the group members (Hugé and Mukherjee, 2018). The process of individual and collective reflection and co-production of knowledge among participants in NGT allows for a depolarising approach to the study and management of contentious issues (Hugé and Mukherjee, 2018). NGT interactions also fare well in satisfaction ratings by participants, considerably outranking Delphi and PMs and beating FTF by a small margin (Graefe and Armstrong, 2011).

Note that studies that compare the performance of consensus group judgement with a straight mathematical aggregate of the individual judgments within a group (e.g. the group average) often refer to this control as the 'nominal group' (e.g. Schultze et al., 2012). This use of the term *nominal* is different to its use in NGT. The NGT steps are outlined in Box 5.3.

Box 5.3 Process for running a Nominal Group Technique

1. The facilitator frames the problem, challenge or question(s).
2. Participants silently and independently generate ideas.
3. Group members take turns to each present their views (recorded by the facilitator — no discussion at this point).
4. Each contribution is clarified and discussed in turn. The original contributor need not defend or explain the idea (anyone can do this).
5. The facilitator outlines the process and criteria for prioritising ideas. Each participant privately prioritises each submission (e.g. scoring, voting).
6. The final group judgement is typically the highest-rated idea (by the group).

5.5.3 IDEA and related protocols

IDEA — Investigate, Discuss, Estimate and Aggregate (Burgman, 2015; Hanea et al., 2017) — is a structured expert elicitation protocol designed to help elicit more accurate quantitative and probabilistic judgements (along with rationales) from experts. Akin to the Delphi technique it involves the essential steps of allowing individuals to make a private individual estimate, before viewing other people's anonymised judgements and rationales, and revising their estimates. It is designed explicitly for the elicitation of quantities and probabilities (facts, rather than qualitative opinions); in contrast to Delphi technique it is more prescriptive in its guidance for making these estimates (e.g. using three-step and four-step interval estimates; Burgman, 2015), and uses guided social interactions to avoid the biassing elements of group deliberation and behavioural aggregation (Hanea et al., 2017). Final round judgements are usually aggregated by taking equal-weighted means. Performance weighting may also be used, but requires that calibration questions are seeded in the survey from which to develop weights (e.g. see Hemming et al., 2020b). See Box 5.4 for an outline of the basic steps.

Two other notable protocols for eliciting expert judgements that are also applied in ecology include the Sheffield Elicitation Framework (SHELF) (Gosling, 2018), which is also akin to the Delphi technique but which relies on behavioural aggregation (Fitzgerald et al., 2021; O'Hagan, 2019), and the Classical Model (also known as Cooke's Method), which typically only allows a single estimate from experts and uses performance-weighted aggregation (Cooke, 1991). Recent applications of the IDEA protocol have adopted the performance-weighted aggregations of the Classical Model, that is, giving greater weight to the judgements of those who performed better on test questions (Barons and Aspinall, 2020; Hanea et al., 2021a).

Box 5.4 Process for running an IDEA group

1. Recruit a diverse group of 8–12 individuals (to account for attrition). Diversity should be represented in domain expertise, experience, and demographic diversity. Experts simply need to know enough to understand the questions being asked.
2. Ask experts to investigate the question being asked, and provide an initial, private, individual estimate. Typically experts are asked to provide their estimates as single event probabilities with upper and lower bounds (three-step question format), or a best estimate with upper and lower bounds, and a level of confidence (four-step question format), but IDEA is flexible and continuous probabilities and other question formats can be used.
3. If you have elicited four-step uncertainty intervals (where participants assign different confidences), these will need to be transformed to a common level of confidence for aggregation and display, usually 80% or 90%. To do this, you'll need to make an assumption about the underlying distribution. A linear transformation is straightforward and minimises assumptions.
4. Once estimates and rationales are obtained, collate the responses into a feedback document or platform and present the results, including plots of judgements with uncertainty intervals, where possible, for the experts to review (e.g. Figure 5.1).
5. Facilitate a discussion to promote counterfactual and 'consider the opposite' reasoning.
6. Allow experts to revise their initial estimates if they have reason to do so, making clear that consensus is not the goal.
7. Aggregate estimates, most often using an equal-weighted aggregation (quantile or linear pooling). Performance weighting may also be used.

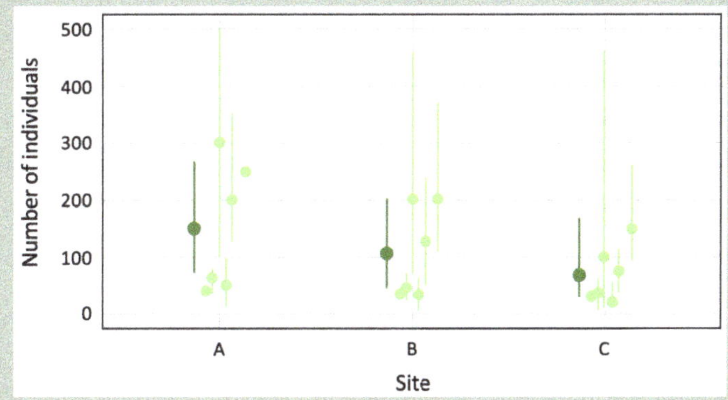

Figure 5.1 Example feedback plot of anonymised expert estimates of the number of individuals of a given species present at three different sites, elicited using the IDEA protocol. Error bars reflect elicited four-step intervals transformed to a 90% confidence level. Light green denotes estimates from individual participants; dark green denotes the aggregate of these estimates. Individual estimates may be labelled with anonymous participant IDs if desired. (*Source*: authors)

Testing of the protocol has provided compelling evidence that: 1) entrusting an equal-weighted aggregate of a diverse group of individuals typically provides a better judgement than a single well-credentialled individual (Burgman et al., 2011; Hemming et al., 2020a, 2018b); 2) discussion after providing an initial, private estimate typically improves group judgements of individuals and groups (Wintle et al., 2021) — sometimes this is simply because linguistic ambiguity and mistakes made in the first round are addressed, other times it is because critical evidence has been introduced, and diverse opinions discussed; and 3) the way questions are asked can help experts to better communicate their judgements, for example, a four-step question format by Speirs-Bridge et al. (2010) helps overcome overconfidence. The countless case studies that have applied the protocol show it is also practical to apply regardless of whether experts are to be convened remotely (e.g. McBride et al., 2012), in a face-to-face workshop (e.g. Wintle et al., 2021), or a hybrid approach (e.g. Hanea et al., 2017), and regardless of whether estimation and discussion take place synchronously or asynchronously.

5.5.4 Prediction Markets

Prediction Markets (PMs) allow individuals to trade on the outcome of future events (Wolfers and Zitzewitz, 2004). They create contracts that pay a fixed amount if an event occurs, and then allow people to trade on the contract by buying or selling in a manner comparable to financial markets. For example, a contract may pay $1 if an event occurs by a prespecified date (e.g. will average oil prices be higher at midnight the last day of this month than at 00.01am on the first day?), and $0 otherwise. Participants are incentivised to buy shares when they think the current estimate is too low, and to sell when they think it is too high. The price at which the contract trades at a given time can then be interpreted, with some caveats (Manski, 2006), as the market's collective forecast of the probability of the event occurring.

Similar to the Delphi technique, PMs are typically anonymous. They differ from the Delphi technique in that they are continuously updated by participants, who typically do not share information with each other. Rather, they respond directly to the price signal. These features mean that PMs often correct, rather than amplify, the effects of individual errors by creating powerful incentives to disclose, rather than to conceal, privately held views. PMs have been successfully applied in domains such as politics, sports and business, and have been found to generate relatively accurate forecasts, in many cases outperforming the statistical aggregation of prior individual estimates (e.g. Graefe and Armstrong, 2011; Dreber et al., 2015), but not always (e.g. see Atanasov et al., 2017). A drawback is that to work they require a large enough number of active traders with different opinions, and participants' satisfaction with them is usually low (Graefe and Armstrong, 2011; Kerr and Tindale, 2011). The PM steps are outlined in Box 5.5.

5.5.5 Superforecasters

The notion of superforecasting (Mellers et al., 2014, 2015; Tetlock and Gardner, 2016) arose from a large forecasting tournament funded by the US government (IARPA) to overcome

Box 5.5 Process for running a Prediction Market

1. Find an online platform to host your Prediction Market.
2. Formulate forecasting questions (usually yes/no) and provide question background and resolution information where relevant. Questions should have a verifiable answer (e.g. if asking about the future price of oil, you should be able to verify whether the price exceeded a given value by a given date).
3. Include comments/justifications, text boxes and discussion threads for participants to provide reasoning, links and information if desired.
4. Invite participants to trade, and prompt them to update forecasts.
5. Measure market accuracy (usually Brier scores) and relevant metrics when the outcome is known, and notify participants of performance across questions and account balances.

some of the forecasting challenges faced by the US intelligence community. Elite teams, comprising the top 2% of performers from the first year of the forecasting tournament — the superforecasters — worked together in the second year, outperforming all the other teams by a wide margin for two years in a row, rather than regressing to the mean. Mellers et al. (2014) argue that team communication, in particular, the exchange of rationales, news articles and other information, and debating differences of opinion, produced 'enlightened cognitive altruism' among group members. This, combined with psychological interventions, including training on probability and scenarios, proved key to the success of their team's forecasting. The researchers suggest that simply receiving the 'exalted' title of *superforecasters* substantially boosted effort and engagement from team members. See Box 5.6 for an outline of the basic steps.

5.6 Practical Methods for Improving Routine Judgements

The last section described five structured approaches that warrant much greater use. The aim of this section is to recognise that there are occasions when these methods might be too complex or time consuming to adopt comprehensively, but that some features can usually be included to improve the rigour of judgements. This section illustrates how some of these features might be applied to a series of those routine judgements. We envisage that the organiser of the job selection panel or the group deciding which projects to fund will introduce these more rigorous processes.

Regardless of the issue, there are a series of general considerations that apply to all methods. Adding any of these components will improve the rigour. Section 5.4 describes options for improving further.

Box 5.6 Process for running a Superforecasting group

1. Test participants on their forecasting abilities (based on Brier scores).
2. High-performing forecasters are assembled into teams and given training, e.g. in probability and calibration, avoiding overconfidence, and considering base rates.
3. Forecasters make individual predictions.
4. Forecasters are allowed to interact in an online chat room, encouraged to share and discuss information, and to update predictions until the event deadline passes.
5. Final individual probability estimates are transformed and combined using statistical algorithms.

1. Decide on the precise issue that is being evaluated.
2. Determine whether this is about making a choice (e.g. employing someone, deciding which projects to fund) or considering a fact (the veracity of an observation or estimating a parameter).
3. Decide whether to break down the issue into intermediate stages, such as the probability of rats colonising an island, and the conditional probability of various reptile species going extinct if they do. This helps decompose the problem.
4. Clearly define the exact questions to avoid ambiguity.
5. Decide on a scoring process for eliciting responses. A wider score range, say 1–10 or 1–20 reduces the number of ties; a 1–5 range tends to result in many rather uninformative 3s and 4s. Where there is a threshold of acceptability, or you wish to avoid fence-sitting, you might use, say, 1–10, where 1 = definitely not; 5 = on balance, no; 6 = on balance; yes, 10 = definitely yes.
6. Taking the median reduces the impact of extreme values or those wishing to manipulate the outcome.
7. Decide how to present evidence, such as giving reasons for and against and elicit further evidence from the group.
8. Decide how to enhance the independence and ideally anonymity of judgements. A key element is that the participants do not state their views prior to scoring, to retain independence. The minimum approach is for everyone to write down their score but not show others and then state this in turn. A better strategy is one where everyone writes their score on paper or cards, which are collated (e.g. can just be laid out on the table — providing identical pens increases anonymity). Better still is

to score online using a template, survey or software, so anonymous scores can be easily compiled.

9. Decide on the scoring sequence. The minimum is discussion, then scoring. Introducing an initial round of scoring, then discussion, then scoring again, improves judgements.
10. Decide on how the discussions will be run, including ground rules for participation (e.g. steps to avoid exposing who produced which judgements).

The following methods describe minimal ways of improving three of the most common types of expert elicitation. Each of these are inferior to adopting more comprehensive methods described above and in other chapters but superior to conventional practice.

5.6.1 Judging the veracity of a statement

Evidence is very often expressed as a statement. This could be a claim to have seen a species, the assertion that a species is locally extinct, or the conclusion that the species usually nests in old woodpecker holes. Such a process could be adopted for record committees, such as bird recorders deciding which records they accept. It can be part of the decision-making processes described in Chapter 8 or used when embedding evidence into processes, such as management plans, policies or models, as described in Chapter 9. Courts could be changed so that jurors have to apply a similar process in deciding whether or not the accused is guilty. Box 5.7 outlines a simple process for undertaking such a task.

Box 5.7 A simple process for judging the veracity of a statement

1. Decide upon the main statement being considered (e.g. salmon bred on the site last year; the bears have low breeding success due to inbreeding).
2. Decide on a scoring system, such as 1–10 for each criterion where 1 = definitely not, 5 = on balance not, 6 = on balance yes, 10 = definitely yes.
3. Consider the range of elements to the discussion (e.g. whether observations of salmon spawning are accurate, whether some salmon records are actually trout, whether the young salmon found dead could have hatched elsewhere).
4. Collate and present the evidence for each element.
5. Ask participants not to indicate their overall view of the main statement or the scores given.
6. Discuss each element including considering arguments for, and against, before discussing the main statement.
7. Each participant scores the main statement privately.
8. Display the results anonymously. Discuss reasons for high or low scores.
9. Rescore and take the median score.

If making repeated decisions, one option to improve efficiency is to accept the statement if everyone gives a high score above, say 7 or above, and reject if everyone gives a low score, say, 3 or lower. Thus a panel deciding upon bird records can concentrate on those records that all are uncertain about, and those where there is disagreement.

5.6.2 Selecting options

Ranking items is a regular component of decision making in which there are a series of options of which some need to be selected, for example deciding which research projects to support or which species or areas are prioritised for conservation action. Sometimes the task is to identify a subset placed in rank order, for example, judging the most suitable job applicant and three backups.

A common practice is to score a range of criteria and then add to give a total sum. However, there are various problems with this, including how to weight each criteria, whether they are additive, and whether scores can be considered linear and equidistant (is the distance between a 4 and 5 the same as the distance between a 9 and 10? Do scores of 10, 10, 2 beat 8, 7, 7?). It is now accepted that adding scores is deeply flawed and better replaced by decision science methods (Klein et al., 2014), as described in Section 8.5. This process entails removing options that are below the minimum threshold, those that are inferior across all criteria, i.e. 'dominated alternatives', and also removing criteria that are no longer informative, i.e. 'redundant criteria', for example, criteria on which all remaining options score about the same, or all options are at a sufficient level. Applying methods from decision science prompts us to explicitly grapple with trade-offs, for example, whether the more impressive research project justifies the increased cost or whether to support the project with lower expected conservation gains but exciting opportunities for education.

Box 5.8 A simple process for selecting options

1. Clarify what criteria are important to the ranking. For selecting projects, this could include the number of threatened species present, the project cost and the fit to the organisation's objectives.
2. Ask participants not to indicate their overall views.
3. Decide on the criteria and the performance measures to be used to evaluate options. These may be quantitatively measured or estimated on a natural scale (e.g. number of species present, cost) or judged on a constructed or proxy scale (e.g. organisational fit). If using indirect scales, decide upon a scoring system, such as 1–10, where 1 = definitely unsuitable, 5 = on balance unsuitable, 6 = on balance suitable, 10 = definitely suitable. Decide on minimal criteria.

4. Collect, present and discuss evidence for those criteria that need to be judged, including conflicting evidence, and then estimate or score each, e.g. in a consequence table (see 8.5.1).
5. Discuss estimates/scores of each criterion, and reasons why they might be high or low. Revise estimates/scores in table.
6. Tabulate median or mean scores for each option and criteria
7. Exclude options that do not satisfy any minimum criteria (unacceptable options). Look for a dominant alternative (i.e. one that outperforms all others across all criteria). If there's no clear winner, exclude those that perform worse on all criteria (dominated alternatives). Remove criteria that are no longer informative (redundant criteria).
8. Discuss and decide on preferences based on trade-offs. Is the more effective project worth the extra cost?
9. If necessary, discuss which options are acceptable, for example, which applicants are considered employable if the preferred candidate turns down the offer.

5.6.3 Estimating a numeric value

Estimating a quantity or probability is often called for, but the evidence may be unclear, complicated, or disputed. This could be the population size of a species, the change in population size over the last decade, or parameters for a model. The process outlined below is effectively the same as the IDEA or Delphi technique protocols.

Box 5.9 A simple process for estimating numeric values

1. Clearly define the exact value that is sought to avoid ambiguity. For example, if estimating the current population size, be clear if you are referring to individuals or breeding pairs, what time of year to focus on, and the exact location or area to consider. For predictions, be especially clear about the time frame to consider, e.g. what's the probability that a well-specified event will occur within 2 years, or 50 years?
2. Examine relevant evidence, including contradictory evidence.
3. Decide whether to first decompose the question into different parts. For example, in estimating a population of a species, what is the likely geographic

distribution of that species? What are the densities in different areas? How large might the non-breeding population be? Ideally, these are judged independently before a judgement on the overall value.

4. Make independent estimates, ideally look at results presented anonymously, discuss and revise estimates.
5. Combine the final estimates, e.g. by taking the mean or median.

References

Anderson, J.R. 1981. *Cognitive Skills and Their Acquisition* (Psychology Press), https://doi.org/10.4324/9780203728178.

Armstrong, J.S. 2001. Role playing: A method to forecast decisions. In: *Principles of Forecasting*, International Series in Operations Research & Management Science, Vol. 30, ed. by J.S. Armstrong (Boston, MA: Springer), pp.15–30, https://doi.org/10.1007/978-0-306-47630-3_2.

Ashton, R.H. 1992. Effects of justification and a mechanical aid on judgment performance. *Organizational Behavior and Human Decision Processes* 52: 292–306, https://doi.org/10.1016/0749-5978(92)90040-e.

Aspinall, W.P. and Cooke, R.M. 2013. Quantifying scientific uncertainty from expert judgement elicitation. In: *Risk and Uncertainty Assessment for Natural Hazards*, ed. by J. Rougier, et al. (Cambridge: Cambridge University Press), pp.64–99, https://doi.org/10.1017/CBO9781139047562.005.

Atanasov, P., Rescober, P., Stone, E. et al. 2017. Distilling the wisdom of crowds: Prediction markets vs. prediction polls. *Management Science* 63: 691–706, https://doi.org/10.1287/mnsc.2015.2374.

Azmitia, M. and Perlmutter, M. 1989. Social influences on children's cognition: State of the art and future directions. *Advances in Child Development and Behavior* 22: 89–144, https://doi.org/10.1016/S0065-2407(08)60413-9.

Bar-Hillel, M. 1980. The base-rate fallacy in probability judgments. *Acta Psychologica* 44: 211–33, https://doi.org/10.1016/0001-6918(80)90046-3.

Barnett, A.G., van der Pols, J.C,. and Dobson, A.J. 2005. Regression to the mean: What it is and how to deal with it. *International Journal of Epidemiology* 34: 215–20, https://doi.org/10.1093/ije/dyh299.

Baron, J., Mellers, B.A., Tetlock, P.E., et al. 2014. Two reasons to make aggregated probability forecasts more extreme. *Decision Analysis* 11: 133–45, https://doi.org/10.1287/deca.2014.0293.

Barons, M.J. and Aspinall, W. 2020. Anticipated impacts of Brexit scenarios on UK food prices and implications for policies on poverty and health: A structured expert judgement approach. *BMJ Open* 10: e032376, https://doi.org/10.1136/bmjopen-2019-032376.

Benson, P.G and Önkal, D. 1992. The effects of feedback and training on the performance of probability forecasters. *International Journal of Forecasting* 8: 559–73, https://doi.org/10.1016/0169-2070(92)90066-I.

Berthet, V. 2022. The impact of cognitive biases on professionals' decision-making: A review of four occupational areas. *Frontiers in Psychology,* 12: Article 802439, https://doi.org/10.3389/fpsyg.2021.802439.

Bilalic, M., Langner, R., Erb, M., et al. 2010. Mechanisms and neural basis of object and pattern recognition: A study with chess experts. *Journal of Experimental Psychology* 139: 728–42, https://doi.org/10.1037/a0020756.

Bonner, B.L. and Baumann, M.R. 2008. Informational intra-group influence: The effects of time pressure and group size. *European Journal of Social Psychology* 38: 46–66, https://doi.org/10.1002/ejsp.400.

Bourne, L., Jr, Kole, J. and Healy, A. 2014. Expertise: Defined, described, explained. *Frontiers in Psychology* 5: Article 186, https://doi.org/10.3389/fpsyg.2014.00186.

Brier, G.W. 1950. Verification of forecasts expressed in terms of probability. *Monthly Weather Review* 78: 1–3, https://doi.org/10.1175/1520-0493(1950)078<0001:VOFEIT>2.0.CO;2.

Burgman, M.A. 2005. *Risks and Decisions for Conservation and Environmental Management,* Ecology, Biodiversity, and Conservation (New York: Cambridge University Press), https://doi.org/10.1017/CBO9780511614279.

Burgman, M.A. 2015. *Trusting Judgements: How to Get the Best out of Experts* (Cambridge: Cambridge University Press), https://doi.org/10.1017/CBO9781316282472.

Burgman, M.A., McBride, M., Ashton, R., et al. 2011. Expert status and performance. *PLoS ONE* 6: e22998., https://doi.org/10.1371/journal.pone.0022998.

Canessa, S., Taylor, G., Clarke, R.H., et al. 2020. Risk aversion and uncertainty create a conundrum for planning recovery of a critically endangered species. *Conservation Science and Practice* 2: e138, https://doi.org/10.1111/csp2.138.

Caputo, A. 2013. A literature review of cognitive biases in negotiation processes. *International Journal of Conflict Management* 24: 374–98, https://doi.org/10.1108/IJCMA-08-2012-0064.

Charness, N., Tuffiash, M., Krampe, R., et al. 2005. The role of deliberate practice in chess expertise. *Applied Cognitive Psychology* 19: 151–65, https://doi.org/10.1002/acp.1106.

Clemen, R. 1989. Combining forecasts: A review and annotated bibliography. *International Journal of Forecasting* 5: 559–83, https://doi.org/10.1016/0169-2070(89)90012-5.

Clemen, R. and Winkler, R.L. 1985. Limits for the precision and value of information from dependent sources. *Operations Research* 33: 427–42, https://doi.org/10.1287/opre.33.2.427.

Clemen, R. and Winkler, R.L. 1999. Combining probability distributions from experts in risk analysis. *Risk Analysis: An International Journal* 19: 187–203, https://doi.org/10.1111/j.1539-6924.1999.tb00399.x.

Colson, A.R. and Cooke, R.M. 2017. Cross validation for the classical model of structured expert judgment. *Reliability Engineering & System Safety* 163: 109–20, https://doi.org/10.1016/j.ress.2017.02.003.

Cooke, R. 1991. *Experts in Uncertainty: Opinion and Subjective Probability in Science* (New York: Oxford University Press on Demand).

Cooper, G.S. and Meterko, V. 2019. Cognitive bias research in forensic science: A systematic review. *Forensic Science International* 297: 35–46, https://doi.org/10.1016/j.forsciint.2019.01.016.

Coskun, H. 2011. The effects of group size, memory instruction, and session length on the creative performance in electronic brainstorming groups. *Kuram Ve Uygulamada Egitim Bilimleri* 11: 91–95, https://files.eric.ed.gov/fulltext/EJ919891.pdf.

Dawes, R.M. 1994. *House of Cards: Psychology and Psychotherapy Built on Myth* (New York: Free Press).

De Dreu, C.K.W., Koole, S.J. and Steinel, W. 2000. Unfixing the fixed pie: A motivated information-processing approach to integrative negotiation. *Journal of Personality and Social Psychology* 79: 975–87, https://doi.org/10.1037//0022-3514.79.6.975.

Dougherty, D. 1992. Interpretive barriers to successful product innovation in large firms. *Organization Science* 3: 179–202, https://doi.org/10.1287/orsc.3.2.179.

Dreber, A., Pfeiffer, T., Almenberg, J., et al. 2015. Using prediction markets to estimate the reproducibility of scientific research. *Proceedings of the National Academy of Sciences* 112: 15343-7, https://doi.org/10.1073/pnas.1516179112.

Drescher, M. and Edwards, R.C. 2018. A systematic review of transparency in the methods of expert knowledge use. *Journal of Applied Ecology* 56: 436–49, https://doi.org/10.1111/1365-2664.13275.

Ehrlinger, J., Gilovich, T. and Ross, L. 2005. Peering into the bias blind spot: Peoples' assessments of bias in themselves and others. *Personality and Social Psychology Bulletin* 31: 680–92, https://doi.org/10.1177/0146167204271570.

Elder, L. and Paul, R. 1998. The role of Socratic questioning in thinking, teaching, and learning. *The Clearing House* 71: 297–301, https://doi.org/10.1080/00098659809602729.

Ericsson, K.A., Krampe, R.T. and Tesch-Roemer, C. 1993. The role of deliberate practice in the acquisition of expert performance. *Psychological Review* 100: 363–406, https://doi.org/10.1037/0033-295X.100.3.363.

Ericsson, K.A. and Lehmann, A.C. 1996. Expert and exceptional performance: Evidence of maximal adaptation to task constraints. *Annual Review of Psychology* 47: 273–305, https://doi.org/10.1146/annurev.psych.47.1.273.

Failing, L., Gregory, R. and Harstone, M. 2007. Integrating science and local knowledge in environmental risk management: A decision-focused approach. *Ecological Economics* 64: 47–60, https://doi.org/10.1016/j.ecolecon.2007.03.010.

Fischer, G.W. 1982. Scoring-rule feedback and the overconfidence syndrome in subjective probability forecasting. *Organizational Behavior and Human Performance* 29: 352–69, https://doi.org/10.1016/0030-5073(82)90250-1.

Fitzgerald, D.B., Smith, D.R., Culver, D.C., et al. 2021. Using expert knowledge to support endangered species act decision-making for data-deficient species. *Conservation Biology* 35: 1627–38, https://doi.org/10.1111/cobi.13694.

Fournier, A.M.V, White, E.R. and Heard, S.B. 2019. Site-selection bias and apparent population declines in long-term studies. *Conservation Biology* 33: 1370–79, https://doi.org/10.1111/cobi.13371.

French, S. 2011. Aggregating expert judgement. *Revista de La Real Academia de Ciencias Exactas, Fisicas y Naturales. Serie A. Matematicas* 105: 181–206, https://doi.org/10.1007/s13398-011-0018-6.

French, S. 2012. Expert judgment, meta-analysis, and participatory risk analysis. *Decision Analysis* 9: 119–27, https://doi.org/10.1287/deca.1120.0234.

Froggatt, W.W. 1936. The introduction of the great Mexican toad *Bufo Marinus* into Australia. *The Australian Naturalist* 9: 163–64.

Galesic, M., Bruine de Bruin, W., Dumas, M., et al. 2018. Asking about social circles improves election predictions. *Nature Human Behaviour* 2: 187–93, https://doi.org/10.1038/s41562-018-0302-y.

Galinsky, A. and Kray, L. 2003. From thinking about what might have been to sharing what we know: The effects of counterfactual mind-sets on information sharing in groups. *Journal of Experimental Social Psychology* 40: 606–18, https://doi.org/10.1016/j.jesp.2003.11.005.

Galinsky, A.D., Magee, J.C., Gruenfeld, D.H., et al. 2008. Power reduces the press of the situation: Implications for creativity, conformity, and dissonance. *Journal of Personality and Social Psychology* 95: 1450–66, https://doi.org/10.1037/a0012633.

Gigerenzer, G., Gaissmaier, W., Kurz-Milcke, E., et al. 2008. Helping doctors and patients make sense of health statistics. *Psychological Science in the Public Interest*: 53–96, https://doi.org/10.1111/j.1539-6053.2008.00033.x.

Gosling, J.P. 2018. SHELF: The Sheffield Elicitation Framework. In: *Elicitation. International Series in Operations Research & Management Science, vol CCLXI*, ed. by L. Dias, et al. (Cham, Switzerland: Springer), https://doi.org/10.1007/978-3-319-65052-4_4.

Graber, M. 2005. Diagnostic errors in medicine: A case of neglect. *The Joint Commission Journal on Quality and Patient Safety* 31: 106–13, https://doi.org/10.1016/S1553-7250(05)31015-4.

Graefe, A. and Armstrong, J.S. 2011. Comparing face-to-face meetings, nominal groups, Delphi and prediction markets on an estimation task. *International Journal of Forecasting* 27: 183–95, https://doi.org/10.1016/j.ijforecast.2010.05.004.

Gregory, R, Failing, L., Harstone, M., et al. 2012. *Structured Decision Making: A Practical Guide to Environmental Management Choices* (Chichester, West Sussex: Wiley-Blackwell), https://doi.org/10.1002/9781444398557.

Gregory, R. and Keeney, R.L. 2017. A practical approach to address uncertainty in stakeholder deliberations. *Risk Analysis* 37: 487–501, https://doi.org/10.1111/risa.12638.

Hanea, A.M., McBride, M.F., Burgman, M.A., et al. 2016. Classical meets modern in the IDEA protocol for structured expert judgement. *Journal of Risk Research* 21: 417–33, https://doi.org/10.1080/13669877.2016.1215346.

Hanea, A.M., McBride, M.F., Burgman, M.A., et al. 2017. Investigate Discuss Estimate Aggregate for structured expert judgement. *International Journal of Forecasting* 33: 267–79, https://doi.org/10.1016/j.ijforecast.2016.02.008.

Hanea, A.M., Hemming, V. and Nane, G.F. 2021a. Uncertainty quantification with experts: Present status and research needs. *Risk Analysis* 42: 254–63, https://doi.org/10.1111/risa.13718.

Hanea, A.M., Wilkinson, D.P., McBride, M., et al. 2021b. Mathematically aggregating experts' predictions of possible futures. *PLoS ONE* 16: e0256919, https://doi.org/10.1371/journal.pone.0256919.

Hemming, V., Armstrong, N., Burgman, M.A., et al. 2020a. Improving expert forecasts in reliability: Application and evidence for structured elicitation protocols. *Quality and Reliability Engineering International* 36: 623–41, https://doi.org/10.1002/qre.2596.

Hemming, V., Burgman, M.A., Hanea, A.M., et al. 2018a. A practical guide to structured expert elicitation using the IDEA protocol. *Methods in Ecology and Evolution* 9: 169–80, https://doi.org/10.1111/2041-210X.12857.

Hemming, V., Camaclang, A.E., Adams, M.S., et al. 2022. An introduction to decision science for conservation. *Conservation Biology* 36: e13868, https://doi.org/10.1111/cobi.13868.

Hemming, V., Hanea, A., Walshe, T., et al. 2020b. Weighting and aggregating expert ecological judgements. *Ecological Applications* 30: e02075, https://doi.org/10.1002/eap.2075.

Hemming, V., Walshe, T.V., Hanea, A.M., et al. 2018b. Eliciting improved quantitative judgements using the IDEA protocol: A case study in natural resource management. *PloS ONE* 13: e0198468, https://doi.org/10.1371/journal.pone.0198468.

Herzog, S.M. and Hertwig, R. 2009. The wisdom of many in one mind: Improving individual judgments with dialectical bootstrapping. *Psychological Science* 20: 231–37, https://doi.org/10.1111/j.1467-9280.2009.02271.x.

Hoch, S.J. 1985. Counterfactual reasoning and accuracy in predicting personal events. *Journal of Experimental Psychology-Learning Memory and Cognition* 11: 719–31, https://doi.org/10.1037//0278-7393.11.1-4.719.

Hogarth, R.M. 1978. A note on aggregating opinions. *Organizational Behavior and Human Performance* 21: 40–46, https://doi.org/10.1016/0030-5073(78)90037-5.

Hollnagel, E. and Fujita, Y. 2013. The Fukushima disaster–Systemic failures as the lack of resilience. *Nuclear Engineering and Technology* 45: 13–20, https://doi.org/10.5516/NET.03.2011.078.

Hong, L. and Page, S.E. 2004. Groups of diverse problem solvers can outperform groups of high-ability problem solvers. *Proceedings of the National Academy of Sciences of the United States of America* 101: 16385–89, https://doi.org/10.1073/pnas.0403723101.

Hong, L. and Page, S.E. 2020. The contributions of diversity, accuracy, and group size on collective accuracy. *DecisionSciRN: Other Human Behavior & Game Theory (Topic)*, October 15 2020, https://doi.org/10.2139/ssrn.3712299.

Hora, S.C. 2004. Probability judgments for continuous quantities: Linear combinations and calibration. *Management Science* 50: 597–604, https://doi.org/10.1287/mnsc.1040.0205.

Hugé, J. and Mukherjee, N. 2018. The nominal group technique in ecology and conservation: Application and challenges. *Methods in Ecology and Evolution* 9: 33–41, https://doi.org/10.1111/2041-210X.12831.

Janis, I.L. 1982. *Groupthink*, Second edition (Boston: Houghton Mifflin).

Jasanoff, S. 2006. Transparency in public science: Purposes, reasons, limits. *Law and Contemporary Problems* 69: 21–45, https://scholarship.law.duke.edu/cgi/viewcontent.cgi?referer=&httpsredir=1&article=1385&context=lcp.

Jehn, K.A., Northcraft, G.B. and Neale, M.A. 1999. Why differences make a difference: A field study of diversity, conflict, and performance in workgroups. *Administrative Science Quarterly* 44: 741–63, https://doi.org/10.2307/2667054.

Johnson, J. and Bruce, A.C. 2001. Calibration of subjective probability judgments in a naturalistic setting. *Organizational Behavior and Human Decision Processes* 85: 265–90, https://doi.org/10.1006/obhd.2000.2949.

Kahneman, D. 1991. Judgment and decision making: A personal view. *Psychological Science* 2: 142–45, https://doi.org/10.1111/j.1467-9280.1991.tb00121.x.

Kahneman, D. and Klein, G. 2009. Conditions for intuitive expertise: A failure to disagree. *American Psychologist* 64: 515–26, https://doi.org/10.1037/a0016755.

Kahneman, D. and Tversky, A. 1972. Subjective probability: A judgment of representativeness. *Cognitive Psychology* 3: 430–54, https://doi.org/10.1016/0010-0285(72)90016-3.

Kahneman, D. and Tversky, A. 1979. Prospect theory: Analysis of decision under risk. *Econometrica* 47: 263–91, https://doi.org/10.2307/1914185.

Katzenstein, G. 1996. The debate on structured debate: Toward a unified theory. *Organizational Behavior and Human Decision Processes* 66: 316–32, https://doi.org/10.1006/obhd.1996.0059.

Kent, S. 1964. Words of estimative probability. *Studies in Intelligence* 8: 49–65, https://www.cia.gov/static/0aae8f84700a256abf63f7aad73b0a7d/Words-of-Estimative-Probability.pdf.

Keogh, L. 2011. Introducing the cane toad. *Queensland Historical Atlas,* 25 March 2011, https://www.qhatlas.com.au/introducing-cane-toad.

Kerr, N. and Tindale, R.S. 2011. Group-based forecasting?: A social psychological analysis. *International Journal of Forecasting* 27: 14–40, https://doi.org/10.1016/j.ijforecast.2010.02.001.

Klein, C.J., Jupiter, S and Possingham, H.P. 2014. Setting conservation priorities in Fiji: Decision science versus additive scoring systems *Marine Policy* 48: 204–05, https://doi.org/10.1016/j.marpol.2014.03.008.

Kluger, A.N. and DeNisi, A. 1996. The effects of feedback interventions on performance: A historical review, a meta-analysis, and a preliminary feedback intervention theory. *Psychological Bulletin* 119: 254–84, https://doi.org/10.1037//0033-2909.119.2.254.

Kopelman, R.E. 1986. Objective feedback. In: *Generalizing From Laboratory to Field Settings,* ed. by E.A. Locke (Lexington, MA: Lexington Books), pp.119–45.

Koriat, A., Lichtenstein, S. and Fischhoff, B. 1980. Reasons for confidence. *Journal of Experimental Psychology-Human Learning and Memory* 6: 107–18, https://doi.org/10.1037/0278-7393.6.2.107.

Kunda, Z. 1990. The case for motivated reasoning. *Psychological Bulletin* 108: 480–98, https://doi.org/10.1037/0033-2909.108.3.480.

Kuran, T. and Sunstein, C.R. 1999. Availability cascades and risk regulation. *Stanford Law Review* 51: 683–768, https://doi.org/10.2307/1229439.

Larkin, J., McDermott, J., Simon, D.P., et al. 1980. Expert and novice performance in solving physics problems. *Science* 208: 1335–42, https://doi.org/10.1126/science.208.4450.1335.

Larrick, R.P. and Soll, J.B. 2006. Intuitions about combining opinions: Misappreciation of the averaging principle. *Management Science* 52: 111–27, https://doi.org/10.1287/mnsc.1050.0459.

Lerner, J.S and Tetlock, P.E. 1999. Accounting for the effects of accountability. *Psychological Bulletin* 125: 255–75, https://doi.org/10.1037/0033-2909.125.2.255.

Lichtenstein, S. and Fischhoff, B. 1980. Training for calibration. *Organizational Behavior and Human Performance* 26: 149–71, https://doi.org/10.1016/0030-5073(80)90052-5.

Lin, S.W. and Bier, V.M. 2008. A study of expert overconfidence. *Reliability Engineering & System Safety* 93: 711–21, https://doi.org/10.1016/j.ress.2007.03.014.

Linstone, H.A. and Turoff, M. 1975. *The Delphi Method: Techniques and Applications* (Reading, MA: Addison-Wesley Publishing Company), https://web.njit.edu/~turoff/pubs/delphibook/index.html.

Mahoney, M.J. 1977. Publication prejudices: An experimental study of confirmatory bias in the peer review system. *Cognitive Therapy and Research* 1: 161–75, https://doi.org/10.1007/BF01173636.

Manski, C.F. 2006. Interpreting the predictions of prediction markets. *Economics Letters* 91: 425–29, https://doi.org/10.1016/j.econlet.2006.01.004.

Margalida, A., Martínez, J.M., Gómez de Segura, A., et al. 2017. Supplementary feeding and young extraction from the wild are not a sensible alternative to captive breeding for reintroducing bearded vultures *Gypaetus barbatus*. *Journal of Applied Ecology* 54: 334–40, https://doi.org/10.1111/1365-2664.12541.

Martin, T.G., Burgman, M.A., Fidler, F., et al. 2012. Eliciting expert knowledge in conservation science. *Conservation Biology* 26: 29–38, https://doi.org/10.1111/j.1523-1739.2011.01806.x.

McBride, M., Garnett, S.T., Szabo, J.K., et al. 2012. Structured elicitation of expert judgments for threatened species assessment: A case study on a continental scale using email. *Methods in Ecology and Evolution* 3: 906–20, https://doi.org/10.1111/j.2041-210X.2012.00221.x.

McKenzie, C., Liersch, M. and Yaniv, I. 2008. Overconfidence in interval estimates: What does expertise buy you? *Organizational Behavior and Human Decision Processes* 107: 179–91, https://doi.org/10.1016/j.obhdp.2008.02.007.

Mellers, B., Stone, E., Murray, T., et al. 2015. Identifying and cultivating superforecasters as a method of improving probabilistic predictions. *Perspectives On Psychological Science: A Journal Of The Association For Psychological Science* 10: 267–81, https://doi.org/10.1177/1745691615577794.

Mellers, B., Ungar, L., Baron, J., et al. 2014. Psychological strategies for winning a geopolitical forecasting tournament. *Psychological Science* 25: 1106–15, https://doi.org/10.1177/0956797614524255.

Mercier, H. and Claidière, N. 2022. Does discussion make crowds any wiser? *Cognition* 222: 104912, https://doi.org/10.1016/j.cognition.2021.104912.

Mercier, H. and Sperber, D. 2011. Why do humans reason? Arguments for an argumentative theory. *Behavioral and Brain Sciences* 34: 57–74, https://doi.org/10.1017/S0140525X10000968.

Mohanani, R., Salman, I., Turhan, B., et al. 2018. Cognitive biases in software engineering: A systematic mapping study. *IEEE Transactions on Software Engineering* 46: 1318–39, https://doi.org/10.1109/TSE.2018.2877759.

Mojzisch, A. and Schulz-Hardt, S. 2010. Knowing others' preferences degrades the quality of group decisions. *Journal of Personality and Social Psychology* 98: 794–808, https://doi.org/10.1037/a0017627.

Moore, D.A. and Healy, P.J. 2008. The trouble with overconfidence. *Psychological Review* 115: 502–17, https://doi.org/10.1037/0033-295x.115.2.502.

Morgan, M.G. 2014. Use (and abuse) of expert elicitation in support of decision making for public policy. *Proceedings of the National Academy of Sciences* 111: 7176–84, https://doi.org/10.1073/pnas.1319946111.

Mukherjee, N., Hugé, J., Sutherland, W.J., et al. 2015. The Delphi technique in ecology and biological conservation: Applications and guidelines. *Methods in Ecology and Evolution* 6: 1097–109, https://doi.org/10.1111/2041-210x.12387.

Mukherjee, N., Zabala, A., Hugé, J., et al. 2018. Comparison of techniques for eliciting views and judgements in decision-making. *Methods in Ecology and Evolution* 9: 54–63, https://doi.org/10.1111/2041-210X.12940.

Murphy, A.H. and Winkler, R.L. 1977. Can weather forecasters formulate reliable probability forecasts of precipitation and temperature? National Weather Digest 2: 2–9, http://nwafiles.nwas.org/digest/papers/1977/Vol02No2/1977v002no02-MurphyWinkler.pdf.

Navajas, J., Niella, T., Garbulsky, G., et al. 2017. Aggregated knowledge from a small number of debates outperforms the wisdom of large crowds. *Nature Human Behaviour* 2: 126–32, https://doi.org/10.1038/s41562-017-0273-4.

Newell, B.R., Weston, N.J., Tunney, R.J., et al. 2009. The effectiveness of feedback in multiple-cue probability learning. *Quarterly Journal Of Experimental Psychology* 62: 890–908, https://doi.org/10.1080/17470210802351411.

Nickerson, R.S. 1998. Confirmation bias: A ubiquitous phenomenon in many guises. *Review of General Psychology* 2: 175–220, https://doi.org/10.1037/1089-2680.2.2.175.

O'Hagan, A. 2019. Expert knowledge elicitation: Subjective but scientific. *The American Statistician* 79: 69–81, https://doi.org/10.1080/00031305.2018.1518265.

O'Reilly, C. and Chatman, J. 1986. Organizational commitment and psychological attachment: the effects of compliance, identification, and internalization on pro-social behavior. *Journal of Applied Psychology* 71: 492–99, https://doi.org/10.1037/0021-9010.71.3.492.

Oskamp, S. 1965. Overconfidence in case-study judgments. *Journal of Consulting Psychology* 29: 261–65, https://doi.org/10.1037/h0022125.

Page, S. 2007. *The Difference: How the Power of Diversity Creates Better Groups, Firms, Schools, and Societies*, New Edition (Princeton, NJ: Princeton University Press), https://doi.org/10.2307/j.ctt7sp9c.

Palan, S., Huber, J., and Senninger, L. 2020. Aggregation mechanisms for crowd predictions. *Experimental Economics* 23: 788–814, https://doi.org/10.1007/s10683-019-09631-0.

Perneger, T.V. and Agoritsas, T. 2011. Doctors and patients' susceptibility to framing bias: A randomized trial. *Journal of General Internal Medicine* 26: 1411–17, https://doi.org/10.1007/s11606-011-1810-x.

Postmes, T., Spears, R. and Cihangir, S. 2001. Quality of decision making and group norms. *Journal of Personality and Social Psychology* 80: 918–30, https://doi.org/10.1037//0022-3514.80.6.918.

Puncochar, J.M. and Fox, P.W. 2004. Confidence in individual and group decision making: When 'two heads' are worse than one. *Journal of Educational Psychology* 96: 582–91, https://doi.org/10.1037/0022-0663.96.3.582.

Regan, H., Colyvan, M. and Burgman, M. 2002. A taxonomy and treatment of uncertainty for ecology and conservation biology. *Ecological Applications* 12: 618–28, https://doi.org/10.1890/1051-0761(2002)012[0618:ATATOU]2.0.CO;2.

Rothschild, D. and Wolfers, J. 2011. Forecasting elections: Voter intentions versus expectations. *SSRN Electronic Journal*, https://doi.org/10.2139/ssrn.1884644.

Rowe, G. and Wright, G. 1999. The Delphi technique as a forecasting tool: Issues and analysis. *International Journal of Forecasting* 15: 353–75, https://doi.org/10.1016/S0169-2070(99)00018-7.

Saposnik, G., Redelmeier, D., Ruff, C.C., et al. 2016. Cognitive biases associated with medical decisions: A systematic review. *BMC Medical Informatics and Decision Making* 16: Article 138, https://doi.org/10.1186/s12911-016-0377-1.

Satopää, V.A., Baron, J., Foster, D.P., et al. 2014. Combining multiple probability predictions using a simple logit model. *International Journal of Forecasting* 30: 344–56, https://doi.org/10.1016/j.ijforecast.2013.09.009.

Scholten, L., van Knippenberg, D., Nijstad, B.A., et al. 2007. Motivated information processing and group decision-making: Effects of process accountability on information processing and decision quality. *Journal of Experimental Social Psychology* 43: 539–52, https://doi.org/10.1016/j.jesp.2006.05.010.

Schultze, T., Mojzisch, A. and Schulz-Hardt, S. 2012. Why groups perform better than individuals at quantitative judgment tasks: Group-to-individual transfer as an alternative to differential weighting. *Organizational Behavior and Human Decision Processes* 118: 24–36, https://doi.org/10.1016/j.obhdp.2011.12.006.

Schulz-Hardt, S., Brodbeck, F.C., Mojzisch, A., et al. 2006. Group decision making in hidden profile situations: Dissent as a facilitator for decision quality. *Journal of Personality and Social Psychology* 91: 1080–93, https://doi.org/10.1037/0022-3514.91.6.1080.

Schwenk, C.R. 1990. Effects of devil's advocacy and dialectical inquiry on decision-making: A meta-analysis. *Organizational Behavior and Human Decision Processes* 47: 161–76, https://doi.org/10.1016/0749-5978(90)90051-a.

Shanteau, J. 1992. Competence in experts: The role of task characteristics. *Organizational Behavior and Human Decision Processes* 53: 252–66, https://doi.org/10.1016/0749-5978(92)90064-E.

Shanteau, J., Weiss, D.J., Thomas, R.P., et al. 2003. How can you tell if someone is an expert? Performance-based assessment of expertise. In: *Emerging Perspectives on Judgment and Decision Research*, ed.by S.L. Schneider and J. Shanteau (Cambridge: Cambridge University Press), pp.620–42, https://doi.org/10.1017/CBO9780511609978.021.

Siegel-Jacobs, K. and Yates, J.F. 1996. Effects of procedural and outcome accountability on judgment quality. *Organizational Behavior and Human Decision Processes* 65: 1–17, https://doi.org/10.1006/obhd.1996.0001.

Silver, I., Mellers, B.A. and Tetlock, P.E. 2021. Wise teamwork: Collective confidence calibration predicts the effectiveness of group discussion. *Journal of Experimental Social Psychology* 96: 104157, https://doi.org/10.1016/j.jesp.2021.104157.

Simon, H.A. 1977. The logic of heuristic decision making. In: *Models of Discovery*, Boston Studies in the Philosophy of Science, Vol LIV (Dordrecht: Springer), pp.154–75, https://doi.org/10.1007/978-94-010-9521-1_10.

Sniezek, J.A. 1992. Groups under uncertainty: An examination of confidence in group decision making. *Organizational Behavior and Human Decision Processes* 52: 124–55, https://doi.org/10.1016/0749-5978(92)90048-C.

Sniezek, J.A., Paese, P.W. and Switzer, F.S., III. 1990. The effect of choosing on confidence in choice. *Organizational Behavior and Human Decision Processes* 46: 264–82, https://doi.org/10.1016/0749-5978(90)90032-5.

Soll, J.B. and Klayman, J. 2004. Overconfidence in interval estimates. *Journal of Experimental Psychology: Learning Memory and Cognition* 30: 299–314, https://doi.org/10.1037/0278-7393.30.2.299.

Soll, J.B. and Larrick, R.P. 2009. Strategies for revising judgment: How (and how well) people use others' opinions. *Journal of Experimental Psychology: Learning, Memory, and Cognition* 35: 780–805, https://doi.org/10.1037/a0015145.

Speirs-Bridge, A., Fidler, F., McBride, M., et al. 2010. Reducing overconfidence in the interval judgments of experts. *Risk Analysis: An International Journal* 30: 512–23, https://doi.org/10.1111/j.1539-6924.2009.01337.x.

Stasser, G. and Stewart, D. 1992. Discovery of hidden profiles by decision-making groups — Solving a problem versus making a judgment. *Journal of Personality and Social Psychology* 63: 426–34, https://doi.org/10.1037/0022-3514.63.3.426.

Stasser, G., Taylor, L.A. and Hanna, C. 1989. Information sampling in structured and unstructured discussions of 3-person and 6-person groups. *Journal of Personality and Social Psychology* 57: 67–78, https://doi.org/10.1037//0022-3514.57.1.67.

Stasser, G. and Titus, W. 1985. Pooling of unshared information in group decision-making: Biased information sampling during discussion. *Journal of Personality and Social Psychology* 48: 1467–78, https://doi.org/10.1037/0022-3514.48.6.1467.

Stasser, G. and Titus, W. 2003. Hidden profiles: A brief history. *Psychological Inquiry* 14: 304–13, https://doi.org/10.1080/1047840X.2003.9682897.

Steiner, I.D. 1972. *Group Process and Productivity*. Social Psychology (New York: Academic Press).

Stewart, D.D., Billings, R.S. and Stasser, G. 1998. Accountability and the discussion of unshared, critical information in decision making groups. *Group Dynamics: Theory, Research and Practice* 2: 18–23, https://doi.org/10.1037//1089-2699.2.1.18.

Stewart, T.R., Roebber, P.J. and Bosart, L.F. 1997. The importance of the task in analyzing expert judgment. *Organizational Behavior and Human Decision Processes* 69: 205–19, https://doi.org/10.1006/obhd.1997.2682.

Surowiecki, J. 2004. *The Wisdom of Crowds: Why the Many Are Smarter than the Few* (London: Abacus).

Sutherland, W.J. and Burgman, M. 2015. Policy advice: Use experts wisely. *Nature* 526: 317–18, https://doi.org/10.1038/526317a.

Taber, C.S. and Lodge, M. 2006. Motivated skepticism in the evaluation of political beliefs. *American Journal of Political Science* 50: 755–69, https://doi.org/10.1111/j.1540-5907.2006.00214.x.

Teigen, K.H. and Jørgensen, M. 2005. When 90% confidence intervals are 50% certain: On the credibility of credible intervals. *Applied Cognitive Psychology* 19: 455–75, https://doi.org/10.1002/acp.1085.

Tetlock, P.E. 2005. *Expert Political Judgment* (Princeton, NJ: Princeton University Press).

Tetlock, P.E. and Gardner, D. 2016. *Superforecasting: The Art and Science of Prediction* (New York: Random House).

Thomas, E.J. and Fink, C.F. 1961. Models of group problem solving. *Journal of Abnormal and Social Psychology* 63: 53–63, https://doi.org/10.1037/h0040512.

Thorndike, E.L. 1920. A constant error in psychological ratings. *Journal of Applied Psychology* 4: 25–29, https://doi.org/10.1037/h0071663.

Trumbo, D., Adams, C., Milner, M., et al. 1962. Reliability and accuracy in the inspection of hard red winter wheat. *Cereal Science Today* 1: 62–71.

Tversky, A. and Kahneman, D. 1971. Belief in the law of small numbers. *Psychological Bulletin* 76: 105–10, https://doi.org/10.1037/h0031322.

Tversky, A. and Kahneman, D. 1973. Availability: A heuristic for judging frequency and probability. *Cognitive Psychology* 5: 207–32, https://doi.org/10.1016/0010-0285(73)90033-9.

Tversky, A. and Kahneman, D. 1974. Judgment under uncertainty: Heuristics and biases: Biases in judgments reveal some heuristics of thinking under uncertainty. *Science* 185: 1124–31, https://doi.org/10.1126/science.185.4157.1124.

Tversky, A. and Kahneman, D. 1983. Extensional versus intuitive reasoning: The conjunction fallacy in probability judgment. *Psychological Review* 90: 293–315, https://doi.org/10.1037/0033-295X.90.4.293.

Tversky, A. and Kahneman, D. 1985. The framing of decisions and the psychology of choice. In: *Behavioral Decision Making* (Boston, MA: Springer), pp.25–41, https://doi.org/10.1007/978-1-4613-2391-4_2.

Valacich, J.S., Jessup, L.M., Dennis, A.R., et al. 1992. A conceptual framework of anonymity in group support systems. *Group Decision and Negotiation* 1: 219–41, https://doi.org/10.1007/BF00126264.

Van de Ven, A. and Delbecq, A.L. 1971. Nominal versus interacting group processes for committee decision making effectiveness. *Academy of Management Journal* 14: 203–12, https://doi.org/10.5465/255307.

Vul, E. and Pashler, H. 2008. Measuring the crowd within: Probabilistic representations within individuals. *Psychological Science* 19: 645–47, https://doi.org/10.1111/j.1467-9280.2008.02136.x.

Wattanacharoensil, W. and La-ornual, D. 2019. A systematic review of cognitive biases in tourist decisions. *Tourism Management* 75: 353–69, https://doi.org/10.1016/j.tourman.2019.06.006.

Williams, J.J. and Mandel, D.R. 2007. Do evaluation frames improve the quality of conditional probability judgment? In: *Proceedings of the 29th Annual Meeting of the Cognitive Science Society*, ed. by D.S. McNamara and J.G. Trafton (Mahwah, NJ: Erlbaum), pp.1653–58, https://escholarship.org/uc/item/8vc9s8sm.

Winkler, R.L. 1967. Assessment of prior distributions in Bayesian analysis. *Journal of the American Statistical Association* 62: 776–800, https://doi.org/10.2307/2283671.

Winkler, R.L. and Murphy, A.H. 1968. 'Good' probability assessors. *Journal of Applied Meteorology* 7: 751–58, https://doi.org/10.1175/1520-0450(1968)007<0751:PA>2.0.CO;2.

Winman, A., Hansson, P. and Juslin, P. 2004. Subjective Probability intervals: How to reduce overconfidence by interval evaluation. *Journal of Experimental Psychology-Learning Memory and Cognition* 30: 1167–75, https://doi.org/10.1037/0278-7393.30.6.1167.

Wintle, B.C., Fidler, F., Vesk, P., et al. 2013. Improving visual estimation through active feedback. *Methods in Ecology and Evolution* 4: 53–62, https://doi.org/10.1111/j.2041-210x.2012.00254.x.

Wintle, B.C., Fraser, H., Wills, B.C., et al. 2019. Verbal probabilities: Very likely to be somewhat more confusing than numbers. *PloS ONE* 14: e0213522, https://doi.org/10.1371/journal.pone.0213522.

Wintle, B., Mody, F., Smith, E., et al. 2021. Predicting and reasoning about replicability using structured groups. *MetaArXiv*, 4 May 2021, https://doi.org/10.31222/osf.io/vtpmb.

Wolfers, J. and Zitzewitz, E. 2004. Prediction markets. *Journal of Economic Perspectives* 18: 107–26, https://doi.org/10.1257/0895330041371321.

Worchel, S., Wood, W. and Simpson, J.A. 1992. *Group Process and Productivity* (Newbury Park, CA: Sage Publications).

Yang, Y.-T.C. 2008. A catalyst for teaching critical thinking in a large university class in Taiwan: Asynchronous online discussions with the facilitation of teaching assistants. *Educational Technology Research and Development* 56: 241–64, https://doi.org/10.1007/s11423-007-9054-5.

Yaniv, I. 2011. Group diversity and decision quality: Amplification and attenuation of the framing effect. *International Journal of Forecasting* 27: 41–49, https://doi.org/10.1016/j.ijforecast.2010.05.009.

Yaniv, I. and Choshen-Hillel, S. 2012a. When guessing what another person would say is better than giving your own opinion: Using perspective-taking to improve advice-taking. *Journal of Experimental Social Psychology* 48: 1022–28, https://doi.org/10.1016/j.jesp.2012.03.016.

Yaniv, I. and Choshen-Hillel, S. 2012b. Exploiting the wisdom of others to make better decisions: Suspending judgement reduces egocentrism and increases accuracy. *Journal of Behavioral Decision Making* 25: 427–34, https://doi.org/10.1002/bdm.740.

Yaniv, I, Choshen-Hillel, S. and Milyavsky, M. 2009. Spurious consensus and opinion revision: Why might people be more confident in their less accurate judgments? *Journal of Experimental Psychology-Learning Memory and Cognition* 35: 558–63, https://doi.org/10.1037/a0014589.

PART III
MAKING AND APPLYING DECISIONS

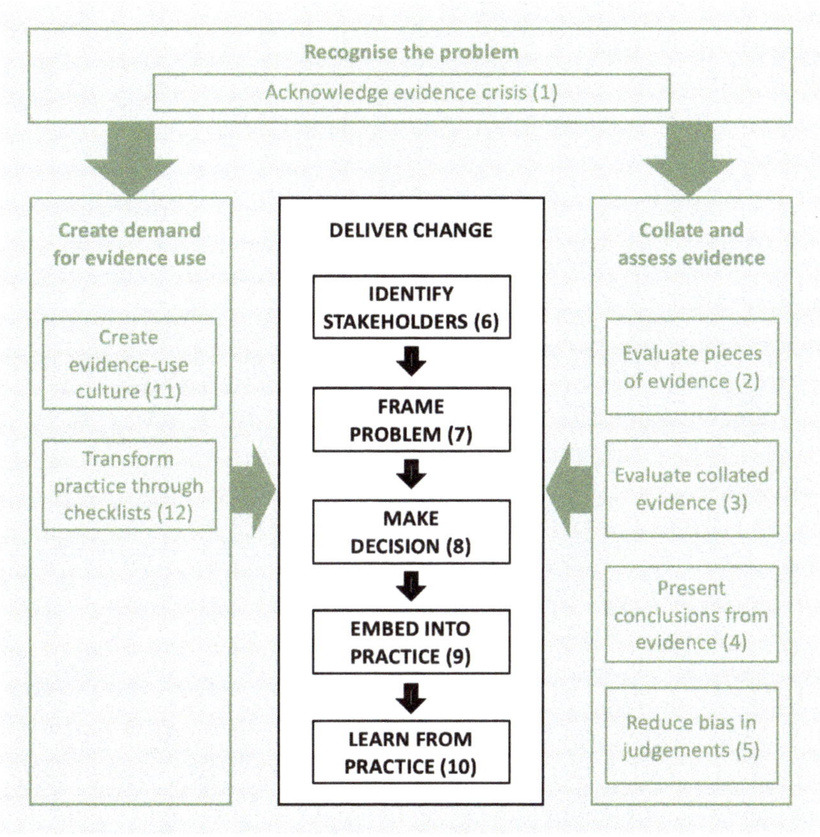

6. Identifying Stakeholders and Collaborating with Communities

Marina Best[1], Robyn Irvine[2], Longji Bako[3], Geoffroy Citegetse[4], Alison Field[5], Dilys Roe[6], David Rose[7], Jonathan Spencer[5]

Working with communities, including local and Indigenous communities, is fundamental to most successful conservation practice. Key elements include determining the appropriate level of engagement, identifying the key stakeholders, identifying appropriate means of collaborating with different stakeholders, creating and maintaining trust, and collaborating to deliver the objectives.

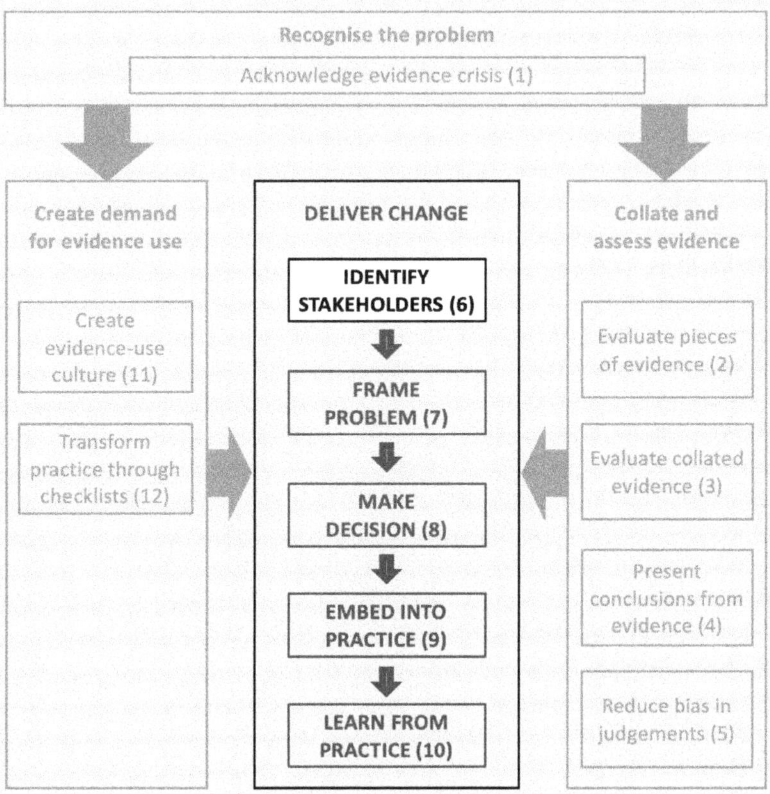

https://doi.org/10.11647/OBP.0321.06

[1] Parks Canada Agency, Indigenous Conservation, Indigenous Affairs and Cultural Heritage, 30 rue Victoria, Gatineau, Quebec, Canada
[2] Parks Canada Agency, Ecosystem Conservation, Protected Areas Establishment and Conservation, 30 rue Victoria, Gatineau, Quebec, Canada
[3] The A.P. Leventis Ornithological Research Institute (APLORI), University of Jos Biological Conservatory, P.O. Box 13404, Laminga, Jos-East LGA, Plateau State, 930001, Nigeria
[4] BirdLife International, Conservation of Migratory Birds and Habitats, Mermoz Pyrotechnie, Lot 23 Rue MZ 56, Dakar, Senegal
[5] Field and Forest, Land Management, 2 Beech Copse, Winchester, UK
[6] International Institute for Environment and Development, 235 High Holborn, London, UK
[7] School of Water, Energy and the Environment, Cranfield University, College Road, Cranfield, Bedford, UK

Contents

6.1 The Benefits of Community-Working

1. The natural environment is embedded within the cultural–political–economic systems in which communities live their lives. These communities are affected by conservation decisions, and in turn, their decisions, responses and actions often determine conservation outcomes. There are thus several reasons why such communities should be involved in conservation decision making.
2. Moral and legal: local communities and Indigenous peoples have a moral and legal right (embodied in international human rights law) to be involved in decision making about the land, waters, and ice that they may have lived on and stewarded for generations. Despite legal rights, these communities have often been, and continue to be, evicted from their homes and may face diluted land rights in the name of conservation (Adams and Mulligan, 2002; Sandlos, 2005; Borras et al. 2011).
3. Improved evidence for decision making: communities often hold considerable knowledge about their traditional and ancestral territories with positive associated conservation outcomes (Schuster et al. 2019); similarly, in landscapes of intensive agriculture and urbanised populations, communities may hold important knowledge of the locale and its history. Such knowledge should be respectfully included by scientists who could learn from lay, place-based stories of what works for conservation, and what does not; for example, Australian Aborigines possess sophisticated knowledge of the consequences of different types of fires (Pascoe and Gammage, 2021). Conversely, the loss of traditional knowledge, or the disappearance of traditional practices, can lead to impoverishment and decline in the quality of semi-natural habitats surviving as relics in modern, transformed landscapes; knowledge of their past management can be key to their successful future management. The United Nations (UN) Convention on Biological Diversity (CBD) specifically recognises the value of Indigenous and local knowledge and highlights the need to protect it.
4. Building support for conservation: interventions made in the name of conservation should benefit, or at least not negatively impact, local and Indigenous communities, for example by not restricting traditional sources of food and income or violating spiritual and cultural practices. It is then far more likely that projects will work and be durable, with ongoing community support and ownership, and will serve to begin the process of repairing relationships that have not necessarily been developed in honourable ways in the past (Wong et al. 2020).

Box 6.1 summarises key principles for collaborating with local and Indigenous communities.

Box 6.1 Principles and methods for working with local and Indigenous communities

Local and Indigenous peoples lead lives that are particularly intertwined with the natural environment. Indeed, the worldview of Indigenous peoples is that they are part of nature not separate from it (United Nations, 2009).

Define a clear purpose for engaging with communities

Making conservation decisions in a participatory fashion with communities is important (Section 6.1). However, participation is a spectrum of forms of engagement from the close and equal sharing of power associated with co-production and co-design down to exploitative and tokenistic consultation, education, or coercion of those with the least power (Arnstein, 1969; Bell and Reed, 2021). A clearly defined and communicated engagement plan can help identify potential problems and offer opportunities to correct them.

Engage appropriately to build local ownership

Ensuring local and Indigenous communities have a sense of ownership over a project or intervention is a prerequisite to working successfully with them. Engagement should take place well before the planning stage of a project and needs to happen throughout each stage of its conceptualisation, from implementation and monitoring to evaluation (Hunt, 2013). Approach engagement on the community's terms. Let them determine who is at the table and understand that this could look different from community to community, as each group will offer diverse perspectives and provide opportunities for capacity building. Developing and selecting topics of interest should ideally be done in collaboration, as projects of interest for Indigenous and local communities can look very different from those of interest to scientists.

The UN and international legal framework of Free Prior Informed Consent (FPIC) have core principles of the policy of self-determined development and respect for the knowledge of Indigenous peoples, their cultures and traditions, which are valuable to the conservation and sustainable use of natural resources. The community should be engaged before starting the project to foster shared objectives and share decision-making power.

Understand 'the community' and the power relationships within it

There is no such thing as a homogeneous community. Every community is different and every person within a community is different. It is critical to ensure that community engagement strategies reflect an understanding of these differences and the power relationships that affect who has a place at the table and who does not. Effective engagement most likely means identifying locally relevant stakeholder groups within the community (which may be socially stratified by age, wealth, caste, gender, or occupation) and ensuring separate discussions are held with each group so that all voices can be heard and taken into account.

Use participatory approaches

Making use of participatory approaches (interviews, focus groups, discussions and workshops) helps to deliver successful project outcomes (Hou-Jones et al., 2021). The specific approaches used should be tailored to the cultural norms of the community.

Develop trust and manage conflict effectively

The chosen spaces for discussion and collaboration need to be safe and foster trust (Ermine, 2007; Bell and Reed, 2021) to create lasting participatory engagement (Madden and McQuinn, 2014; National Coalition for Dialogue and Deliberation, 2010). An attitude open to community collaboration is required, with sufficient time allowed to build trust, depending on the starting point of the various relationships at the project's outset (Madden and McQuinn, 2014). For collaborative meetings, it is recommended to use facilitators who are Indigenous or who adequately understand Indigenous ways of knowing; they also need to be fluent in relevant languages or be able to communicate via a translator, if required.

Foster two-way dialogue and transparency

Cultural factors, gender, and income inequality, the digital divide, amongst other factors, can result in an unequal representation of views (Bell and Reed, 2021; Hurley et al., 2022). The ability to travel to meetings or to engage online, to discuss matters in a common language, and to find the time to engage, will not be equal among participants.

Trust and transparency can also be fostered by creating a two-way reflexive dialogue. This means letting communities know how their knowledge and opinions have affected decision making and creating the feedback loops necessary to show that they have been listened to. Language is important to consider, particularly in areas where the Indigenous or local languages have historically been suppressed by authorities. Consideration should also be given to the preferred method for communication of information; highly technical written material should not be prioritised if local communities have an oral tradition.

6.2 Types of Community Engagement

The benefits of community working, whether with Indigenous communities exercising traditional rights and practices or local communities in often intensively managed landscapes, essentially emerge from a wider understanding and deeper appreciation of views, interests, rights and responsibilities of the parties involved. The appropriate type of community engagement depends on the impact of actions and extent of involvement (Figure 6.1). A proposal to create a wildflower area behind the churchyard may simply justify mentioning in local media while a river restoration project that overlaps land with Indigenous rights will usually be expected to have deep collaboration and to only proceed with the consent of the community.

Engagement ↓ / Impact →	Minor	Moderate	Considerable
Minor	Mention	Discussion	Discussion
Moderate	Discussion	Consultation	Consultation
Considerable	Discussion	Consultation	Joint decision making

Figure 6.1 Appropriate type of community involvement according to the likely impact caused by the proposal and the extent of community engagement in the site. (*Source*: authors)

Table 6.1 lists the different levels of engagement ranging from incorporating information from the community to providing information to aid a community's decision. The lower levels of involvement apply where there are few external responsibilities. Resources and timelines for conservation projects may practically influence the level of engagement that is possible (White et al., 2022). Of course, the level of engagement has to reflect the risks or likelihood of adverse impact; for instance, in the UK the consultation and negotiation associated with discussions on the reintroduction of long-extinct beavers (with concerns over flooding and tree damage) or lynx (with concerns for sheep and pets) far exceed that involved with the return of locally extinct butterflies or orchids.

Stakeholder analysis at the outset can be used to frame both these levels of concern and the level of appropriate response. Partnerships on all sides need to consider what the project aims to achieve, and what level of engagement is appropriate.

Table 6.1 Types of interactions with communities.

Type of interaction	Description
Finding and including information	Using existing information provided by the community
Requesting information	Asking community members for relevant information
Consulting	Speaking to individuals or groups prior to decision making
Involving	Individuals or representatives present and consulted during decision making but not final decision makers
Co-assessing	Collaboratively look at the full range of evidence and assess
Co-operation	Community members or representatives work on and in the projects

Type of interaction	Description
Collaboration	Individuals or representatives fully involved as equal in decision making
Co-production	Community knowledge and expertise brought together with technical conservation knowledge and expertise to jointly create and use information produced
Co-decision	Joint decision with community and conservationists
Informing	Providing information to aid communities in making their own decisions

6.3 Identifying Who to Collaborate With

Human communities are complex. The groups and sub-groups that will be impacted by the proposed intervention, and in turn that the intervention will be impacted by, need to be identified at the initial stages of project development. These could be community groups, Indigenous communities, non-governmental organisations, or associations representing different interest groups. A point of contact with each identified group should be established at this stage as the impacts on and by the groups will differ. It is also important to recognise that there may be differences *within* groups and efforts should be made to understand these differences and take them into account where necessary. Table 6.2 lists the groups that need to be identified and acknowledged accordingly.

Table 6.2 Groups that may be impacted by interventions and other key figures.

Group	Description
Concerned groups	Communities, including governmental and non-governmental agencies, with specific concerns about the management decisions and who have obligations to manage parts or all of the important resources.
Dependent groups	Those whose livelihoods may be at stake due to their dependence on the resources under consideration.
Groups with claims	Communities with territory or resource claims or any form of traditional or legal rights, claims or entitlements.
Holders of knowledge and skills	Who are the most knowledgeable individuals or groups in the area? Does local, valuable knowledge pertain to the conservation question?
Impacted groups	Made up of those who live in close proximity to an intervention site and who may be physically, culturally or economically affected directly by the intervention.
Impacting groups	Groups with members whose activities may be impacting the area or its natural resources, legal or otherwise.
Managers and users	Are there individuals or groups that currently manage the area and its resources or have done so in the past?

Group	Description
National authorities	Are there any national authorities with the mandate to develop and implement policies and rules regarding the area and its resources?
Neighbours	Who are the individuals or communities living near the resources?
Potential investors	Who are the individuals or groups who may be willing to invest human and/ or other capital resources in the area and resources at stake?
Special circumstances	Perhaps the resource use and dependence by the group in question are affected by seasons (e.g. are there seasonal migration patterns or any seasonal events that have important impacts on the area and its resources?) or other factors.
Traditional authorities	Who are the traditional authorities in the area at stake?
Trusted individuals	Are there individuals or groups who are particularly trusted, e.g. as being skilled in conflict management, liaison, and facilitation?

A key stage in this process is to undertake a stakeholder mapping and analysis exercise (see Box 6.2) to determine who should be included, keeping in mind that this list may change due to various reasons, such as capacity or interest.

6.4 Initiating Contact

Prior to establishing contact with a community group, it is important to be clear on why changes to an existing situation are necessary. Then outline the anticipated level of engagement required (see Table 6.1). A consultation programme might start by assessing existing levels of evidence and resources (Sutherland et al., 2017), and researching any past history of challenge or conflict so that clear expectations can be set when first interacting. It is useful at this stage to find out what has, or has not, worked well for previous conservation projects in that (or a similar) community and the reasons why. This can provide a framework for the consultation

Box 6.2 Stakeholder mapping and analysis

Stakeholder mapping is a simple visual technique that enables the depiction of all the stakeholders of a project on one diagram and leads to stakeholder analysis. This example is for a proposal to restore a section of canalised river.

1. Articulate the focus of the mapping.
2. What exactly does the mapping hope to achieve? This may be a subset of the overall programme.
3. Identify stakeholders.

4. Decide the key groups that pertain to the focus of the map (see Table 6.2). In a group, this may be done with post-it notes or added to a flipchart.
5. Identify criteria.
6. Decide upon the key criteria that affect who is useful to consult, such as influence and interest.
7. Analyse stakeholders.
8. Discuss and ask questions to determine where individuals are placed along the criteria. Place in categories usually either low/high or low/medium/high, for instance on power (might they affect the decisions or not) and interest (are they showing considerable interest or not).
9. Map stakeholders.
10. Place each stakeholder within a matrix of the two criteria (Figure 6.2). If a third criterion is used, such as interest in evidence, then this can be represented by font size, colour or whether bold/normal/italic.
11. Use the power/interest matrix to guide the engagement strategy.

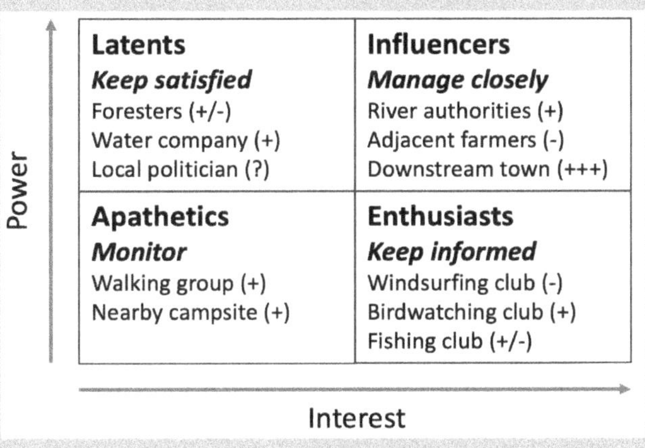

Figure 6.2 Stakeholder mapping classifying them into four groups depending on their interest and power. Support shown in brackets. (*Source*: authors)

12. Prioritise stakeholders.

Stakeholder analysis (Table 6.3) is used to assess the potential role of each stakeholder. This is then used as the basis for determining those stakeholders with the most influence on a project, i.e. which stakeholders to concentrate efforts on and the appropriate engagement approach for each.

Stakeholder mapping thus allows project managers to understand who their stakeholders are, helps identify those who may have the greatest impact on the success of an initiative, and provides the foundations for an engagement/communications plan. This information can then be tabulated in more detail, as in the following stakeholder analysis. Critically, it is important to bear in mind that the most vociferous of voices are not always the ones to most closely adhere to – they may be loud but ill-informed or unrepresentative of the community.

Table 6.3 An example of a stakeholder analysis.

Name and contact details	**Impact: low, medium, high**	**Influence: low, medium, high**	**What is important to them**	**How could they contribute?**	**How could they block the project?**	**Engagement strategy**
River authorities Alex Lamprey a.lamprey@gmail.com	High	High	Reduced flood risk	Provide funding	Not fund if consider not cost effective or influenced by objections	Agree to collaborate on hydrological models
Fishing community spokesperson Maria Sturgeon Mobile: 03145 926535	High	Medium	Fish population and suitable fishing areas	Could help maintain site	Might object if consider detrimental to fish or trees block fishing	Take to meet fishing community on other restored sites. Agree on retaining tree-free locations
Local bird club (based at the museum)	Medium	Low	Good bird watching along the river with places to watch from	Could monitor changes	Unlikely but might complain if vegetation prevents seeing river	Keep informed and agree on access points to birdwatch

There is a range of software that can help with the creation of stakeholder mapping and analyses.

process. Following local protocols is important, including, where appropriate, the provision of gifts or recompensation for time and effort.

6.5 Creating and Maintaining Trust

Trust is the basis for almost everything we do (Frei and Morriss, 2020). Effective community relationships are dependent upon trust; for example, Lachapelle and McCool (2012) suggested that the lack of trust with conservation agencies was often the fundamental barrier to the negotiation and construction of natural resource management plans. More positively, Young et al. (2016) showed how increased trust, through fair processes, makes conflict resolution more likely.

The elements of trust can be classified as Contractual trust (promises upheld, commitments and expectations explicit, participants can rely on each other), Communication trust (key information provided appropriately and important material not withheld), Competency trust (collaborators will deliver knowledgeably and effectively), and Caring trust (support for diversity of needs and understanding when needed).

Frei and Morriss (2020) suggest that people are trusted when others think they are interacting with the real person (authenticity), when others have faith in their judgement and competence (logic), and when others believe that they are cared about (empathy). Achieving and maintaining trust takes time (but can be lost quickly) and requires an approach including being honest about objectives, showing integrity, listening to concerns from a broad community, asking questions with genuine curiosity, showing humility, sharing knowledge effectively, being helpful, delivering on promises, admitting mistakes, giving credit, and providing praise.

6.6 Collaborating

The beginning of any project never starts with a blank sheet. There are always existing interests and resource uses that need to be understood. Critical questions at the outset of a project consultation include the role of individuals and communities engaged in the consultation, the extent to which existing activity or usage can be relocated elsewhere, what the current position is (including the veracity of assumptions), the range of views, and what means are required to find out the answers. Initial or pre-engagement consultations benefit from a sense of direction, even if it takes time to complete, to encourage focus in discussions.

Some projects never find such a compromise: consensus does not always lead to an agreed or successful outcome. Even reaching a consensus can be difficult or impossible in wide consultations, where some participants may have opinions that are difficult to reconcile. Nevertheless, the discussions and meetings, proposals and counter proposals should have led all participants to understand why the decision was ultimately chosen and what the level or distribution of support for it was.

Bringing together different communities can be challenging, with the potential for significant disagreement and conflict. There are often skilled intermediaries or trusted facilitators in communities who can help to reduce biases and power inequalities as well as manage conflicts. For collaborative initiatives with Indigenous peoples, it is recommended to use facilitators who are Indigenous or who adequately understand Indigenous worldviews and ways of knowing.

For projects in developed countries with large urban populations or nationally-based interest groups, facilitators from more neutral or widely respected organisations are more appropriate. In all cases it is key that a facilitator can speak the local language and understands local cultural norms. Once trust and understanding have been gained between the parties involved, the further services of such facilitators may not be necessary.

For interventions that involve research, there are often ethical standards that need to be observed. Most universities and research institutes will have their own ethics procedures but, in the absence of these, external codes of conduct can be used such as the code of ethics of the Society for Ethnobiology (https://ethnobiology.org/about-society-ethnobiology/ethics).

Everything is more complicated than it seems. Successful projects acknowledge this truth. Their hallmark is a complex mix of engagement with communities and stakeholders from the outset, leading to compromise and an agreed common vision of what needs to be achieved and how.

Relationship building is key and usually goes well beyond the scope of a single project. This element is foundational to the success of a truly co-developed and collaborative project. Building a solid relationship takes considerable time, commitment, and continued nurturing. Ideally, effort should be made to build the relationship with key partners, specifically local people, local interest groups and Indigenous communities, and should be pursued well in advance of any specific project. Building such relationships outside of the pressures of meeting objectives or ties to any project goals helps to show the sincerity and commitment of an organisation to a particular area or community. This often means, when possible, ethical, frequent and accepted participation in a community's events and activities, creating bonds with community members, and hosting meetings and discussions on a broad range of topics. For example, in remote locations, an individual's role as a representative of a conservation organisation remains even outside working hours. Getting involved in community activities outside of work hours helps foster stronger and more meaningful relationships. This may mean participating in cultural celebrations, weddings, funerals, festivals, or feasts, or other events that might fall well outside of the normal scope of one's organisational duties.

Community meetings benefit from an open and flexible approach that allows for the scope of discussion to change depending on the priorities and issues that are raised by community members. Moreover, it is important, especially with Indigenous communities, to conduct meetings in places that make sense to them, which typically means hosting a gathering on the lands, ice, and waters in question. Keeping a balance between the players involved is critical, and Madden and McQuinn (2014) provide some advice on how to balance the 'cards' participants are holding. Examples of this balancing can range from things like using various facilitation and structural methods to ensure all voices are heard in meetings or ensuring that data sovereignty is rigorously upheld.

Prior to any project getting underway, it is important to determine how a community's knowledge and information will be used and stored. For example, First Nations in Canada have principles of ownership, control, access, and possession (or OCAP™), which assert their stewardship of their information and data. Courses and workshops are available for researchers and are highly recommended before conducting any work. Moreover, Inuit in Canada have also put together the National Inuit Strategy on Research (Inuit Tapiriit Kanatami, 2018), which aims to advance Inuit governance, capacity, and access to research processes in Inuit Nunangat. Ensuring that researchers follow guidance from the communities will help build trust and foster the relationship, while at the same time working to advance projects in a manner that respects the communities' involvement.

The Healthy Country Planning approach to the Open Standards for the Practice of Conservation co-develops conservation projects with Indigenous partners (Carr et al., 2017). Adopted by many conservation planners around the world, this approach provides a foundation for 'Two-Eyed Seeing' (or *Etuaptmumk*), coined by Mi'kmaq Elders Albert and Murdena Marshall, this is the principle of bringing together Indigenous and western ways of knowing through seeing the strengths of each perspective, viewing the world with the 'two eyes' (i.e. from both sides) and advancing in a collaborative space (Reid et al. 2020). The approach allows cultural and socio-economic objectives to parallel and overlap the ecological and quantitative ones. This enables the co-creation of conservation projects and sharing decision making.

Box 6.3 gives a range of examples of conservation projects for which community engagement was fundamental.

Box 6.3 Examples of community engagement

Chiixuu Tll iinasdll — kelp forest restoration project

This project, aimed to enable the recovery of an area of kelp forest through the removal of hyperabundant sea urchins (guudingaay) (Bellis et al., 2019), at Gwaii Haanas National Park Reserve, Canada, is an example of approaching engagement on a community's terms. The Gwaii Haanas Gina 'Waadluxan KilGuhlGa (Talking about Everything) Land–Sea–People Management Plan (Council of the Haida Nation and Canada, 2018) was organised on the six Haida ethics and values (respect, responsibility, interconnectedness, balance, seeking wise council, and giving and receiving). These principles underpinned the planning and delivery of the project, which was co-developed with the Haida Nation and the Haida Fisheries Program. A delivery of sea urchins to local communities several times through the project was an important communication approach that enabled engagement and dialogue over food. Monitoring suggests this programme has been beneficial for kelp species, kelp cover and northern abalone and this approach has acted as a model for collaborative working elsewhere by establishing enduring working relationships that are respectful of the social and ecological context of each place (Lee et al., 2021).

Sapo National Park Liberia

The Sapo National Park, which provides refuge to many rare and endemic species, was established in 1983 under the military decree. There was no community engagement, with communities moving out of the area. The National Park was further expanded without engaging the community and with further eviction. Though there was not any form of resistance, the communities were aggrieved. Failure to properly engage the community was detrimental and there was a clash between the locals and the rangers that tragically led to a ranger dying. However, reconciliation and making a change to proper community engagement after the tragic incident turned everything around. The community became partners in protecting the National Park. This was seen clearly in the growing enthusiasm and dedication of the community members as they even will go as far as helping to get rid of illegal miners and poachers dwelling within the park. They will arrest the defaulters and take them to the local chief, something that the ranger force was unable to achieve in the past. Sapo National Park has been transformed from being a Park in danger of losing its biodiversity and natural resources to a fully functioning protected area where local communities are deliberately acting to protect their heritage.

New Forest New Future programme

The New Forest (a National Park since 2005) is one of the most important areas of unenclosed pasture, forest and lowland heathland in the UK. The New Forest New Future programme, initiated in 1997, brought together those with an interest in the New Forest's future (foresters, naturalists, statutory bodies, representatives of the Commoners that graze the forest with their animals, and the Verderers that manage the affairs of the Commoners across the Forest, local communities, tourism interests and local economic interests). The eventual objective was to agree on a management plan, specifically for the Forest's timber inclosures (those areas where the management of trees for timber production is permitted under the New Forest Acts) whose management was particularly contentious. The programme comprised meetings and presentations, field visits, discussions, and planning events, at which all parties presented information and insight into their particular interests and concerns. These events ensured that those engaged in the subsequent, more detailed, consultations appreciated the legal, economic, or other aspects of importance in the forest from all other viewpoints. Most of the participants became known to each other, and indeed in many instances, became friends. The programme led to agreement on a management plan, reviewed every five years, which, 25 years on from its inception, continues to deliver an agreed programme of habitat restoration and management that is widely understood and largely endorsed by all the disparate parties with an interest in the future of the New Forest.

Beyond Borders Caribou Workshop series

These events, hosted by Parks Canada for Wapusk National Park and the greater Wapusk Ecosystem (Environment and Climate Change Canada, 2022), brought together Indigenous communities (Cree, Dene, Inuit and Métis), government organisations (Environment and Climate Change Canada, Provincial and Territorial), academic researchers and local communities to share Indigenous and local knowledge and western science perspectives about caribou. The workshops aimed to strengthen and form new relationships, highlight topics of concern, identify gaps in knowledge, and outline actions for effective caribou monitoring and management. An ethical and inclusive space was created that encouraged two-way knowledge sharing and discussion. This was achieved by ensuring everyone was at the table with appropriate resources to enable participation by all groups and equal status in the co-development of this project.

Living on the Edge, communities at the centre of Lac de Mâl conservation

The Lac de Mâl is a permanent lake in a dry area, 65 km south-east of the city of Aleg in Mauritania. It covers about 5,250 ha at the end of the rains, shrinking to 870 ha at the end of the dry season. It is an important site for water birds with at least 35,000 birds congregating in the lake annually. As the only permanent wetland in the region, it is also a key resource for the local people; some 9,000 people live around the lake. In spite of its importance, the area has no official protection status. The main activities around the lake are agriculture, small-scale fishing and cattle breeding. Nature Mauritanie, the BirdLife partner in Mauritania, supported the communities through the European-Union-funded Living on the Edge project. The communities came together and signed a charter with guidelines on the use of the natural resources of the lake by the stakeholders (fishermen, farmers, and livestock breeders). In the past, fuelwood-cutting and overgrazing have reduced the area of woodland around the lake and the cover of herbaceous vegetation, exposing the soil to erosion. The dwindling amount of vegetation cover on surrounding dunes has increased the risk of the dunes shifting and filling in the lake. The current project's activities are centred on tree planting to stabilise the dunes around the lake and the rehabilitation of the dam. Together with the local communities, it developed alternatives and sustainable activities (gardening, solar fish smoking, poultry). The waterbird population has increased and the livelihood of local communities improved (cooperative, income-generating activities). The model of collaboration was the model for the development and management plan of Mâl Commune by the local authorities.

References

Adams, W. and Mulligan, M. 2002. *Decolonizing Nature. Strategies for Conservation in a Post-colonial Era* (London: Taylor and Francis), https://www.taylorfrancis.com/books/mono/10.4324/9781849770927/decolonizing-nature-william-bill-adams-martin-mulligan.

Arnstein, S.R. 1969. A ladder of citizen participation. *Journal of the American Institute of Planners* 35: 216–24, https://doi.org/10.1080/01944366908977225.

Bell, K. and Reed, M. 2021. The tree of participation: A new model for inclusive decision-making. *Community Development Journal* bsab018, https://doi.org/10.1093/cdj/bsab018.

Bellis, K.X.T., Peet, R.T., Irvine, R.L., et al. 2019. Beyond biodiversity: The cultural context of invasive species initiatives in Gwaii Haanas. In: *Island Invasives: Scaling Up to Meet the*

Challenge, ed. by C.R. Veitch, et al. (Occasional Paper SSC No. 62. Gland, Switzerland: IUCN), pp. 494–96, https://doi.org/10.2305/IUCN.CH.2019.SSC-OP.62.en.

Borras, S.M., Jr, Hall, R., Scoones, I., et al. 2011. Towards a better understanding of global land grabbing: an editorial introduction. *Journal of Peasant Studies* 38: 209–16, https://doi.org/10.1080/03066150.2011.559005.

Carr, B., Fitzsimons, J., Holland, N., et al. 2017. CAPitalising on conservation knowledge: Using Conservation Action Planning, Healthy Country Planning and the Open Standards in Australia. *Ecological Management and Restoration* 18: 176–189, https://doi.org/10.1111/emr.12267.

Council of the Haida Nation and the Government of Canada. 2018. *Gwaii Haanas Gina"Waadluxan KilGuhlGa Land-Sea-People Management Plan: 2018* (Gatineau, Quebec:Parks Canada), https://www.pc.gc.ca/en/pn-np/bc/gwaiihaanas/info/consultations/gestion-management-2018.

Environment and Climate Change Canada. 2022. *Species at Risk Act: Annual Report to Parliament for 2020* (Gatineau, QC: Environment and Climate Change Canada), https://wildlife-species.canada.ca/species-risk-registry/virtual_sara/files/reports/Sar-2020-v02-eng.pdf.

Ermine, W. 2007. The ethical space of engagement. *Indigenous Law Journal* 6: 193–234, https://ilj.law.utoronto.ca/sites/ilj.law.utoronto.ca/files/media/ilj-6.1-ermine.pdf.

Frei, F.X. and Morriss, A. 2020. Begin with trust. *Harvard Business Review* 98: 112–21, https://hbr.org/2020/05/begin-with-trust.

Hou-Jones, X, Bass, S. and Roe, D. 2021. *Mainstreaming Biodiversity in Government Decision-Making* (Cambridge: UNEP-WCMC), https://pubs.iied.org/20426g.

Hunt, J. 2013. *Engaging with Indigenous Australia — Exploring the Conditions for Effective Relationships with Aboriginal and Torres Strait Islander Communities*, Issue Paper No. 5, Produced for the Closing the Gap Clearinghouse (Canberra: Australian Institute of Health and Welfare; Melbourne: Australian Institute of Family Studies), https://www.aihw.gov.au/getmedia/7d54eac8-4c95-4de1-91bb-0d6b1cf348e2/ctgc-ip05.pdf.aspx?inline=true.

Hurley, P., Lyon, J., Hall, J., et al. 2022. Co-designing the environmental land management scheme in England: The why, who and how of engaging 'harder to reach' stakeholders. *People and Nature* 4: 744–57, https://doi.org/10.1002/pan3.10313.

Inuit Tapiriit Kanatami. 2018. *National Inuit Strategy on Research* (Ottawa, Ontario, Canada: Inuit Tapiriit Kanatami), https://www.itk.ca/wp-content/uploads/2018/04/ITK_NISR-Report_English_low_res.pdf.

Lachapelle, P.R. and McCool, S.F. 2012. The role of trust in community wildland fire protection planning. *Society and Natural Resources* 25: 321–35, https://doi.org/10.1080/08941920.2011.569855.

Lee, L.C., McNeill, G.D., Ridings, P., et al. 2021. Chiixuu Tll iinasdll: Indigenous ethics and values lead to ecological restoration for people and place in Gwaii Haanas. *Ecological Restoration* 39: 45–51, https://doi.org/10.3368/er.39.1-2.45.

Madden, F. and McQuinn, B. 2014. Conservation's blind spot: The case for conflict transformation in wildlife conservation. *Biological Conservation* 178: 97–106, https://doi.org/10.1016/j.biocon.2014.07.015.

National Coalition for Dialogue and Deliberation. 2010. *Resource Guide on Public Engagement*, https://www.ncdd.org/rc.html DOI:10.1017/S0030605317001296.

Pascoe, B. and Gammage, B. 2021. *Country: Future Fire, Future Farming* (Melbourne: Thames and Hudson Australia Pty Ltd).

Reid, A.J., Eckert, L.E., Lane, J., et al. 2020. 'Two-eyed seeing': An Indigenous framework to transform fisheries research and management. *Fish and Fisheries* 22: 243–61, https://doi.org/10.1111/faf.12516.

Sandlos, J. 2005. Federal spaces, local conflicts: National Parks and the exclusionary politics of the conservation movement in Ontario, 1900–1935. *Journal of the Canadian Historical Association* 16: 293–318, https://doi.org/10.7202/015735ar.

Schuster, R., Germain, R.R., Bennett, J.R., et al. 2019. Vertebrate biodiversity on Indigenous-managed lands in Australia, Brazil, and Canada equals that in protected areas. *Environmental Science and Policy* 101: 1–6, https://doi.org/10.1016/j.envsci.2019.07.002.

Sutherland, W.J., Shackleford, G., Rose, D.C. 2017. Collaborating with communities: Co-production or co-assessment? *Oryx* 51: 569–70, https://doi.org/10.1017/S0030605317001296.

United Nations. 2009. *The State of the World's Indigenous Peoples*, ST/ESA/328 (New York: United Nations Department of Economic and Social Affairs, Secretariat of the Permanent Forum on Indigenous Issues), https://www.un.org/esa/socdev/unpfii/documents/SOWIP/en/SOWIP_web.pdf

White, T.B., Petrovan, S.O., Christie, A.P., et al. 2022. What is the price of conservation? A review of the status quo and recommendations for improving cost reporting. *BioScience* 72: 461–71, https://doi.org/10.1093/biosci/biac007.

Wong, C., Ballegooyen, K., Ignace, L. et al. 2020. Towards reconciliation: 10 Calls to Action to natural scientists working in Canada. *FACETS*. 5: 769–783. https://doi.org/10.1139/facets-2020-0005.

Young, J.C., Searle, K., Butler, A., et al. 2016. The role of trust in the resolution of conservation conflicts. *Biological Conservation* 195: 196–202, https://doi.org/10.1016/j.biocon.2015.12.030.

7. Framing the Problem and Identifying Potential Solutions

Nafeesa Esmail[1,2], Rhys Green[3], Silviu Petrovan[3,4], Nick Salafsky[5], Nigel G. Taylor[3], Jeremy D. Wilson[6]

The creation of effective policy and practice starts by framing the problem to be solved. This requires deciding what is important, identifying the current and potential future threats, diagnosing the actual cause of the problems, and identifying solutions, including innovating to create new ones when required. In this chapter we describe various techniques that can be used to frame the problem including horizon scanning, situation models and theory of change diagrams. These can be used to identify the analytical questions and specific assumptions that underpin the assessment of evidence and decision making.

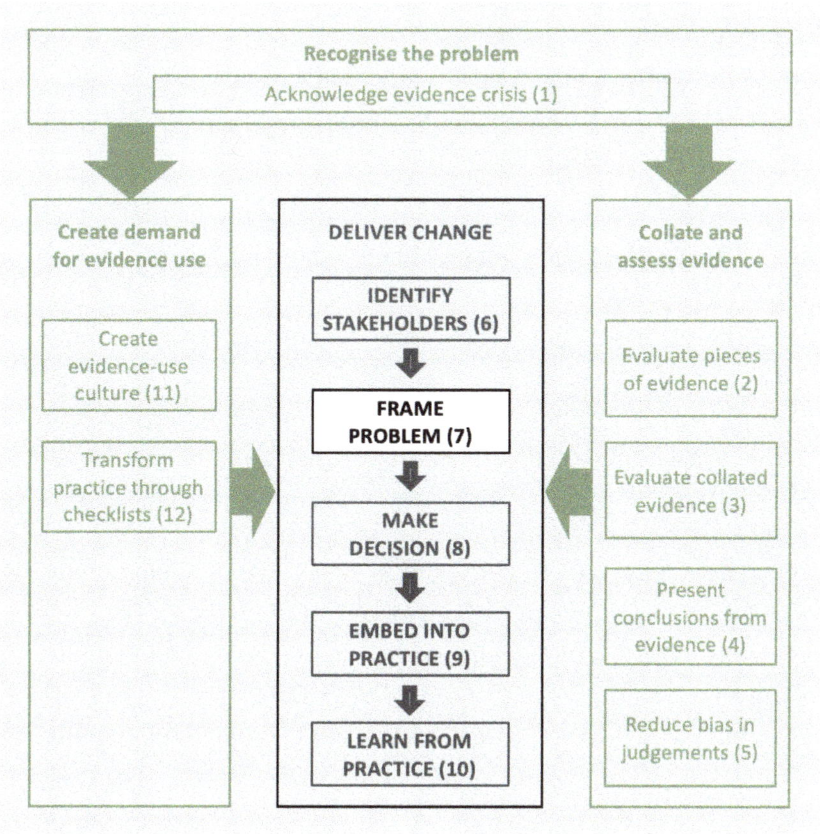

 https://doi.org/10.11647/OBP.0321.07

[1] The Wilder Institute, 1300 Zoo Road NE, Calgary, Alberta, Canada
[2] Women for the Environment Africa (WE Africa) Fellow
[3] Conservation Science Group, Department of Zoology, University of Cambridge, The David Attenborough Building, Pembroke Street, Cambridge, UK
[4] Biosecurity Research Initiative at St Catharine's (BioRISC), St Catharine's College, Cambridge, UK
[5] Foundations of Success, 4109 Maryland Avenue, Bethesda, Maryland, USA
[6] RSPB Centre for Conservation Science, 2 Lochside View, Edinburgh, UK

Contents

7.1 The Approach to Identifying Problems and Potential Solutions

As a basis for this chapter, we are assuming that the practice of conservation (which includes natural resource management) takes place through specific projects and broader programmes (CMP, 2020, Salafsky et al., 2022). Conservation is a process that (i) involves a defined project team understanding the threats, drivers, and opportunities in a given situation, (ii) agreeing on a desired future state for the situation, and then (iii) deciding on, implementing, and managing one or more strategies that involve taking particular actions that are designed to achieve these goals and objectives. This process, which can be applied at any spatial or temporal scale, is implemented through various planning and decision support frameworks (Schwartz et al., 2017). Although it may seem like every organisation has its distinct proprietary framework, at their core, most of these approaches involve a common iterative series of steps that include analysing the situation, identifying and comparing different potential intervention strategies, implementing these strategies, monitoring and evaluating the results of these strategies, and then adaptively managing and learning based on the findings.

A typical process for initiating a project would be:

1. Decide upon the scope of the project and identify features of interest.
2. Collate knowledge about the species, habitat, cultural and environmental features present.
3. Identify current and future threats and opportunities relevant to the project, for example using diagnosis techniques, horizon scanning or scenario planning.
4. Define goals and objectives for the project.
5. Identify possible responses to the threats and opportunities. This may involve solution scanning to list existing options and innovation to develop new options, then using theory of change and assessing evidence of effectiveness to refine the list of options.
6. Identify questions for which further evidence is needed.

These elements seem obvious, but there are numerous examples of projects selecting peculiar goals or objectives, failing to identify the real problem or selecting inefficient options for implementation. For example, many conservation projects undertake mangrove planting as an action, yet this often fails because the goal is framed in terms of the number of trees planted rather than the number surviving, the focus is on planting trees in new areas rather than improving and expanding surviving mangrove forest patches using targeted planting, or because planting is carried out in inappropriate topography or using poorly adapted species for the local conditions (Kodikara et al., 2017).

In contrast, when these elements are followed, they can lead to justifiable, targeted and effective conservation action. Following news of the rapid loss of shorebirds in the east Asian flyway, and the imminent extinction of the spoon-billed sandpiper *Calidris pygmaea*, the IUCN commissioned a situation analysis of the Yellow Sea (MacKinnon et al., 2012). This 70-page IUCN publication documented the evidence for the importance of the Yellow Sea, the evidence of declines in biodiversity and various threats including habitat loss by land reclamation. This was a critical stage in laying down the problem and led to the eventual designation of the Yellow Sea as a World Heritage Site (Crockford, 2018).

This chapter discusses each of these six elements in turn and, by outlining the problem, sets the stage for subsequent chapters that describe how to resolve the problem and deliver change. There will often be some iteration between the steps. For example, although objectives will be based on an initial understanding of the system, they may raise further specific questions to investigate and be refined as these questions are answered. The process will be facilitated by an individual or team, consulting stakeholders and experts as appropriate with workshops, interviews, or questionnaires.

7.2 Defining the Scope of the Project and the Conservation Targets

The team behind a conservation project should start by defining its scope: spatially (e.g. a specific protected area, a geographic region) or thematically (e.g. a specific species, all the economic actors in a given fishery) or often by some combination of these (CMP, 2020). Within this scope, the team then identifies the specific features of interests that represent its goals. These typically include specific species and broader ecosystems for which the area is important, or even cultural features or elements of human wellbeing linked to biodiversity. In identifying features, it is important to take into account global, national, and local listings of endangered and threatened species and ecosystems. For example, a species that is locally abundant, but rare elsewhere, is a likely feature of interest. In the end, however, selecting both a project scope and its features of interest is at least as much a values-based decision as it is a scientific decision but the reasoning and the process should be transparent.

7.3 Understanding the Biological and Human System

7.3.1 Determining feature status

Once features of interest have been identified, it is important to assess their current status. This assessment is typically done by identifying one or more key attributes that help define the viability of the feature (CMP, 2020). Examples of key attributes include water chemistry for aquatic habitats, biological community composition, and access to food resources or recreational opportunities for humans.

Statements of significance identify the important features in the area, such as 'high quality grassland habitat is located within the project site', 'it is predicted that in 10 years the site will be suitable for rainforest ecosystems', or 'the watershed ecosystem is essential for maintaining water quality and regulating flows for the town downstream'. These are all statements that can be backed up by evidence of the status of features, including: species ranges, presence/absence, population size or abundance; habitat presence/absence, quality or distribution; or ecosystem services provided.

Understanding the status of species and habitats is an important step in the design of subsequent actions (Pimm et al., 2014). For example, effectively avoiding and minimising the impact of new energy infrastructure (e.g. power lines, wind turbines) on bird species requires information on the species present in the area, the habitat quality, and their susceptibility to impact before decisions can be made on what actions may be appropriate to undertake (Bennun et al., 2021).

Evidence can also be compiled on the status of ecosystem services in the area of interest, and the uses of various components of the environment by different stakeholders. For example, a claim may be made in the design of a marine conservation project that 100 tons of fishery catch are currently landed per annum from the project area; that the mangrove habitats present are vital for the provision of firewood to local stakeholders; and the mangroves act as nursery habitats that play a major role in the fishery stock. Obtaining information on these uses and ecosystem services can be important to ensure that claims made in the project conception are backed up by appropriate evidence, thus reducing the risk of tackling inappropriate features of interest or having an incorrect theory of change resulting in an ineffective project.

7.3.2 Assessing changes in feature status

Designing a conservation project may also require evidence of changes in status — a key component in identifying the threats and opportunities that the project could address. For example, this may be changes in species populations over time, or changes in habitat quality or integrity.

Considering the change in status of the features of interest usually involves combining wide-scale and local information (Table 7.1) to assess if any changes observed are part of a broader pattern or specific to the site. In some cases, there is regular monitoring that provides the required information. Moussy et al. (2022) provide a database of nearly 1,200 monitoring schemes, ranging in start date from 1800 to 2018. In most cases, however, the evidence comprises a mixture of survey counts, old records and knowledge of local experts. Museum collections, old maps, photos, descriptions, and paintings (which may not be accurate or biologically correct) can all be useful for assessing longer-term changes over decades or centuries. In a particularly innovative study, Belgian researchers assessed changes in tree and shrub phenology over 35 years using archived television footage from the Tour of Flanders bicycle race (De Frenne et al., 2018).

7.3.3 Understanding how the system functions

To correctly identify threats, opportunities and possible actions it is necessary to understand the functioning of systems around the features of interest. For example, the key to determining

Table 7.1 Potential actions concerning wide-scale and local changes.

	Declining widely	Stable/increasing widely
Declining locally	Decide if local action will be effective, or if impossible to reduce or reverse decline due to changes elsewhere	Identify the local problem and consider whether feasible or necessary to resolve
Stable/increasing locally	Identify solutions within the local context and inform others	Monitor

why a species has declined usually lies in a fundamental appreciation of its ecology and local context. This can be achieved by carrying out a literature review and looking at existing empirical studies to understand: (1) the ecological requirements such as habitat preferences, diet, breeding sites, or germination requirements; (2) which external factors affect demographic rates such as reproduction and mortality; and (3) large-scale correlations between distribution and abundance, and key resources or environmental factors. New observation, experiments and analyses may be needed to confirm theories or fill knowledge gaps.

As a specific example, the conservation of the large blue butterfly *Phenagris arion* in Britain required an understanding of the species' larval dependence on a single host-ant species whose own ecological requirements are so specialised that subtle changes in grassland management and vegetation structure could cause its replacement by ecologically similar but unsuitable species. Only when this was understood could appropriate amendments to grassland management be recommended to drive recovery or re-establishment of the butterfly (Thomas et al., 2009). This in turn prompted wider recognition that diagnosis of larval habitat quality and its environmental drivers is central to the evidence-based conservation of many butterflies (Thomas et al., 2011). Section 7.4.2 explores diagnosis of declines in more detail.

Understanding socioeconomic systems can also inform conservation management. This is especially true for invasive alien species, where knowledge of how the species is introduced (the pathways and vectors involved) can highlight areas for intervention. Researchers in North America worked out that shipping activity was a major driver of invasions in the Great Lakes basin, with ballast water release being responsible for around 65% of invasions in the late 20th Century. Canada and the United States subsequently introduced strict ballast water exchange regulations, which reduced the rate of discovery of new non-native species in the Great Lakes by 85% (Ricciardi and MacIsaac, 2022).

7.4 Identifying Threats and Opportunities

7.4.1 Characterising current threats

Building on an understanding of the relevant systems, the project team can identify the threats that are affecting their features of interest. Following CMP (2020), this includes both direct threats (aka pressures) and indirect threats (aka root causes or underlying causes). The aim is to produce a reasonably comprehensive list, including those threats that may not be obvious and not only those that the project team (currently) has the capacity to address.

Direct threats are primarily human activities that immediately degrade a feature of interest (e.g. unsustainable fishing, unsustainable hunting, oil drilling, construction of roads, release of industrial wastewater, or introduction of invasive alien species). Direct threats can also be natural phenomena, perhaps altered by human activities (e.g. increase in extreme storm events or increased evaporation due to global climate change) or, in rare cases, with an impact increased by other human activities (e.g. a potential tsunami that threatens the last remaining population of an Asian rhino already endangered and range-restricted due to poaching). Direct threats can range from obvious and rapid (e.g. deforestation) to subtle and long-term (e.g. changes in grazing) and they often interact and combine to produce cumulative and more complex outcomes. For example, habitat fragmentation, invasive species and climate change can interact to increase impacts from each threat (Macinnis-Ng et al., 2021). Understanding how specific threats are likely to develop enables appropriate planning.

Indirect threats are drivers of direct threats, such as policies around logging or prescribed burning, or market forces that influence fishing pressure. It is usually easier to work backwards from targets, identifying direct threats then considering possible underlying causes for these.

In many cases, there may be a complex series of biophysical factors that link a given threat to a feature of interest. For example, the threat of land use change (e.g. ploughing of former grassland sites) may result in increased sediment input to waterways, which can affect biological features through eutrophication and resulting in changes in oxygen or light availability. It can be useful to explicitly map out these factors to clarify the logic linking threats to targets, and to identify possible intervention options.

7.4.2 Diagnosis: Identifying the likely cause of declines

Based on evidence of the timing and location of change, as well as ecological knowledge of the system in which these changes are taking place, candidate causes of species decline, habitat deterioration or other processes can be listed to inform investigation and determine which causes are most important (Green, 2002; Daszak et al., 2003). If the quality of knowledge is high and causal links are well understood, then this list may be short and arrived at with confidence. If the quality of knowledge is low, then careful judgement is needed: a conservative approach may inadvertently exclude an important cause, whereas a more inclusive approach risks creating an impractically long list. Diagnosis then proceeds by seeking to understand for which of the candidate causes on the list is the strongest evidence that they have driven the observed decline. Sometimes — for instance if a species is declining rapidly — a decision must be made based on the currently available evidence, even if imperfect, then monitored and revised if and when new evidence comes to light.

Here we use a range of species examples to illustrate some of the approaches that can be employed. We first consider approaches that can be used when we are in the fortunate position of being able to call upon spatially and temporally extensive distribution, population and demographic rate data to support diagnosis. Secondly, we consider approaches where such data are unavailable and instead we may need to rely upon primary field studies at one or a small number of sites.

The key issue is usually not whether a threat is causing some change but the magnitude of its impact, whether it is a major contributor to overall status change and how it is changing over time. Claims, such as 'songbird species at our site are severely impacted by overexploitation due to the pet trade' or 'habitat degradation caused by dredging activities outside the marine protected area are a significant threat to marine biodiversity', need estimates of the frequency, magnitude and overall consequence of threatening processes in order to be meaningfully quantified and addressed by actions.

Evidence on threats can come from a variety of sources including scientific studies, databases of threat data or expert and local knowledge. Evidence of threats can also be collected from project sites directly, and during project implementation. Increasingly, new technologies are providing near real-time information on threatening processes that can be used as a source of evidence. For example, GLAD (Global Analysis and Discovery) deforestation alerts are issued for areas where deforestation activities have been detected, Global Fishing Watch offers near real-time monitoring of fishing activity globally, and Firecast provides near real-time alerts of fires occurring in natural habitats. Stephenson and Stengel (2020) provide a synthesis of databases with information on threats to biodiversity.

Lastly, some actors in conservation may also be working with those impacting biodiversity as a result of their actions and thus investigating threats or the likely changes in status in the absence of action. For example, businesses may be developing a new infrastructure project and designing mitigation measures to try and avoid, minimise and compensate for negative impact. Similarly, businesses may wish to enact mitigation measures in their supply chains to minimise impacts on biodiversity. In these instances, their mitigation strategies will be making claims about their impacts, which need to be based on appropriate evidence. Information can be collected on the extent and severity of impacts, so that action taken is appropriate and likely to achieve desired outcomes.

Diagnosis based on large-scale data

In some cases, basic historical information about the population, habitat distribution and trends can reveal so much about causes of decline or endangerment that little further diagnostic work is needed. For example, the historical extirpation of the Eurasian bittern *Botaurus stellaris* in Britain was caused by the drainage of land for agriculture that removed the extensive reedbed habitats on which the species wholly depends. In this case, the cause of loss is clear and recent research has focused on understanding precisely how to restore and manage reedbeds with appropriate habitat structure and fish populations, as the basis for the successful recovery of the bittern (Brown et al., 2012).

Where the loss of carrying capacity is less clear-cut than in the bittern case, human impact may still generate measurable additional mortality or reduction in reproductive success. The effect on carrying capacity can then be estimated using 'counterfactual simulation models'.

These incorporate assumptions about density dependence which compare modelled population sizes with and without the effect of that impact (e.g. see Green et al., 2022, for the impact of lead poisoning on some raptor species across the European continent).

Where data for candidate external causes of decline are available at the same spatial and temporal scales as for trends in the species of interest then the diagnosis can also be assisted by testing for correlations between those causes and species trends. For example, measures of trends in agricultural practices and bird population indices were available for countries across Europe and were used to reveal strong spatial correlations between bird population declines and agricultural intensification (Donald et al., 2001).

Data on demographic rates can also be used as supplementary tests to strengthen the diagnosis. For example, studies of trends found that a decline in survival rates was the demographic mechanism driving population decline for many farmland songbirds (Siriwardena et al., 1999). This implicated reductions in over-winter food supply due to changes in agriculture as a likely external cause, and consequently measures to ensure seed-rich habitats in winter on farmland are now commonplace in European agri-environment schemes. More rigorously, a 'sufficiency test' can be used to test whether the magnitude of the effect of the external factor on a demographic rate is enough to account for an observed population decline. In some cases, this may be a matter of simple arithmetic. For example, the maximum possible reproductive rate of the endangered California condor *Gymnogyps californianus* is so low that the documented high adult mortality rate must have driven the observed species decline (Snyder and Snyder, 2000). Given that lead poisoning was the known cause of the increased mortality rate, then this could be diagnosed with high confidence as the cause of the population decline and thus targeted by remedial actions.

Diagnosis based on field studies

Studies at single sites may struggle to separate causal from irrelevant factors that may be changing simultaneously, and demographic responses may be confounded by the impacts of flows of individuals to and from the surrounding environment (Green, 1995). As Sutherland (2000) put it, '...the main reason for failing to diagnose problems is the myopic concentration on single sites, where factors cannot be unravelled, rather than looking for general patterns across sites'. Wherever possible, diagnostic field studies should seek to be as representative of a species' distribution and range as possible and should try to make comparisons among study sites with differing trends in populations and potential drivers. This favours working collaboratively across networks of sites where the same problem is apparent. Some drivers of population change may be difficult to detect and study (e.g. impacts of pollutants and disease across multiple sites) and, for mobile taxa, some factors may operate over long-distance migration routes or in remote, perhaps even unknown, migration destinations.

Across sites, comparative studies of patterns of occupancy, abundance and demographic rates between places where a species is declining and places where it is not can be especially enlightening (e.g. Peach et al., 2004), as can comparisons of the environment between places where a species persists and those from which it has been lost. For example, by comparing sites where the endangered silver-spotted skipper *Hesperia comma* butterfly persisted with those from which it had become extinct, Thomas et al. (1986) found that persistence was associated with grazing regimes that provided the short, sparse fescue-grass swards that attracted egg-laying females, thus paving the way for a successful conservation intervention based on grazing management. Where possible, it is useful to extend this approach to compare changes over time in environmental variables. For example, Stowe et al. (1993) showed that factors linked to grassland management and mowing distinguished locations that had lost or retained nesting corncrakes *Crex crex*. Repeat surveys then showed that subsequent changes in corncrake numbers were correlated with changes in those same factors (Green and Stowe, 1993), a more powerful basis for conservation action.

Correlative studies like those above may be the only practical option available for most conservation practitioners. However, management interventions suggested by these approaches should ideally be followed up with replicated experimental tests at large sites or networks of sites. Both 'before–after' and 'control–intervention' designs can be used to manipulate a putative cause of the decline and monitor the response. For example, initial studies across an array of lakes showed that threatened common scoter *Melanitta nigra* populations persisted at lakes with shallow water, abundant large invertebrate prey, and few stocked brown trout *Salmo trutta*. These results led to a replicated experiment to test whether the reduction of fish populations through targeted angling can restore lake invertebrate populations as a potential scoter conservation measure (Hancock et al., 2020).

Where they are possible, experimental approaches have the advantage of allowing an iterative, adaptive approach so that a putative diagnosis can be modified — or abandoned in favour of another candidate cause — based on the evidence gathered. In cases where conservation urgency (e.g. due to drastic decline or highly threatened populations) is combined with a lack of a strong pre-existing evidence base, then merging diagnosis and the testing of conservation solutions in this way may be the best course of action, especially if the ideal, replicated experiment is impractical. For example, a simple experiment implementing predator control at one site for three years, with additional collection of reference data for four years before and three years afterwards offered important diagnostic insights in showing that capercaillie *Tetrao urogallus* productivity was highest when predator control coincided with dry weather during the nestling period (Summers et al., 2004). Experiments focused on measuring a demographic rate (e.g. fledgling rate) can be more practical for mobile taxa, where flows of individuals can make population responses (e.g. changes in numbers of adult individuals) difficult to measure and interpret.

Combining approaches

Most attempts at diagnosis ultimately combine approaches. As an example, Box 7.1 summarises the steps taken to diagnose rapid and severe recent declines of three *Gyps* vulture species in Asia. Other case studies are summarised by Sutherland (2000) for the Lord Howe woodhen *Hypotaenidia sylvestris* and Green (2002) for the wandering albatross *Diomedea exulans*.

Box 7.1 Diagnosing declines: vultures on the Indian subcontinent

1. Anecdotal evidence of vulture declines was followed up by repeating a formal 1992 survey: severe and geographically consistent declines were found (Prakash, 1999; Pain et al., 2008).
2. This excluded possible localised threats and showed that cattle carcasses — the main food of vultures, made readily available by humans for cultural reasons — remained abundant, suggesting there was no reduction in the food supply.
3. Post-mortem examination showed that the main cause of vulture excess death was kidney failure and gout, which correlated perfectly with the presence of detectable residues of diclofenac, a non-steroidal anti-inflammatory drug widely used as a veterinary treatment for cattle (Oaks et al., 2004; Shultz et al., 2004).
4. Experiments then showed that low doses of diclofenac rapidly killed captive vultures (Oaks et al., 2004; Swan et al., 2006).
5. Concentrations of diclofenac in tissues of treated cattle were shown to be high enough to kill vultures if consumed within a few days of the cow's death (Green et al., 2007).
6. An interview study found that pharmacies in the Indian subcontinent began to sell veterinary formulations of diclofenac from about 1994 onwards, which coincided with the beginning of the vulture decline (Cuthbert et al., 2014).
7. Surveys of diclofenac concentrations in cattle, combined with toxicity data, were brought together in a population model to estimate the rate of vulture population decline expected from the observed level of exposure to the drug. Observed and predicted rates of population decline matched and no other cause of additional mortality was required to account for the observed decline — the sufficiency test (Green et al., 2007).

This level of diagnostic proof has been sufficient to convince governments to ban veterinary use of diclofenac in India, Pakistan and Nepal, initiate the search for alternative, non-toxic drugs for use on cattle and establish captive breeding programmes to support remaining vulture populations. Population response to these measures has been stronger in Nepal, where the legal ban works, than in India, where low level, illegal diclofenac use continues (Galligan et al., 2020, 2021).

7.4.3 Red team/blue team

The red team vs blue team practice is an approach for challenging systems to discover weaknesses, which may threaten the normal functioning of the system and provide opportunities for proactive intervention. The red team entails an individual or group attempting to challenge a system while the blue team seeks to prevent it. It can either be a real, but carefully planned and controlled physical attempt (e.g. attempting to get an imitation gun through security), or an exercise imagining how the challenge develops. This is widely used in cybersecurity and security, such as tests to challenge airport security. It could also be used to test systems for protecting rare species to identify areas of weakness, identify ways in which a predator could cross a fence protecting sensitive species, or identify ways in which sensitive conservation information could be acquired.

7.4.4 Horizon scanning to explore potential futures

With numerous current urgent problems, it may seem unnecessary or even irresponsible to suggest considering future potential issues. However, appropriate consideration of change is important, in particular for strategic planning. Horizon scanning is the process of identifying trends and phenomena that may gain prominence and that are likely to have a high magnitude of future impact, but for which there is little current awareness (UK Government Office for Science, 2017). This foresight technique helps anticipate unforeseen risks so as to respond efficiently and quickly whilst threats are still manageable as well as capitalise on opportunities by allocating resources to realise novel advances or innovations (Sutherland and Woodroof, 2009).

Horizon scanning is a useful means of identifying forthcoming issues. A review (Sutherland et al., 2019a) undertaken a decade after the first annual conservation scan in 2009 (Sutherland et al., 2010b) showed that a number of the top 15 identified issues subsequently appeared to have considerable salience and impacts (e.g. microplastic pollution, synthetic meat, artificial life, use of mobile-sensing technology for data collection, deoxygenation of the oceans, and invasive Indo-Pacific lionfish). Alongside the horizon scan, data were collected in 2010 as to whether senior representatives of 12 conservation organisations had heard of these issues (Sutherland et al., 2012a). This showed that these issues were on the horizon and not widely known. For instance, although use of mobile-sensing technology for data collection may seem obvious now, it was only known by 69% of these senior conservation leaders in 2010.

Horizon scans have also demonstrated value in informing funding priorities and strategic planning. The Antarctic Horizon Scan and follow-on Antarctic Roadmap Challenges have been used as justification for polar strategic planning efforts worldwide (Wintle et al., 2020) including recent major investment in Australian research.

One key consideration is the scope of the scan, as some horizon scan topics are broad while others are narrow (see Table 7.2 for examples). They also vary in geographical scale, with many global, but some local. For invasive species, for example, horizon scans have been carried out across different geographic areas (Great Britain: Parrott et al., 2009, Roy et al., 2014; western

Europe: Gallardo et al., 2016; global: Ricciardi et al., 2016). Some scans have been designed for identifying emerging public policy issues (Parker et al. 2014). Many others have considered social issues, such as changes affecting forest-linked livelihoods (Oldekop et al., 2020).

Table 7.2 Examples of diversity within environmental horizon scanning.

	Topic scale	
Topic scope	Global	Local or regional
Narrow	Migratory shorebirds (Sutherland et al., 2012b)	Conservation issues for inland waters in Canada (Pérez-Jvostov et al., 2019)
	Illegal wildlife trade (Esmail et al., 2020)	Conservation issues for Mediterranean wetlands (Taylor et al., 2021)
Broad	Conservation (Sutherland et al., 2022)	General conservation issues in Israel (Kark et al., 2016)
	Bioengineering (Kemp et al., 2020)	UK nature conservation (Sutherland et al., 2010a)

Regular scanning can be adopted to ensure continuous monitoring of the landscape, as done for the global horizon scan of conservation that has been carried out annually since 2009 (Sutherland et al., 2010b, 2022), or by periodically updating, as done for bioengineering (Wintle et al., 2017; Kemp et al., 2020).

Horizon scans typically consider either challenges or both challenges and opportunities. Another aspect to determine at the planning stage is whether the scan will include issues that are well-known but not being adequately addressed, as in the scan for forest management in Myanmar (Prescott et al., 2017), or whether it aims to identify topics that most readers are unaware of, as in the global conservation scan (Sutherland et al., 2022). A different approach is to consider all issues that affect a specific subject. For example, Sutherland et al. (2012b) listed all the current and potential threats impacting migratory shorebird conservation. This has the advantage of setting out a reasonably complete set of challenges rather than just the most urgent or least well-known ones. Another advantage of this approach is that it deals with the criticism that horizon scans miss important issues when, in fact, these were just excluded because they are reasonably well known. Taylor et al. (2021) took a combined approach in their horizon scan for the conservation of Mediterranean wetlands, identifying both 'critical' issues (ignoring novelty to focus on the significance of impacts) and 'overlooked' issues (taking both novelty and impact into account).

There are numerous ways to conduct horizon scans across science and policy (Sutherland et al., 2011; Wintle et al., 2020). The core principles of the approach described in Box 7.2 are to consult a wide community and then use a form of structured transparent, anonymised expert elicitation, especially versions of the Delphi Technique (Section 5.5), to produce a final list (Sutherland et al., 2011).

Horizon scanning can consider a range of future types of problems. 'Black swans' was introduced by Taleb (2007) as a term for difficult-to-predict, low-probability events that might be high impact. Black swans (the birds *Cygnus atratus*) were once not considered by Europeans, who had only seen white species, until reported from Australia. Following the popularity of the term, we now have 'black elephants' as well-known subjects that attract insufficient attention (from the term 'the elephant in the room'), such as the overuse of antibiotics and resulting drug resistance. 'Black jellyfish' are well-recognised events that could become more serious or are more complicated than expected (i.e. they could have a nasty sting). COVID-19, for instance, is a zoonotic disease originating from a well-known family of viruses, but which suddenly led to the global pandemic in 2020.

Box 7.2 A widely used approach for horizon scanning

1. Identify the boundaries of the topic and time horizon. This needs to be considered carefully as assessment is best carried out if the identified potential issues are comparable in scale and scope.
2. Identify a group of experts from a wide range of organisations, geographies, demographics, disciplines and interests. Look for diversity including both specialists and broad-thinkers.
3. Each scanner is asked to identify issues from their own knowledge of the topic and from asking colleagues and associates. Some may canvass ideas through social media within their networks, run workshops or simply ask others. This typically entails engaging a few hundred contacts overall for all participants (e.g. Sutherland et al., 2022).
4. Some scans have elicited contributions through an open online survey to increase the diversity of ideas to be considered and the background of contributors, including those who may not normally be considered experts. This approach can reach thousands of people (e.g. Esmail et al., 2020; Community Conservation Horizon Scan Collaboration, 2022). Whilst more inclusive and participatory, this approach is dependent on the effectiveness of dissemination, language translation, and coordination capacity to process and consolidate issues.
5. To be considered for inclusion, contributors are often asked to provide a summary of the issue consisting of a couple of hundred words with links to sources. Typically, the ideas suggested by a wider audience need to then be converted into a convincing paragraph by the scanner or coordinator.
6. Depending on the approach, either the same pre-selected group or a subsequently selected group proceeds to assess issues. One way to assess is

to score between 1 and 1,000 with a higher score deemed more suitable for inclusion in the final list, based on a combined criterion of novelty, plausibility of manifestation, and magnitude of impact. Scores are converted to ranks and aggregated, often through the median value (reducing the influence of individual scores).

7. Assessors are asked to state which issues they have heard of as an indicator of novelty.
8. Issues are then shortlisted, excluding those ranked lowest and potentially also those deemed well known.
9. Assessors may be asked to individually further investigate additional shortlisted issues in preparation for a group discussion.
10. Issues are debated and explored through facilitated dialogue and discussion with others in the assessor group, possibly through a workshop format (online, in person or a combination of both).
11. Issues are further assessed through a second round of ranking against the same 1–1,000 scale and criteria.
12. Descriptions of issues are written up indicating evidence as signals of emergence and projections of future impacts. It is usual to ask someone who did not initially submit the idea to write the text and/or elicit any external reviewers as required. Those whose subjects are included in the final list, but are not authors, are acknowledged in the final paper.

7.4.5 Scenario planning for potential futures

Scenario planning (also known as scenario building or scenario analysis) can be used alongside horizon scanning as a foresight tool that helps explore what the future might be like. It helps with thinking, in advance, about potential future change and uncertainty, providing the basis for strategic thinking about possible responses — both to avoid traps and make the most of opportunities — synthesising knowledge and advancing systems understanding (Peterson et al., 2003). Furthermore, it helps forewarn decision makers of undesirable future impacts of change, support decisions in developing adaptive governance strategies, and discover implications that may come from alternative social-ecological development pathways and policy options (IPBES, 2016). Different types of scenarios are useful in different phases of the policy cycle: exploratory scenarios looking at different possible directions are useful in agenda setting, target-seeking scenarios are useful when designing options, policy-screening scenarios are useful during implementation, and retrospective policy evaluation is useful during the review (Figure 7.1; IPBES, 2016).

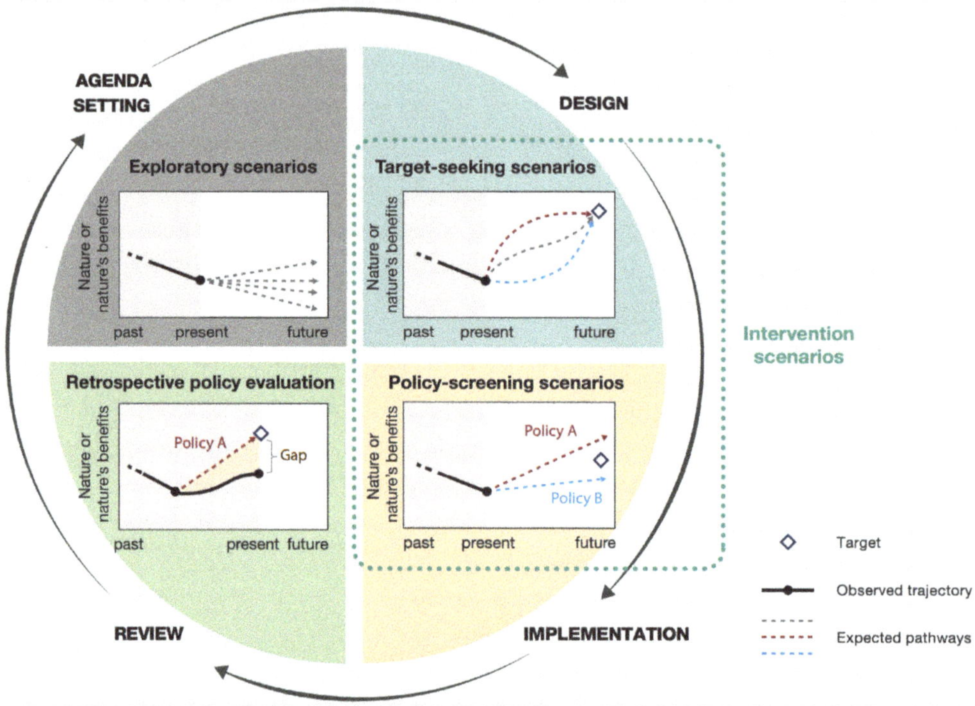

Figure 7.1 Scenarios to support decision making at different phases of the policy phase. (*Source*: © IPBES, 2016, CC-BY-NC-3.0)

There are multiple approaches to scenario planning, but it is generally developed through workshop discussions that allow groups (representing multiple stakeholders and experts) to identify a range of future visions (Box 7.3). Each of the identified scenarios is then detailed further, such as with a narrative or pathway exploring the consequences for existing or potential policy. This is then used as a basis for modelling, or to identify knowledge gaps.

The likely best-known example of scenarios are those created by the Intergovernmental Panel on Climate Change considering a range of societal responses. They then embed these various responses into their climatic models to predict the likely change over time under each scenario (IPCC, 2018). The model outputs have been widely used to explore the possible consequences of climate change, for example for biodiversity.

Scenarios developed for conservation applications range from envisioning a common desirable future for protected areas in Spain (Palomo et al., 2017), to assessing governance reforms for enhancing biodiversity outcomes in Australian landscapes (Mitchell et al., 2016), and developing alternative futures for global biological invasions (Roura-Pascual et al., 2021). Scenarios can provide the basis for deciding what to input into models; for example, models of the dynamics of saiga antelope *Saiga tatarica* parasite infection were underpinned by a range of current and plausible future scenarios of increasing livestock numbers, climate change and antiparasitic treatments (Khanyari et al., 2022).

Box 7.3 Typical process for scenario planning

1. Identify a focal issue or question. Decide on a time frame and spatial scale.
2. Brainstorm trends, drivers and any factors that could affect the future of the focal issue. It can help to consider categories, such as political, economic, social, demographic and environmental.
3. Create a list of critical uncertain outcomes resulting from the action of drivers.
4. Identify a range of possible scenarios. Name each according to its main features.
5. Build out each scenario. This can be done through a narrative that explores potential consequences, or through modelling.
6. Evaluate and test scenarios for consistency by quantification, against stakeholder behaviour, through expert opinion, and against other scenarios.
7. Define the early indicators through policy screening by assessing how existing policies would play out under different scenarios. This allows for the recommendation or update of strategies and policies accordingly. New questions, variables and unknowns may arise and give way to another iteration of the scenario planning process.
8. A backcasting approach may also be useful to unpack a series of management recommendations on how to obtain a desired future, either through one or a combination of scenario pathways.

7.5 Taking Stock

7.5.1 Summarising the problem

A situation model (or situation assessment diagram) is a tool that visually portrays the relationships among the different factors discussed so far, including the targets, direct and indirect threats, opportunities, biophysical stresses, and key actors related to a particular target or targets (Margoluis et al., 2009; CMP, 2020). It can greatly help in defining the project in a standardised and collaborative way. A good model illustrates the main cause-and-effect relationships that exist within the project area or theme. It should include the most important details, yet be as simple as possible. A situation model for an extensive project will need to be at a coarser grain than one for a smaller project. Models can be updated as a project progresses to reassess and re-strategise (Riley et al., 2003). New links can be added, or the specificity of existing links refined, as new evidence comes to light.

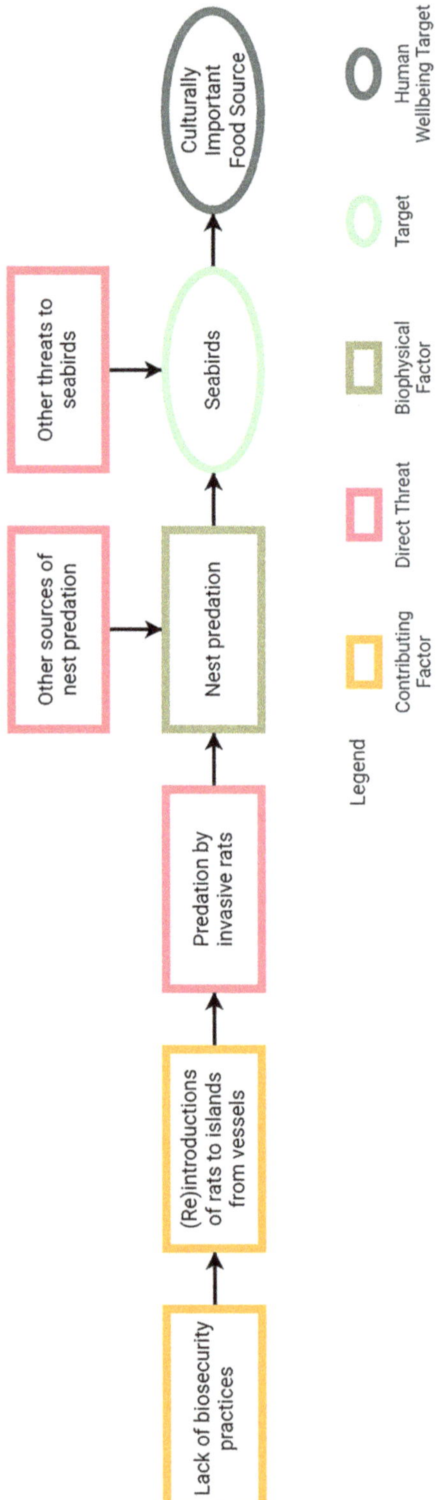

Figure 7.2 Excerpt from a situation model on protecting a seabird colony from invasive rodents. (*Source:* Salafsky et al., 2022, CC-BY-4.0). Note use of the term "Contributing Factor" to describe factors that could be considered indirect threats or opportunities for intervention.

7.5.2 Defining goals and objectives

With a clear understanding of the scope of a conservation project, and its situation, it is possible to determine the project's goals and objectives. These outline the team's desired future status for the features of interest, threats and opportunities (CMP 2020). Goals and objectives should be SMART: specific, measurable, achievable, results-oriented and time-bound. A sensible goal might therefore be: 'Between 2020 and 2030, the available habitat for pond-breeding amphibians in Belize will be doubled through restoration and creation of new pond clusters'.

7.6 Identifying Potential Actions

7.6.1 Solution scanning

To achieve stated goals, conservationists will put in place different actions as part of their projects that aim to protect or restore biodiversity (or other features of interest). For example, in the UK it is common for infrastructure developments to erect bat boxes in an attempt to increase bat abundance in their project area, while farmers often create wildflower strips at the edge of fields to increase bird species richness and pollinator diversity. A reforestation programme may also aim to improve water quality or carbon storage, and reduce soil erosion; an alternative livelihoods programme may hope to improve incomes for local communities, whilst reducing unsustainable harvesting practices. Relevant actions will differ depending on the outcomes desired, the status of the features of interest, and the threatening processes present.

It is often sensible to start by considering all the possible (but reasonable) options for addressing the identified threats and exploiting the identified opportunities. Solution scanning is a more transparent and careful strategy than the traditional approach of selecting a subjective subset of policies using only the experience and beliefs of practitioners (Sutherland et al., 2014). While a complete evidence review for all options would nearly always be preferable, the time and effort required for such reviews mean they are often impractical in practice. The process of solution scanning entails bringing together a wide community to identify possible options for responding to a set of problems or opportunities. This is the initial stage, with no attempt to consider effectiveness. However, this stage is important; one study showed that practitioners had heard of just 57% of the possible 28 interventions (Walsh et al., 2015), suggesting decision makers may be blinkered in the range of available options. Two examples of how solution scanning has recently been applied are a list of possible actions for creating or enhancing ecosystem services (Sutherland et al., 2014) and a list of options for reducing the likelihood that future pandemics of zoonotic origin will emerge (Petrovan et al., 2021). Solution scanning can also be the first stage of subject-wide evidence synthesis, for example during collation, synthesis and assessments of the effectiveness of actions to protect biodiversity (Sutherland et al., 2019b).

The approach adopted for solution scanning (Box 7.4) is to create a list of the challenges, then compile a comprehensive list of possible solutions under each, consult more widely and make the scan available so it can be checked, used and updated. The issue as to whether an action is likely to work in practice is considered at a subsequent stage, when contextual information can be included. Depending on the topic, solution scanning can be carried out at different scales, ranging from quickly listing possible options to an extensive process aiming to be reasonably comprehensive and including structured searching of the literature and consulting a wide range of global experts.

For conservationists, conservationevidence.com and the latest edition of *What Works in Conservation* (Sutherland et al., 2021) provide a range of potential actions for a given threat. Doing nothing (or nothing new) is also an option; in most projects the costs and benefits of this should also be considered.

Box 7.4 Typical process for solution scanning

1. Identify a specific topic with set boundaries.
2. Bring together a core scanning team.
3. Identify the main challenges or threats.
4. The core team thinks through the obvious actions that could reduce these challenges.
5. The core team looks at multiple sources of information that provide suggested solutions.
6. Invite a diverse group of contributors to think of other solutions, aiming to cover the conceptual and geographic scope and including experts in the different aspects of the topic, from theory to practical implementation.
7. Consider conducting additional targeted literature searches where obvious gaps lie or for particularly complex actions.
8. Publish potential solutions with the core team and contributors as authors.

7.6.2 Innovating to devise novel solutions

Innovating, followed by testing, is critical for conservation, whether for novel research methods or techniques, or to improve and create new practical solutions. However, innovation in conservation practice is insufficient in comparison to other disciplines (Game et al., 2014).

Means for encouraging creativity include: experimentation; allocating time to creative side projects; tolerating obviously impractical ideas during initial discussions; embracing a degree of risk and failure; bringing together a group with different knowledge, perspectives and skills; meeting face-to-face; and setting ambitious goals (e.g. by adopting the 10x vs 10% mindset, as they do at Google, i.e. aiming to ambitiously improve something by 10 times, rather than just by 10%). The following paragraphs explore specific tools and approaches to stimulate innovation.

Continual improvement (kaizen)

Kaizen ('doing better' in Japanese) is the principle of continual innovation to improve. Much of the success of Toyota is attributed to kaizen. All departments from construction to accounts and management look at processes and consider means to improve. Similarly, in the 2000s, the British cycling team aimed to improve every component of competitive cycling — from the cleanliness of the maintenance workshop to the cyclists' sleeping position. They suggested that even if each gain is just 1%, these improvements collectively result in large overall gains. Sutherland (2019) suggested that the widespread adoption of kaizen conservation, in which organisations and conservationists look for continual self improvement, would deliver conservation more effectively.

Reflection on the details of the problem

If exploring how to protect a particular species, consider its behavioural and habitat preferences and other requirements, and then what may be attractive and beneficial. For example, the attributes of nesting or hibernation sites give clues as to suitable designs for nest boxes or hibernacula. Equally, understanding the requirements of problematic species can generate innovative ideas for their control, for example by altering the habitat so that it no longer meets those requirements.

Brainstorming

Brainstorm by presenting a problem and encouraging the shouting out of possible solutions (which are documented e.g. on a flip chart) with the rule that no criticism is allowed; building on previous ideas is welcomed. It may be helpful to remove constraints like time, money and even laws of nature to encourage truly creative thinking! At the end of the process, the ideas are reviewed and the useful ones retained.

Physical models for inspiration

In this process materials are provided to make models, such as a hibernaculum or a nesting box that excludes predators, with the aim of generating novel practical design ideas. Building and testing prototypes, for example of wildlife tracking devices, can stimulate novel designs as practical problems are encountered.

Adopt and adapt existing solutions and technologies

Seek out and ask widely how others (including those from other disciplines) solve similar technical problems. In Mauritius, Tatayah et al. (2007) discovered that copper rings deterred African giant land-snails *Achatina* spp. from entering echo parakeet *Psittacula eques* nest cavities, thus preventing the birds from nesting. Newman and Showler (2007) were inspired by this to use copper rings to reduce losses of red helleborine *Cephalanthera rubra* to slug and snail herbivory in the Chiltern Hills, England. As another example, the algorithm used for astronomical pattern matching was used by Arzoumanian et al. (2005) to individually identify whale sharks *Rhincodon typus* by their spots, thus improving the understanding of populations and movements. Sometimes nature itself can inspire innovative design: remodelling the nose of the Japanese Shinkansen bullet train based on the streamlined beak of the kingfisher *Alcedo atthis* increased fuel efficiency and reduced the problem of sonic booms as the train exited tunnels. Bromeliads inspired the design of the *Nucleário* device (https://www.nucleario.com) that provides a reliable water supply to saplings planted for forest restoration.

Hackathons

These were created as a way for computer programmers to collaboratively explore software development with graphic and interface designers, product managers and others, but have since been adapted for conservation solution innovation. Crowdsourced participants are given problem statements and, with limited time and resources, are challenged to come up with practical, often technologically-driven solutions. Teams compete for prizes and prestige using the event to brainstorm ideas, develop prototypes and demonstrate designs. Examples of conservation hackathons are the Zoohackathon for wildlife trafficking (Zoohackathon, 2021) and the Grand Challenges for microfiber pollution, artisanal mining and climate-friendly cooling technologies (ConservationXLabs, 2022). ConservationXLabs winners are subsequently supported to develop and expand their solutions, catalysing innovation for the field and industry.

Gamifying conservation

Games can be used to identify novel possible actions by having teams seeking to solve challenges. In 2010, the World Bank Institute ran the online game EVOKE (https://www.urgentevoke.com/), which encouraged players to develop creative solutions to pressing social problems. Open source and open access platforms and databases lower barriers so multiple people can work on any one problem.

Prizes and awards

Prizes have been awarded to encourage solutions for specific problems; the most famous is the Longitude Prize offered in 1714, which delivered a solution for determining longitude. The annual Earthshot Prize, announced in 2020, has a similar objective of encouraging innovation for the world's most extraordinary eco-solutions. Specific awards for innovative nature conservation include the Ulysses S. Seal Award for Innovation in Conservation, the Pathfinder Award for Innovation in Nature Conservation, and the annual Best Practice (Innovation) award given out by the UK Chartered Institute for Ecology and Environmental Management.

Incubators and accelerators

Incubation is a supportive process of innovators with an impactful idea to help meet investor requirements. Incubators work with innovators to ensure their ideas are relevant, scalable, and impactful. Acceleration is generally the last stage in growing promising ideas (often high-ambition and transformative) into initiatives with further development, validation and financing support. Incubators and accelerators for conservation are emerging and draw upon private-sector models for start-ups to create impactful, de-risked and exciting initiatives for investors and funders: the Luc Hoffmann Institute hosts an ideation–incubation–acceleration programme broadly across biodiversity conservation; the Conservation Finance Alliance Incubator identifies, supports and promotes innovative ideas and solutions to conservation finance challenges; IUCN and partners have set up the Nature+ Accelerator Fund.

7.6.3 Evaluating likely effects of actions

Designing a conservation project involves making claims or assumptions. These will often be about the effect of actions on biodiversity, for example will erecting bat boxes be effective at encouraging bats to roost? Will those bat boxes increase the population of bats in the area? Will the wildflower strips be effective at improving pollinator abundance and species richness? Sometimes assumptions will be about associated socioeconomic effects, for example is there evidence supporting the success of alternative livelihood programmes in increasing local incomes, or evidence showing that carbon storage is enhanced from reforestation programmes? Answering such questions requires evidence on the effectiveness of different actions — what works, and what does not.

It is vital to compile such evidence because sometimes actions presumed to be effective can be ineffective, or at worst harmful. For example, bat gantries have been frequently used in the UK to mitigate the impacts of road developments in areas where bats may frequently cross the infrastructure. Gantries were designed so that bats would use these structures to guide them safely above the road and the height of the incoming traffic. However, once evidence was compiled it showed these gantries were rarely used by bats (Berthinussen et al., 2021), meaning at least £1 million was wasted on ineffective mitigation (Sutherland and Wordley, 2017) and negative traffic threats and impacts for bats remained.

Evidence of the effectiveness of actions can take a range of forms. Effectiveness can be measured both quantitatively and qualitatively, and be from a range of sources (see Section 2.5). Effects can also comprise various measured outcomes, which can be difficult to compare. For example, the effectiveness of wildflower strips for biodiversity conservation is often measured using different metrics (e.g. nesting density, abundance, species richness), and with evidence focused on different species groups (e.g. butterflies, hoverflies, birds, small mammals). Noss (1990) provides a framework of indicators that can be used to measure effects on biodiversity at four organisational levels (genetic, population–species, community–ecosystem, and regional landscape) and including compositional, structural and functional components. A project may wish to compile evidence of the effectiveness of different actions for a range of different metrics. Chapters 2, 3 and 4 explain how to assess and draw conclusions from compiled evidence.

Ideally, evidence of effectiveness should be consulted during the design of a project, but such evidence can also be collected during pilot phases, trials or experiments before full implementation. This may be particularly important where there are large taxonomic or geographical gaps in the published evidence base for the effectiveness of actions (Christie et al., 2021; Junker et al., 2020). Even during implementation, monitoring of observed outcomes can help ensure project success and, if results are shared with the wider community, help build the global evidence base (see Chapter 10).

The effect of a conservation action can often be more complex than just directly impacting a feature of interest and the action may primarily be trying to influence threatening processes or behaviours. For example, a project may be hoping to improve the population outlook of a particular endangered species by reducing rates of hunting through an education programme. In this instance, evidence will also be needed on the impacts of education programmes on rates of hunting in the given context, rather than just evidence on whether reducing hunting benefits the endangered species.

Chapter 13 provides a list of resources which provide evidence of the effects of conservation actions, including some resources focused on biodiversity outcomes, and some focused on other environmental and social outcomes.

7.6.4 Evaluating actions based on other criteria

In practice, conservation actions will also need to be assessed on criteria other than effects on the features of interest or associated threats. The economic costs of actions may be an important consideration (see Section 2.4.4). The most effective action may exceed the budget of a conservation project. If two actions are expected to be similarly effective, one may be preferred because it is less expensive. Local attitudes, values, laws and politics might also restrict what actions can be used (see Section 2.4.5). Public opinion often favours non-lethal over lethal management of problematic species, for example, despite potentially lower effectiveness and higher economic costs (Roberts et al., 2018).

7.6.5 Theory of change

Theories of change are diagrams that illustrate how a given action, or set of actions, is expected to deliver a given outcome based on the available evidence (Figure 7.3). They are popular in many areas of conservation and development and are often expected as part of funding applications. Whereas a situation model (Figure 7.2) shows the system before taking action, the theory of change shows assumptions about how the system will respond after action (Margoluis et al., 2013; CMP, 2020). As such, they are complementary tools.

The example shown in Figure 7.3 illustrates how planned actions are intending to influence the situation expressed in the situation model of Figure 7.2, and which intermediate results and outcomes they are intending to achieve.

7.7 Developing Questions and Assumptions

7.7.1 Developing questions and assumptions in situation models and theories of change

There are many different types of questions and assumptions relevant to conservation policy and practice, each requiring different types of evidence. Some of these may be related to

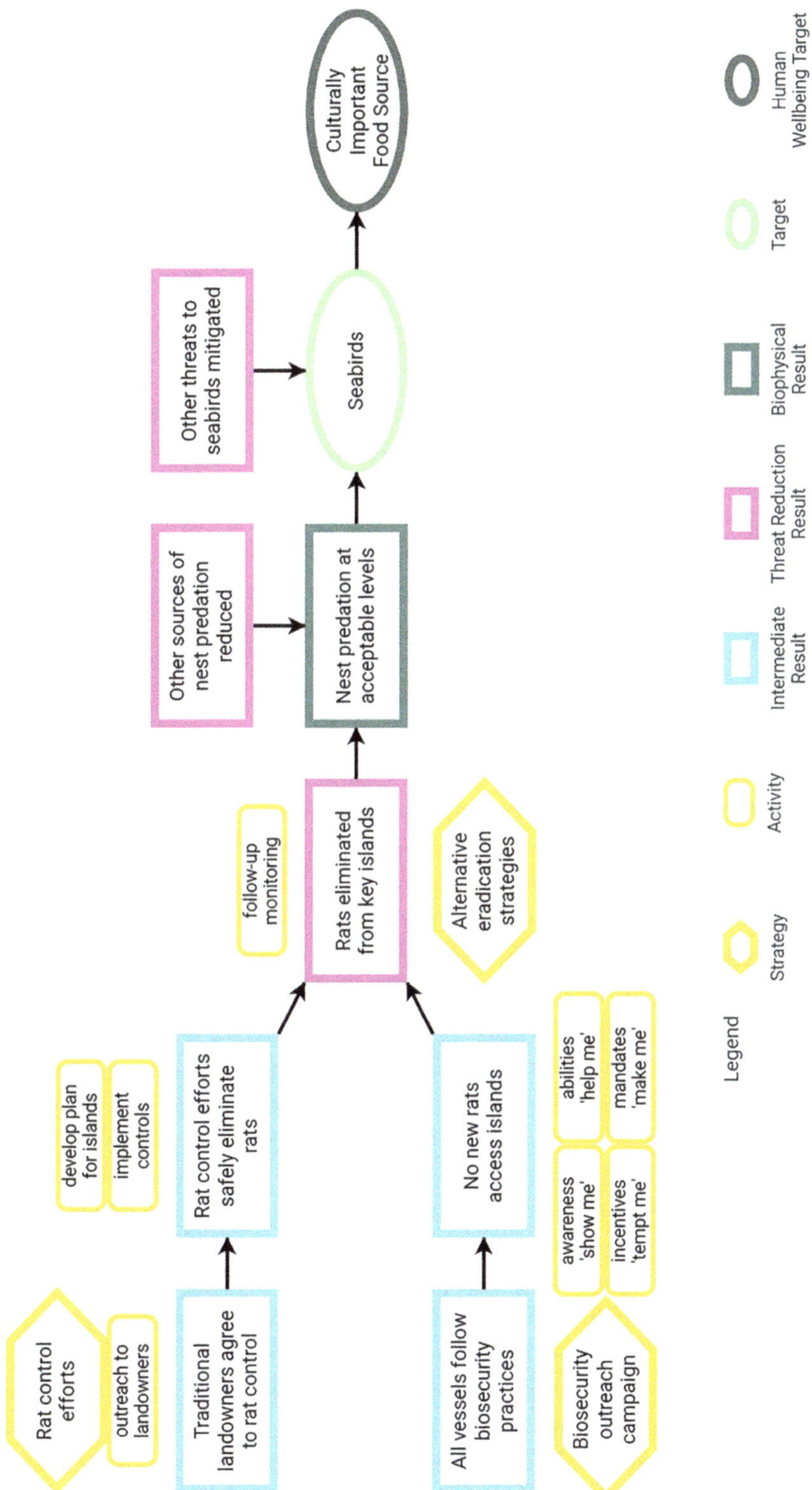

Figure 7.3 Example of a theory of change pathway diagram. (*Source:* Salafsky et al. 2022, CC-BY-4.0)

understanding the situation, such as the status of the target species (Section 7.3) or the cause of a threat (Section 7.4). Others are related to the effectiveness of an action or the conditions under which a given action might be effective (Section 7.6). A large part of the 'art' of evidence-based conservation thus involves understanding the system well enough to figure out the right set of assumptions to consider and the sequence in which these need to be assessed (USAID, 2018). For example, to help conserve seabirds on an island where rat predation might be an issue, the team may need to first confirm that rats are present on the island, then that they are at least a partial cause of seabird nest predation, and finally that poisoning might be an effective action to take to remove the rats given local rainfall patterns and the risks of accidental poisoning for non-target fauna.

The mental models created as part of situation models (Margoluis et al., 2009) (Figure 7.2) and the theory of change pathways (Margoluis et al., 2013) (Figure 7.3) become important tools to generate questions and assumptions. For example, five broad analytical questions need to be answered to justify the proposed rat eradication strategy (Figures 7.4 and 7.5):

1. Do the seabird populations require conservation action?
2. Are rats a major threat to the seabird populations?
3. Is the proposed rat eradication strategy feasible and effective?
4. Are alternative rat control strategies less feasible and/or effective?
5. Can rats be prevented from re-invading the islands?

Questions are easier to answer if specific and well-formulated. For example, rather than the broad question 'Are rats a major threat to our seabird populations?', it is better to formulate a set of specific assumptions such as 'nest predation is a major cause of seabird mortality', 'rats are the primary cause of seabird nest predation', and 'other sources of nest predation are less important'. To this end, it is usually helpful to break down each question into a set of component assumptions as shown in Figures 7.4 and 7.5.

Technically, every factor and every link between factors in situation models and theories of change represents an assumption. Whilst it is usually unrealistic to collect evidence for every assumption, instead, the need is to identify those that are critical as they are more uncertain or have higher risk consequences. The art lies in evaluating which assumptions will benefit from further scrutiny.

7.7.2 Identifying priority questions for policy and practice

Sutherland et al. (2006) created a process by which decision makers and researchers can work together to identify the questions that, if answered, would make the most difference in improving policy and practice. This has led to a wide range of such exercises focusing on subjects such as cetacean conservation (Parsons et al., 2015); habitat restoration (Ockendon et al., 2018); post-2015 development agenda (Oldekop et al., 2016); future of global agriculture (Pretty et al., 2010); microbial ecology (Antwis et al., 2017); Mediterranean wetland conservation

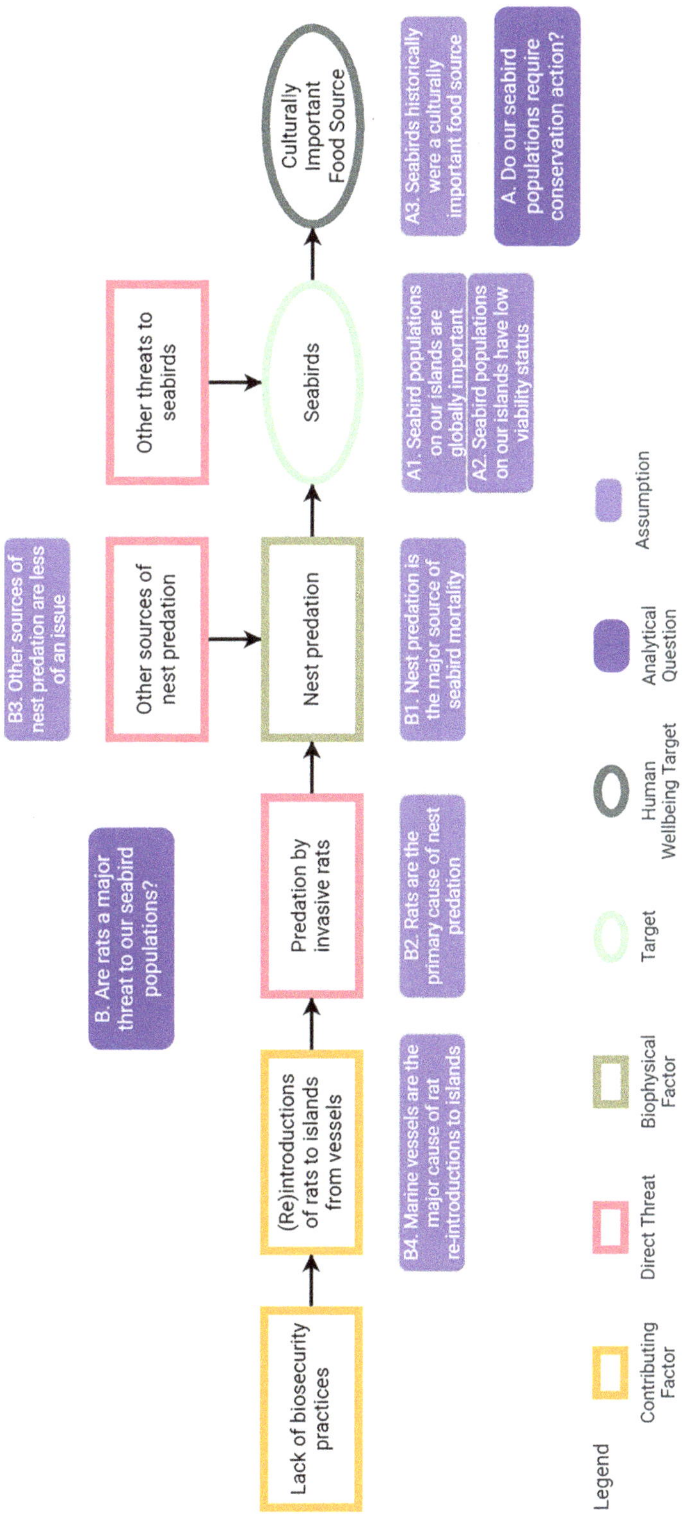

Figure 7.4 Examples of analytical questions and assumptions related to a situation model. (*Source:* Salafsky et al., 2022, CC-BY-4.0)

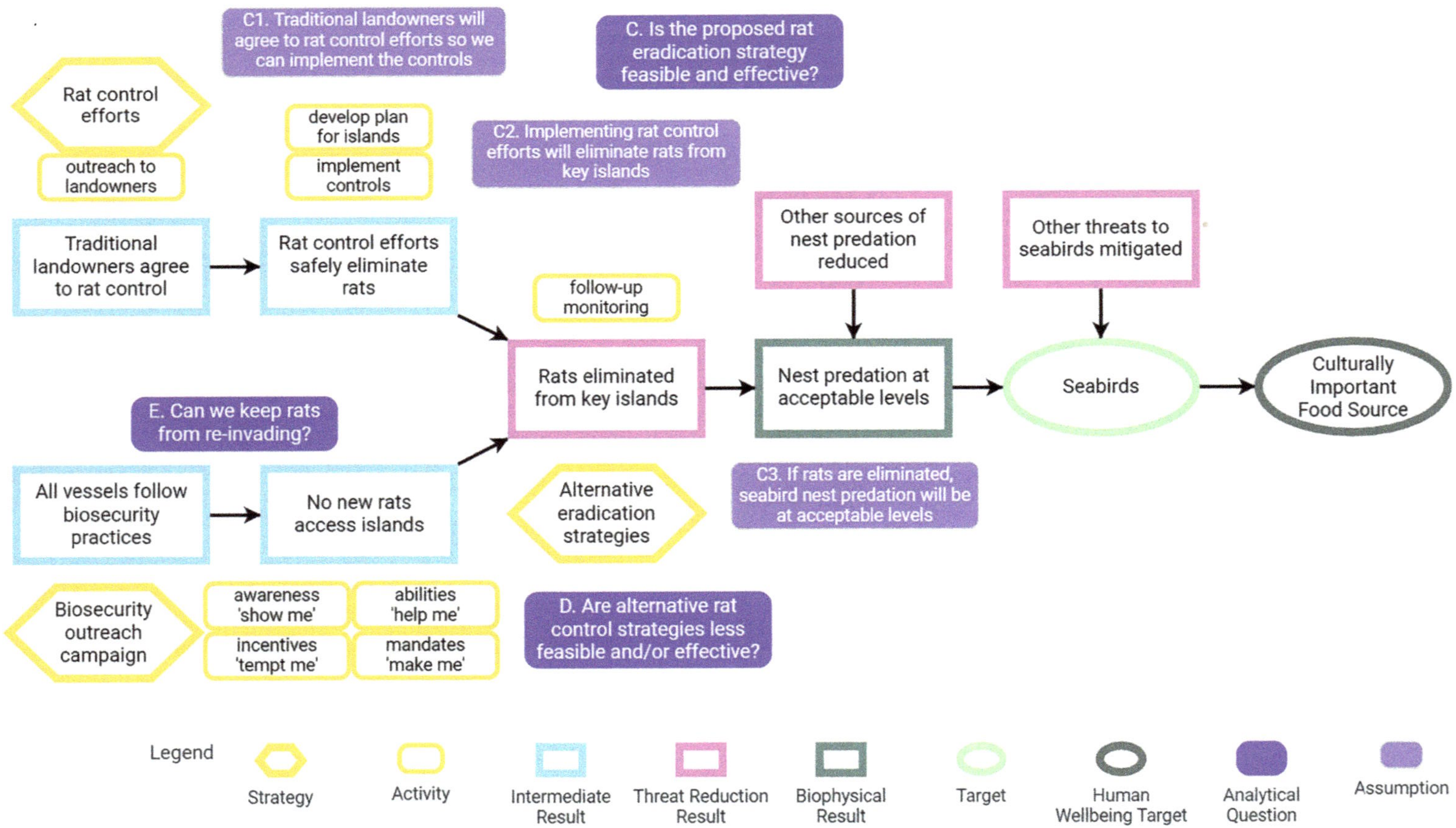

Figure 7.5 Examples of analytical questions and assumptions related to a theory of change. Note that assumptions for Questions D and E are not shown. (*Source:* Salafsky et al., 2022, CC-BY-4.0)

Box 7.5 Creating a research agenda of questions for policy and practice

1. Decide on the precise limits of the subject area, such as seagrass conservation or urban parks in Asia.
2. Decide on the scale of an appropriate question. This should be answerable, such that a research programme with several projects could make substantial progress or could answer the question within a certain time (e.g. 1–5 years).
3. Bring together a group of policymakers, practitioners, and researchers ensuring a diversity of regions, perspectives, and experiences.
4. Decide on the target number of questions. This is often around 20–50, but fewer if the subject area is narrower.
5. Each participant, who may consult their community, provides suggested research questions. Questions could also be solicited through an open online survey.
6. Organiser(s) remove questions clearly out of scope and any repetitions.
7. Questions are grouped by theme or sub-theme. These may often be linked to the number of sessions during the workshop (Point 10). Thus if the initial stage of the workshop comprises two sessions each with three parallel discussions then the questions should be grouped into six themes.
8. Participants then prioritise the top questions (e.g. top 20%) within each thematic grouping. This is usually done remotely and independently.
9. A shortlist of questions is then sent to each participant for consultation.
10. Participants convene in a workshop. The workshop is often divided into multiple stages (ideally three), each gradually reducing the number of questions until the target total is reached.
11. At each session, the number of questions is reduced, particularly by removing those that are unanswerable, not research questions, normative ('should we...') or too broad to fit within the original subject area.
12. During the final workshop session, participants agree on the final list and ensure any similar questions are removed.

(Taylor et al., 2021); Antarctic science (Kennicutt et al., 2015); UK poverty reduction (Sutherland et al., 2013); and UK biosecurity (Kemp et al., 2021). This process of creating research agendas is a promising way to identify what is needed to address societal challenges and risks, and can thus usefully inform policy and government strategies.

An analysis (Jucker et al., 2018) of the outputs of an exercise to identify priority questions for the conservation of global biological diversity a decade earlier (Sutherland, 2009) showed that, as of July 2016, seventy documents cited the exercise specifically to justify research on topics it highlighted. They also identified 21 questions that met their criteria for knowledge gaps and so needed further work. This shows such exercises are indeed used to generate new research and can be used to record progress in addressing priority questions.

Box 7.5 outlines how to adopt this approach. Our experience is that this process needs to combine the strengths of practitioners and policymakers, who know which knowledge gaps are important for practice and policy, and researchers, who can convert general interests into specific research questions.

7.7.3 Unpacking questions

The processes described above often identify gaps in knowledge across a range of subjects or topics, and questions that are critical to delivering action. Many questions asked as part of the process of decision making are general and need unpacking to make them more specific. For questions relating to actions or interventions, the PICO format is often used (Box 7.6).

Box 7.6 Designing PICO (population, intervention, comparison, outcome) questions

For interventions it is common that questions follow the PICO format:

P = Population — What is the target of the intervention, or the group you are interested in studying?

Do bumblebees in Lithuania...

I = Intervention — What main intervention is being considered?

...when provided with headlands seeded with wildflower mix...

C = Comparison — Is there an alternative to compare with the intervention?

...compared to naturally regenerated headlands...

O = Outcome — What is the objective or response being measured?

...increase more in abundance?

Do bumblebees in Lithuania increase more in abundance when provided with headlands seeded with wildflower mix compared to naturally regenerated headlands?

An equivalent approach applies to other questions (EFSA, 2010; James et al., 2016). Questions about threats or impacts can be framed in a PECO format (population, exposure, comparison, outcome), for example 'Do bumblebees with varroa mites have a lower reproductive output than bumblebees without varroa mites?' Questions about the accuracy of tests or monitoring methods can be framed in a PIT format (population, index test, target condition), for example

'Can eDNA sampling of flowers be used to detect the presence of endangered bees?' Descriptive questions about prevalence, occurrence or incidence can be framed in a PO format (population, outcome), for example 'What proportion of fish in UK rivers show signs of disease?'

As discussed in the previous section, it is often helpful to narrow down broad questions into specific ones, so that it is clear what data are needed to answer the question. So rather than asking 'Are seabirds successfully nesting in Eastern Bay?', the question could be 'Are there at least 100 breeding pairs of ruby-crested puffins who have fledged an average of at least one chick during each of the last five breeding seasons in Eastern Bay?'

References

Antwis, R.E., Griffiths, S.M., Harrison, X.A. et al. 2017. Fifty important research questions in microbial ecology. *FEMS Microbiology Ecology* 93: fix044, https://doi.org/10.1093/femsec/fix044.

Arzoumanian, Z., Holmberg, J. and Norman, B. 2005., An astronomical pattern-matching algorithm for computer-aided identification of whale sharks *Rhincodon typus*. *Journal of Applied Ecology* 42: 999–1011, https://doi.org/10.1111/j.1365-2664.2005.01117.x.

Bennun, L., van Bochove, J., Ng, C., et al. 2021. *Mitigating Biodiversity Impacts Associated with Solar and Wind Energy Development: Guidelines for Project Developers* (Gland, Switzerland: IUCN; Cambridge, UK: The Biodiversity Consultancy), https://doi.org/10.2305/IUCN.CH.2021.04.en.

Berthinussen, A., Richardson, O.C. and Altringham, J.D. 2021. *Bat Conservation: Global Evidence for the Effects of Interventions* (2021 Edition) (Cambridge: University of Cambridge), https://www.conservationevidence.com/synopsis/pdf/32.

Brown, A., Gilbert, G. and Wotton, S. 2012. Bitterns and bittern conservation in the UK. *British Birds* 105: 58–87.

Christie, A.P., Amano, T., Martin, P.A., et al. 2021. The challenge of biased evidence in conservation. *Conservation Biology* 35: 249–62, https://doi.org/10.1111/cobi.13577.

CMP. 2020. *Conservation Standards,* https://conservationstandards.org/about/.

Community Conservation Horizon Scan Collaboration. 2022. 2021–2022 Global Horizon Scan for Community Conservation: informing future priorities and decision-making (Calgary, Canada: The Wilder Institute/Calgary Zoo), https://cchorizonscan.org/updates-outputs/.

ConservationXLabs. 2022. *Grand Challenges for Conservation,* https://conservationxlabs.com/grand-challenges.

Crockford, N. 2018. More game-changing good news on coastal wetland conservation — a policy perspective: from Yellow Sea to global conservation OR how an IWSG conference talk can catalyse conservation action... *Wader Study* 125: 158–61, https://doi.org/10.18194/ws.00128.

Cuthbert, R.J., Taggart, M.A., Prakash, V., et al. 2014. Avian scavengers and the threat from veterinary pharmaceuticals. *Philosophical Transactions of the Royal Society B* 369: 20130574, https://doi.org/10.1098/rstb.2013.0574.

Daszak, P., Cunningham, A.A. and Hyatt, A.D. 2003. Infectious disease and amphibian population declines. *Diversity & Distributions* 9: 141–50, https://doi.org/10.1046/j.1472-4642.2003.00016.x.

De Frenne, P., Van Langenhove, L., Van Driessche, A., et al. 2018. Using archived television video footage to quantify phenology responses to climate change. *Methods in Ecology and Evolution* 9: 1874–82, https://doi.org/10.1111/2041-210X.13024.

Donald, P.F., Green, R.E. and Heath, M.F. 2001. Agricultural intensification and the collapse of Europe's farmland bird populations. *Proceedings of the Royal Society of London B*. 268: 25–29, https://doi.org/10.1098/rspb.2000.1325.

EFSA. 2010. Application of systematic review methodology to food and feed safety assessments to support decision making: EFSA guidance for those carrying out systematic reviews. *EFSA Journal* 8: 1637, https://doi.org/10.2903/j.efsa.2010.1637.

Esmail, N., Wintle, B.C., Sas-Rolfes, M.T., et al. 2020. Emerging illegal wildlife trade issues: A global horizon scan. *Conservation Letters* 13: e12715, https://doi.org/10.1111/conl.12715.

Gallardo, B., Zieritz, A., Adriaens, T., et al. 2016. Trans-national horizon scanning for invasive non-native species: A case study in western Europe. *Biological Invasions* 18: 17–30, https://doi.org/10.1007/s10530-015-0986-0.

Galligan, T.H., Bhusal, K.P., Paudel, K., et al. 2020. Partial recovery of critically endangered *Gyps* vulture populations in Nepal. *Bird Conservation International* 30: 87–102, https://doi.org/10.1017/S0959270919000169.

Galligan, T.H., Mallord, J.W., Prakash, V.M., et al. 2021. Trends in the availability of the vulture-toxic drug, diclofenac, and other NSAIDs in South Asia, as revealed by covert pharmacy surveys. *Bird Conservation International* 31: 337–53, https://doi.org/10.1017/S0959270920000477.

Game, E.T., Meijaard, E., Sheil, D., et al. 2014. Conservation in a wicked complex world; challenges and solutions. *Conservation Letters* 7: 271–77, https://doi.org/10.1111/conl.12050.

Green, R.E. 1995. Diagnosing causes of bird declines. *Ibis* 137(Suppl. 1): 47–55, https://doi.org/10.1111/j.1474-919X.1995.tb08457.x.

Green, R.E. 2002. Diagnosing causes of population declines and selecting remedial actions. In: *Conserving Bird Biodiversity: General Principles and their Application*, Conservation Biology, ed. by K. Norris and D.J. Pain (Cambridge: Cambridge University Press), pp. 139–56, https://doi.org/10.1017/CBO9780511606304.008.

Green, R.E., Pain, D.J. and Krone, O. 2022. The impact of lead poisoning from ammunition sources on raptor populations in Europe. *Science of the Total Environment* 823: 154017, https://doi.org/10.1016/j.scitotenv.2022.154017.

Green, R.E. and Stowe, T.J. 1993. The decline of the corncrake *Crex crex* in Britain and Ireland in relation to habitat change. *Journal of Applied Ecology* 30: 689–95, https://doi.org/10.2307/2404247.

Green, R.E., Taggart, M.A., Senacha, K.R., et al. 2007. Rate of decline of the Oriental white-backed vulture population in India estimated from a survey of diclofenac residues in carcasses of ungulates. *PLoS ONE* 2: e686, https://doi.org/10.1371/journal.pone.0000686.

Hancock, M.H., Robson, H.J., Smith, T.D., et al. 2020. From a research study to a conservation partnership: Developing approaches for restoring common scoter populations. *Aquatic Conservation: Marine and Freshwater Ecosystems* 30: 1770–74, https://doi.org/10.1002/aqc.3414.

IPBES. 2016. *Summary for Policymakers of the Methodological Assessment of Scenarios and Models of Biodiversity and Ecosystem Services of the Intergovernmental Science-Policy Platform on Biodiversity and Ecosystem Services (IPBES)*, ed. by S. Ferrier, et al. (Bonn: Secretariat of the Intergovernmental Science-Policy Platform on Biodiversity and Ecosystem Services), https://ipbes.net/sites/default/files/downloads/pdf/SPM_Deliverable_3c.pdf.

IPCC. 2018. *Global Warming of 1.5°C. An IPCC Special Report on the impacts of global warming of 1.5°C above pre-industrial levels and related global greenhouse gas emission pathways, in the context of strengthening the global response to the threat of climate change, sustainable development, and efforts to eradicate poverty*, ed. by V. Masson-Delmotte, et al. (Cambridge: Cambridge University Press), https://doi.org/10.1017/9781009157940.

IUCN. 2021. *The IUCN Red List of Threatened Species. Version 2021–3*, https://www.iucnredlist.org.

James, K.L., Randall, N.P. and Haddaway, N.R. 2016. A methodology for systematic mapping in environmental sciences. *Environmental Evidence* 5: Article 7, https://doi.org/10.1186/s13750-016-0059-6.

Jucker, T., Wintle, B., Shackelford, G., et al. 2018. Ten-year assessment of the 100 priority questions for global biodiversity conservation. *Conservation Biology* 32: 1457–63, https://doi.org/10.1111/cobi.13159.

Junker, J., Petrovan, S.O., Arroyo-RodrÍguez, V., et al. 2020. A severe lack of evidence limits effective conservation of the world's primates. *BioScience 70*: 794–803, https://doi.org/10.1093/biosci/biaa082.

Kark, S., Sutherland, W.J., Shanas, U., et al. 2016. Priority questions and horizon scanning for conservation: A comparative study. *PLoS ONE* 11(1): e0145978, https://doi.org/10.1371/journal.pone.0145978.

Kemp, L., Adam, L., Boehm, C.R., et al. 2020. Bioengineering horizon scan 2020. *eLife* 9: e54489, https://doi.org/10.7554/eLife.54489.

Kemp, L., Aldridge, D.C., Booy, O., et al. 2021. 80 questions for UK biological security. *PLoS ONE* 16: e.0241190, https://doi.org/10.1371/journal.pone.0241190.

Kennicutt, II, M.C., Chown, S.L., Cassano, J.J., et al. 2015. A roadmap for Antarctic and Southern Ocean science for the next two decades and beyond. *Antarctic Science* 27: 3–18, https://doi.org/10.1017/S0954102014000674.

Khanyari, M., Milner-Gulland, E.J., Oyanedel, R., et al. 2022. Investigating parasite dynamics of migratory ungulates for sustaining healthy populations: Application to critically-endangered saiga antelopes *Saiga tatarica*. *Biological Conservation* 266: 109465, https://doi.org/10.1016/j.biocon.2022.109465.

Kodikara, K.A.S., Mukherjee, N., Jayatissa, L.P., et al. 2017. Have mangrove restoration projects worked? An in-depth study in Sri Lanka. *Restoration Ecology* 25: 705–16, https://doi.org/10.1111/rec.12492.

Macinnis-Ng, C., Mcintosh, A.R., Monks, J.M., et al. 2021. Climate-change impacts exacerbate conservation threats in island systems: New Zealand as a case study. *Frontiers in Ecology and the Environment* 19: 216–24, https://doi.org/10.1002/fee.2285.

MacKinnon, J., Verkuil, Y.I. and Murray, N. 2012. *IUCN Situation Analysis on East and Southeast Asian Intertidal Habitats, with Particular Reference to the Yellow Sea (Including the Bohai Sea)*. Occasional Paper of the IUCN Species Survival Commission No. 047 (Gland, Switzerland; Cambridge: IUCN), https://portals.iucn.org/library/sites/library/files/documents/SSC-OP-047.pdf.

Margoluis, R., Stem, C., Salafsky, N., et al. 2009. Using conceptual models as a planning and evaluation tool in conservation. *Evaluation and Program Planning* 32: 138–47, https://doi.org/10.1016/j.evalprogplan.2008.09.007.

Margoluis, R., Stem, C., Swaminathan, V., et al. 2013. Results chains: A tool for conservation action design, management, and evaluation. *Ecology and Society* 18: Article 22, https://doi.org/10.5751/ES-05610-180322.

Mitchell, M., Lockwood, M., Moore, S.A., et al. 2016. Using scenario planning to assess governance reforms for enhancing biodiversity outcomes. *Land Use Policy* 50: 559–72, https://doi.org/10.1016/j.landusepol.2015.10.020.

Moussy, C., Burfield, I. J., Stephenson, P. J., et al. 2022. A quantitative global review of species population monitoring. *Conservation Biology* 36: e13721. https://doi.org/10.1111/cobi.13721.

Newman, R.D. and Showler, A.J. 2007. The use of copper rings to reduce losses of red helleborine *Cephalanthera rubra* to slug and snail herbivory in the Chiltern Hills, Buckinghamshire, England. *Conservation Evidence* 4: 66–68, https://conservationevidencejournal.com/reference/pdf/2263.

Noss, R.F., 1990. Indicators for monitoring biodiversity: a hierarchical approach. *Conservation Biology* 4: 355–64, https://doi.org/10.1111/j.1523-1739.1990.tb00309.x.

Oaks, J.L., Gilbert, M., Virani, M.Z., et al. 2004. Diclofenac residues as the cause of vulture population decline in Pakistan. *Nature* 427: 630–33, https://doi.org/10.1038/nature02317.

Ockendon, N., Thomas, D.H.L., Cortina, J., et al. 2018. One hundred priority questions for landscape restoration in Europe. *Biological Conservation* 221: 198–208, https://doi.org/10.1016/j.biocon.2018.03.002.

Oldekop, J.A., Fontana, L.B., Grugel, J., et al. 2016. 100 key research questions for the post-2015 development agenda. *Development Policy Review* 34: 55–82, https://doi.org/10.1111/dpr.12147.

Oldekop, J.A., Vang Rasmussen, L., Agrawal, A., et al. 2020. Forest-linked livelihoods in a globalized world. *Nature Plants* 6: 1400–07, https://doi.org/10.1038/s41477-020-00814-9.

Pain, D.J., Bowden, C.G.R., Cunningham, A.A., et al. 2008. The race to prevent the extinction of South Asian vultures. *Bird Conservation International* 18: S30–S48, https://doi.org/10.1017/S0959270908000324.

Palomo, I., Múgica, M., Piñeiro, C., et al. 2017. Envisioning protected areas through participatory scenario planning: Navigating coverage and effectiveness challenges ahead. *Parks* 23: 29–44, https://doi.org/10.2305/IUCN.CH.2017.PARKS-23-1IP.en.

Parker, M., Acland, A., Armstrong, H.J., et al. 2014. Identifying the science and technology dimensions of emerging public policy issues through horizon scanning. *PLoS ONE* 9 e96480, https://doi.org/10.1371/journal.pone.0096480.

Parrott, D., Roy, S., Baker, R., et al. 2009. *Horizon Scanning for New Invasive Non-Native Animal Species in England.* Natural England Commissioned Report NECR009 (Sheffield: Natural England), http://publications.naturalengland.org.uk/publication/43005.

Parsons, E.C.M., Baulch, S., Bechshoft, T., et al. 2015. Key research questions of global importance for cetacean conservation. *Endangered Species Research* 27: 113–18, https://doi.org/10.3354/esr00655.

Peach, W.J., Denny, M., Cotton, P.A., et al. 2004. Habitat selection by song thrushes in stable and declining farmland populations. *Journal of Applied Ecology* 41: 275–93, https://doi.org/10.1111/j.0021-8901.2004.00892.x.

Pérez-Jvostov, F., Sutherland, W.J., Barrett, R.D.H., et al. 2019. Horizon scan of conservation issues for inland waters in Canada. *Canadian Journal of Fisheries and Aquatic Sciences* 77: 869–81, https://doi.org/10.1139/cjfas-2019-0105.

Peterson, G.D., Cumming, G.S. and Carpenter, S.R. 2003. Scenario planning: A tool for conservation in an uncertain world. *Conservation Biology* 17: 358–66, https://doi.org/10.1046/j.1523-1739.2003.01491.x.

Petrovan, S.O., Aldridge, D.C., Bartlett, H., et al. 2021. Post COVID-19: A solution scan of options for preventing future zoonotic epidemics. *Biological Reviews* 96: 2694–715. https://doi.org/10.1111/brv.12774.

Pimm, S.L., Jenkins, C.N., Abell, R., et al. 2014. The biodiversity of species and their rates of extinction, distribution, and protection. *Science* 344: 1246752, https://doi.org/10.1126/science.1246752.

Prakash, V. 1999. Status of vultures in Keoladeo National Park, Bharatpur, Rajasthan, with special reference to population crash in *Gyps* species. *Journal of the Bombay Natural History Society* 96: 365–78.

Prescott, G.W., Sutherland, W.J., Aguirre, D., et al. 2017. Political transition and emergent forest-conservation issues in Myanmar. *Conservation Biology* 31: 1257–70, https://doi.org/10.1111/cobi.13021.

Pretty, J.J., Sutherland, W.J., Ashby, J. et al. 2010. The top 100 questions of importance to the future of global agriculture. *International Journal of Agricultural Sustainability* 8: 219–36, https://doi.org/10.3763/ijas.2010.0534.

Ricciardi, A., Blackburn, T.M., Carlton, J.T., et al. 2016. Invasion science: A horizon scan of emerging challenges and opportunities. *Trends in Ecology and Evolution* 32: 464–74, https://doi.org/10.1016/j.tree.2017.03.007.

Ricciardi, A. and MacIsaac, H.J. 2022. Vector control reduces the rate of species invasion in the world's largest freshwater ecosystem. *Conservation Letters* 15: e12866, https://doi.org/10.1111/conl.12866.

Riley, S., Siemer, W., Decker, D., et al. 2003. Adaptive Impact Management: An integrative approach to wildlife management. *Human Dimensions of Wildlife* 8: 81–95, https://doi.org/10.1080/10871200304301.

Roberts, M., Cresswell, W. and Hanley, N. 2018. Prioritising invasive species control actions: Evaluating effectiveness, costs, willingness to pay and social acceptance. *Ecological Economics* 152: 1–8, https://doi.org/10.1016/j.ecolecon.2018.05.027.

Roura-Pascual, N., Leung, B., Rabitsch, W., et al. 2021. Alternative futures for global biological invasions. *Sustainability Science* 16: 1637–50, https://doi.org/10.1007/s11625-021-00963-6.

Roy, H., Peyton, J., Aldridge, D.C., et al. 2014. Horizon-scanning for invasive alien species with the potential to threaten biodiversity in Great Britain. *Global Change Biology* 20: 3859–71, https://doi.org/10.1111/gcb.12603.

Salafsky, N., Irvine, R., Boshoven, J., et al. 2022. A practical approach to assessing existing evidence for specific conservation strategies. *Conservation Science and Practice* 4: e12654, https://doi.org/10.1111/csp2.12654.

Schwartz, M.W., Cook, C.N., Pressey, R.L. et al. 2018. Decision Support Frameworks and Tools for Conservation. *Conservation Letters* 11: e12385. https://doi.org/10.1111/conl.12385.

Shultz, S., Baral, H.S., Charman, S., et al. 2004. Diclofenac poisoning is widespread in declining vulture populations across the Indian subcontinent. *Proceedings of the Royal Society of London B. (Suppl.)* 271: S458–S460, https://doi.org/10.1098/rsbl.2004.0223.

Siriwardena, G.M., Baillie, S.R. and Wilson, J.D. 1999. Temporal variation in annual survival rates of six granivorous birds with contrasting population trends. *Ibis* 141: 621–36, https://doi.org/10.1111/j.1474-919X.1999.tb07370.x.

Snyder, N. and Snyder, H. 2000. *The California Condor: A Saga of Natural History and Conservation* (San Diego: Academic Press).

Stephenson, P.J. and Stengel, C., 2020. An inventory of biodiversity data sources for conservation monitoring. *PLoS ONE 15*: e0242923, https://doi.org/10.1371/journal.pone.0242923.

Stowe, T.J., Newton, A.V., Green, R.E., et al. 1993. The decline of the corncrake *Crex crex* in Britain and Ireland in relation to habitat. *Journal of Applied Ecology* 30: 53–62, https://doi.org/10.2307/2404270.

Summers, R.W., Green, R.E., Proctor, D., et al. 2004. An experimental study of the effects of predation on the breeding productivity of capercaillie and black grouse. *Journal of Applied Ecology* 41: 513–25, https://doi.org/10.1111/j.0021-8901.2004.00891.x.

Sutherland, W.J. 2000. *The Conservation Handbook. Research, Management and Policy* (Oxford: Wiley-Blackwell).

Sutherland, W.J. 2019. Kaizen conservation? *Oryx* 53: 397–98, https://doi.org/10.1017/S0030605319000619.

Sutherland, W.J., Adams, W.M., Aronson, R.B., et al. 2009. One hundred questions of importance to the conservation of global biological diversity. *Conservation Biology* 23: 557–67, https://doi.org/10.1111/j.1523-1739.2009.01212.x.

Sutherland, W.J., Albon, S.D., Allison, H., et al. 2010a. The identification of priority opportunities for UK nature conservation policy. *Journal of Applied Ecology* 47: 955–65, https://doi.org/10.1111/j.1365-2664.2010.01863.x.

Sutherland, W.J., Allison, H., Aveling, R., et al. 2012a. Enhancing the value of horizon scanning through collaborative review. *Oryx* 46: 368–74, https://doi.org/10.1017/S0030605311001724.

Sutherland, W.J., Alves, J.A., Amano, T., et al. 2012b. A horizon scanning assessment of current and potential future threats to migratory shorebirds. *Ibis* 154: 663–79, https://doi.org/10.1111/j.1474-919X.2012.01261.x.

Sutherland, W.J., Armstrong-Brown, S., Armsworth, P.R., et al. 2006. The identification of 100 ecological questions of high policy relevance in the UK. *Journal of Applied Ecology* 43: 617–27, https://doi.org/10.1111/j.1365-2664.2006.01188.x.

Sutherland, W.J., Atkinson, P.W., Butchart, S.H.M., et al. 2022. A horizon scan of global biological conservation issues for 2022. *Trends in Ecology and Evolution* 37: 95–104, https://doi.org/10.1016/j.tree.2021.10.014.

Sutherland, W.J., Clout, M., Côté, I.M., et al. 2010b. A horizon scan of global conservation issues for 2010. *Trends in Ecology and Evolution* 25: 1–7, https://doi.org/10.1016/j.tree.2009.10.003.

Sutherland, W.J., Dicks, L.V., Petrovan, S.O., et al. (eds.). 2021. *What Works in Conservation 2021* (Cambridge: Open Book Publishers), https://doi.org/10.11647/obp.0267.

Sutherland, W.J., Fleishman, E., Clout, M., et al. 2019a. Ten years on: A review of the first global conservation horizon scan. *Trends in Ecology and Evolution* 34: 139–53, https://doi.org/10.1016/j.tree.2018.12.003.

Sutherland, W.J., Fleishman, E., Mascia, M.B., et al. 2011. Methods for collaboratively identifying research priorities and emerging issues in science and policy. *Methods in Ecology and Evolution* 2: 238–47, https://doi.org/10.1111/j.2041-210X.2010.00083.x.

Sutherland, W.J., Gardner, T., Bogich, T.L., et al. 2014. Solution scanning as a key policy tool: Identifying management interventions to help maintain and enhance regulating ecosystem services. *Ecology and Society* 19: Article 3, https://doi.org/10.5751/ES-06082-190203.

Sutherland, W.J., Goulden, C., Bell, K., et al. 2013. 100 questions: identifying research priorities for poverty prevention and reduction. *Journal of Poverty and Social Justice* 21: 189–205, https://doi.org/10.1332/175982713X671210.

Sutherland, W.J., Taylor, N.G., MacFarlane, D., et al. 2019b. Building a tool to overcome barriers in research-implementation spaces: The Conservation Evidence database. *Biological Conservation* 238: 108199, https://doi.org/10.1016/j.biocon.2019.108199.

Sutherland, W.J. and Woodroof, H.J. 2009. The need for environmental horizon scanning. *Trends in Ecology and Evolution* 24: 523–27, https://doi.org/10.1016/j.tree.2009.04.008.

Sutherland, W.J. and Wordley, C.F. 2017. Evidence complacency hampers conservation. *Nature Ecology and Evolution* 1: 1215–16, https://doi.org/10.1038/s41559-017-0244-1.

Swan G.E., Cuthbert, R., Quevedo, M., et al. 2006. Toxicity of diclofenac to *Gyps* vultures. *Biology Letters* 2: 279–82, https://doi.org/10.1098/rsbl.2005.0425.

Taleb, N. 2007. *The Black Swan: The Impact of the Highly Improbable* (New York: Random House).

Tatayah, R.V.V., Malham, J. and Haverson, P. 2007. The use of copper strips to exclude invasive African giant land-snails *Achatina* spp. from echo parakeet *Psittacula eques* nest cavities, Black River Gorges National Park, Mauritius. *Conservation Evidence* 4: 6–8, https://conservationevidencejournal.com/reference/pdf/2245.

Taylor, N.G., Grillas, P., Al Hreisha, H., et al. 2021. The future for Mediterranean wetlands: 50 key issues and 50 important conservation research questions. *Regional Environmental Change* 21: Article 33, https://doi.org/10.1007/s10113-020-01743-1.

Thomas, J.A., Simcox, D.J. and Clarke, R.T. 2009. Successful conservation of a threatened *Maculinea* butterfly. *Science* 325: 80–83, https://doi.org/10.1126/science.1175726.

Thomas, J.A., Simcox, D.J. and Hovestadt, T. 2011. Evidence based conservation of butterflies. *Journal of Insect Conservation* 15: 241–48, https://doi.org/10.1007/s10841-010-9341-z.

Thomas, J.A., Thomas, C.D., Simcox, D.J., et al. 1986. Ecology and declining status of the silver-spotted skipper butterfly (*Hesperia comma*) in Britain. *Journal of Applied Ecology* 23: 365–80, https://doi.org/10.2307/2404023.

UK Government Office for Science. 2017. *The Futures Toolkit: Tools for Futures Thinking and Foresight Across UK Government*, Edition 1.0 (London: Government Office for Science), https://assets.publishing.service.gov.uk/government/uploads/system/uploads/attachment_data/file/674209/futures-toolkit-edition-1.pdf.

USAID. 2018. *Evidence in Action: Using and Generating Evidence About Effectiveness in Biodiversity Programming. Unit 1: Understanding an Evidence-Based Approach* (Washington DC: USAID Bureau for Economic Growth, Education, and the Environment Office of Forestry and Biodiversity), https://pdf.usaid.gov/pdf_docs/PA00SXVJ.pdf.

Walsh, J.C., Dicks, L.V. and Sutherland, W.J. 2015. The effect of scientific evidence on conservation practitioners' management decisions. *Conservation Biology* 29: 88–98, https://doi.org/10.1111/cobi.12370.

Wintle, B., Boehm, C., Rhodes, C., et al. 2017. A transatlantic perspective on 20 emerging issues in biological engineering. e-*Life* 6: e30247, https://doi.org/10.7554/eLife.30247.

Wintle, B.C., Kennicutt, M.C. and Sutherland, W.J. 2020. Scanning horizons in research, policy and practice. In: *Conservation Research, Policy and Practice,* Ecological Reviews, ed. by W.J. Sutherland, et al. (Cambridge: Cambridge University Press), pp. 29–47, https://doi.org/10.1017/9781108638210.003.

Zoohackathon. 2021. *Zoohackathon,* https://www.state.gov/zoohackathon-2021/.

8 Making Decisions for Policy and Practice

Carly N. Cook[1], Mark Burgman[2], Libby Rumpff[3], Ann Thornton[4]

Having collaborated with the community to decide upon the objectives, identified the major threats to be addressed, considered the possible options and assessed the evidence, the next stage is to decide what to do. Many decisions can be decided easily, as either obvious or trivial, thus requiring no additional assessment of different stages of the process. The harder decisions then require assessing the likely consequences of options and determining the preferred trade off. There is a range of approaches for making each of these stages more rigorous, reducing the likelihood of making inefficient decisions. These approaches are described with an account of the situations under which is most appropriate to address the conservation problem.

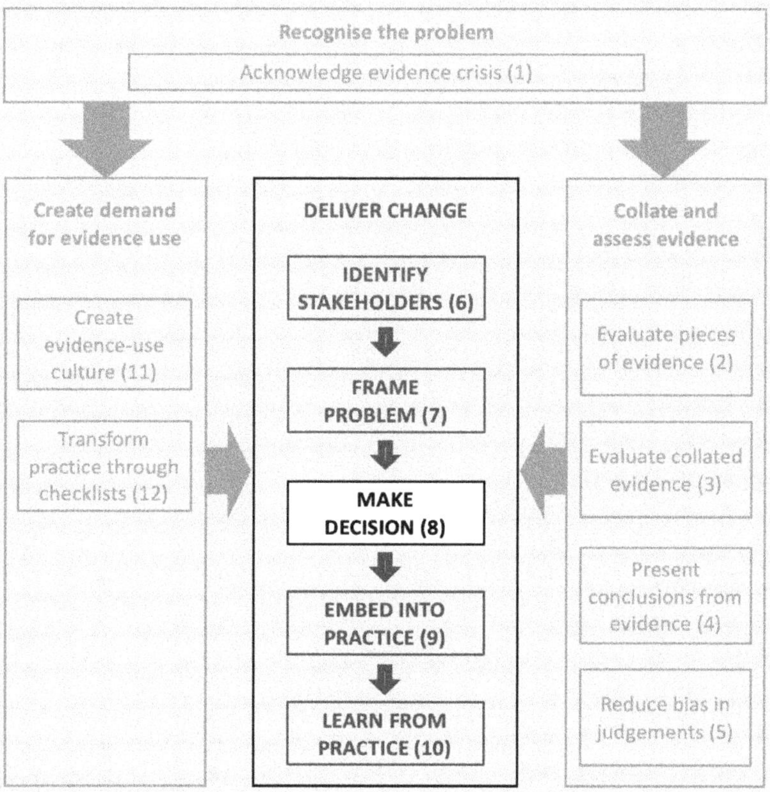

 https://doi.org/10.11647/OBP.0321.08

[1] Monash University, School of Biological Sciences, Monash University, Clayton, VIC 3800, Australia
[2] Centre for Environmental Policy, Imperial College, Princes Gardens, South Kensington, London, UK
[3] School of BioSciences, University of Melbourne, Parkville 3052, Melbourne, Australia
[4] Conservation Science Group, Department of Zoology, University of Cambridge, The David Attenborough Building, Pembroke Street, Cambridge, UK

Contents

8.1 What is a Structured Approach to Decision-Making?

The previous chapters have described the processes involved in preparing to make a decision: understanding the context for the decision (Chapters 6 and 7), clearly identifying the targets (Chapter 7), and compiling all the information required to make an informed choice (Chapters 2, 3, 4 and 5). Almost any conservation decision will need to use elements from these chapters.

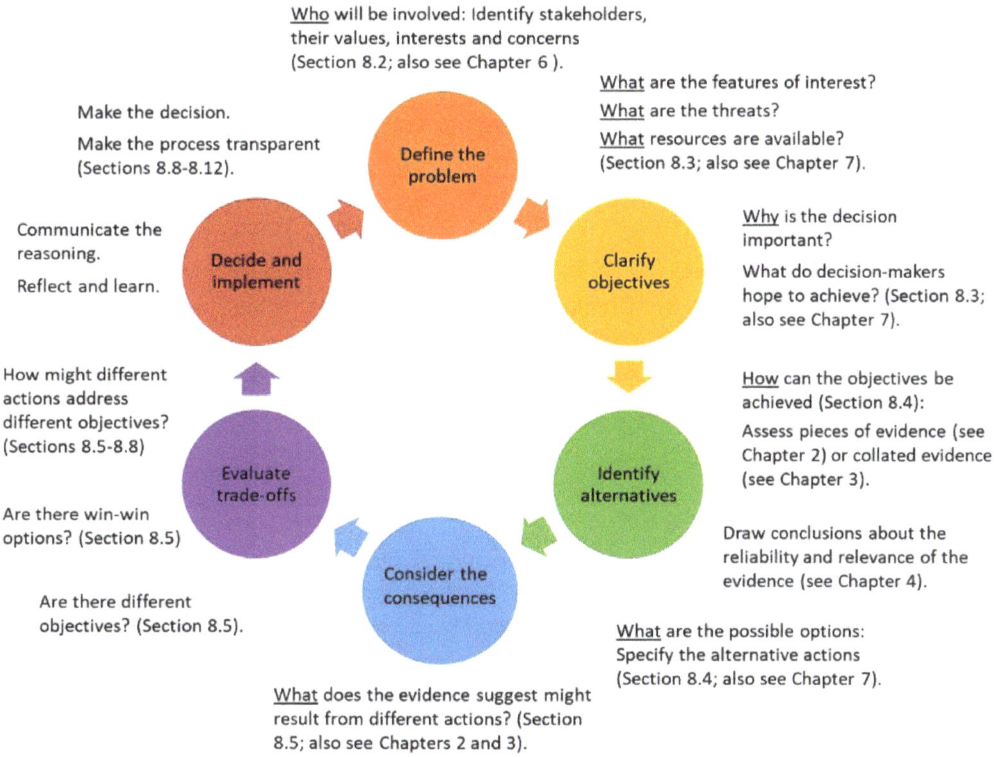

Figure 8.1 The stages in a structured decision process are represented by coloured circles, alongside text suggesting questions to consider during the process. While many of these considerations can be made unconsciously, a structured process helps decision makers to explicitly consider *who, what, why* and *how,* even if done quickly and simply (Box 8.1). (*Source*: authors, adapted from Gregory et al., 2012)

This chapter describes the various stages in the decision process (Figure 8.1; Box 8.1) and provides tools to support the decision maker to integrate relevant information when considering potential alternatives, assessing trade-offs and deciding how to act. The same process can be adopted whether an individual is thinking through a decision alone or as part of a group. Using a structured process to consider the essential elements of the decision offers decision makers the opportunity to document the process involved in reaching their decision, which has three important advantages. First, decision makers can communicate the rationale behind the choice to stakeholders who can then decide whether any concerns they raised were considered (Moon et al., 2019). Second, it can ensure the decision-making process is transparent and repeatable, thus helping to identify where knowledge about the system needs to be improved (Hemming et al., 2022). Third, it provides an opportunity to scrutinise the process and potentially improve future decisions (Gregory et al., 2012; Schwartz et al., 2018). Capturing the rationale and evidence supporting a decision can be a form of knowledge transfer, making sure that future decision makers understand why past decisions were made (Christie et al., 2022). Implementing the decision then offers an opportunity to learn, update our understanding of the system, and work through the process again to refine our management (Gregory et al., 2012).

8.1.1 The process of decision making

We make decisions all the time, consciously or not. Some are simple matters that do not require much thought; deciding not to do anything is also a decision. Others are complex and multi-faceted with potentially irreversible consequences for which a structured process and careful consideration could guard against disaster. Regardless of how simple or complex the decision is, most are improved by, at least briefly, considering the different elements of the problem using the information currently available (Figure 8.1).

Box 8.1 outlines the general decision-making process with an example.

8.1.2 Summary of tools to help structure decisions

In this chapter we discuss tools that can support decision makers to work through the decision-making process. These can be used selectively, depending on the type of decision, the needs of the decision maker, and the time and resources available. The tools and the processes they promote can help decision makers be transparent about, and communicate the rationale behind, their decisions (Hemming et al., 2022). This helps engage stakeholders and promotes learning and improvement. Table 8.1 summarises the various tools, and provides examples of how each can be used and when it might be appropriate.

Box 8.1. The decision-making process

(see Keeney, 2004; Gregory et al., 2012)

Regardless of the time or resources available to make a decision there is an underlying logic to the process (Figure 8.1). Decisions can be made quickly, and it is worthwhile moving through these steps rapidly to identify whether there is a preferred solution, or whether iteration is needed (Garrard et al., 2017). When multiple stakeholders are involved, the following steps are worked through together, to get a shared understanding of the problem, and preferred solution(s).

- Identify who will be consulted or involved in the decision making process.
- Frame the decision – collectively agree on the problem to solve and the subsequent decision to be made, and brainstorm what you are fundamentally trying to achieve, or avoid, with your decision (these form your objectives).

- Consider the range of options available to achieve the objectives (Chapter 7). Start with a long list of possible options that focus on meeting all objectives. Narrow the list by excluding actions that are impossible to implement (e.g. too expensive, too technical, not socially acceptable), do not work in the relevant context (e.g. good for low-lying areas but not for uplands) or are irrelevant (e.g. focused on mammals, not birds). This generally leaves a much shorter list of relevant options.
- Consider which options are likely to give the best outcomes for all objectives, using the available evidence (Chapters 2 and 3). This step often involves trade-offs between competing objectives (see Section 8.5); these could relate to cost (e.g. an option may be preferred because it is cheaper and can be used over a larger area) or social constraints (e.g. it may be more important to keep the community on-side by selecting a less cost-effective action). We often make trade-offs unconsciously but discussion of preferred alternatives is critical, and if required, decision tools can support choices.
- Make a choice, or refine the preferred options. This choice may be based solely on the evidence for the most effective action. Usually, it will be a compromise, incorporating value judgements, and economic, social or political constraints. Multiple stakeholders often will have different preferred options that a decision maker must weigh up.
- Present draft conclusions to stakeholders for consultation, or iterate (e.g. including new options) where needed.

For example, I need to decide how to reduce the impact of roads on an arboreal mammal.

1. Managers and other stakeholders want possums to be able to cross the road safely, to connect populations that have become separated.
2. I consult local ecologists and others with local ecological knowledge, look at the Conservation Evidence database for actions to increase connectivity across roads, and reports produced by road safety agencies in other parts of the country. Based on this research, I create a list of options (in consultation with stakeholders) — building an overpass, creating tunnels, installing rope bridges, installing glider poles or doing nothing. I can quickly reduce these options because those with local knowledge tell me that possums rarely come to the ground and won't use tunnels. I do not have the money to build an overpass. So I must consider glider poles or rope bridges if I want to improve connectivity.
3. The local ecologist shows me a draft paper revealing that rope bridges are twice as effective as poles in increasing connectivity for this species. I consider the cost of the two options. Rope bridges cost four times more than glider poles

so our budget will only enable three rope bridges to be installed along a 20 km stretch of road, whereas we can install 12 glider poles along the same area. Glider poles will, therefore, increase connectivity more efficiently.

4. Given the trade-off between cost and effectiveness, I decide to install glider poles.
5. When discussing this choice with stakeholders, they point out that, on bends in the road, headlights illuminate the forest thereby discouraging the possums. They suggest selecting straight stretches of road to install the glider poles.

These tools can be carried out with different levels of complexity. They can be adapted for a decision (such as which vehicle to hire next season) in a process that takes a few minutes involving a few colleagues. On the other extreme, for a difficult and contentious issue, a series of workshops may bring together a broad community, allowing the creation of models to support decision making.

Table 8.1 Summary of tools described in this chapter.

Tool	Section	Description	Uses
Just doing	8.2	Act (but occasionally reflect)	Suitable for self evident or low stakes problems
Decision sketching (or rapid prototyping)	8.3.1	Outline the components of the decision	Allows relatively quick decisions
Benefit-cost/Cost-effectiveness analyses	8.3.2	Assess the change that is expected to occur when considering the cost of each action; suitable when objectives can be distilled into a single measure	Fundamental for any decision with a limited budget and a single measure of utility.
Clarifying objectives	8.3.3	Identifying fundamental objectives of the stakeholders and decision makers	Identifying objectives underpins the decision and understanding differences in objectives illuminates disputes
Means-ends networks	8.3.4	Sketch ways of achieving end given means, much like theory of change	Clarifies the problem: separate what you want to achieve from how to achieve it.
Multi-criteria decision making/structured decision making	8.5	Assess the outcome of different options against the key criteria/objectives, negotiate trade-offs	Complex decisions where win-win outcomes are unlikely, and trade-offs are involved

Tool	Section	Description	Uses
Consequence tables	8.5.1-8	Compare the performance of different alternatives against multiple objectives	Simplifies a decision context by ruling out redundant objectives, identifying dominant or dominated alternatives, identifying unavoidable trade-offs or win-win alternatives.
Revealing hidden values	8.5.3	Identifying values that have not been stated but influence decision choices	Helps understand decisions and reasons for variation in decisions.
Strategy Table	8.6	Organises alternative actions into packages 'strategies' or series of actions	When need to consider which combination of measures to adopt
Classifying decisions	8.7	Linking set of decisions	When there is a hierarchy of decisions and sub-decisions
Decision trees	8.8	Presenting a range of different options and consequences	When decisions involve a series of smaller, related decisions
Models	8.9	Identify potential actions and make predictions about potential outcomes	Essential for quantifying many changes, to populate
Achieving consensus and dealing with conflict	8.10	Bringing individuals together to find mutually acceptable solution	When stakeholders hold different objectives

8.2 Filter Easy Decisions: Deciding Whether to Invest in Decision Making

It is a common misconception that a structured approach to decision making is a time consuming and complicated process reserved for the most challenging decisions. This does not need to be the case. Figure 8.2 illustrates appropriate levels of analysis for an imaginary 10,000 decisions. The main point is that most decisions can be dealt with almost instantaneously because the best course of action is self-evident, or the alternative actions have such small consequences that make little difference. Many others can be dealt with by simple processes. Detailed consideration necessary for only a small proportion of complex decisions in which there are competing objectives, and the actions may result in substantial differences.

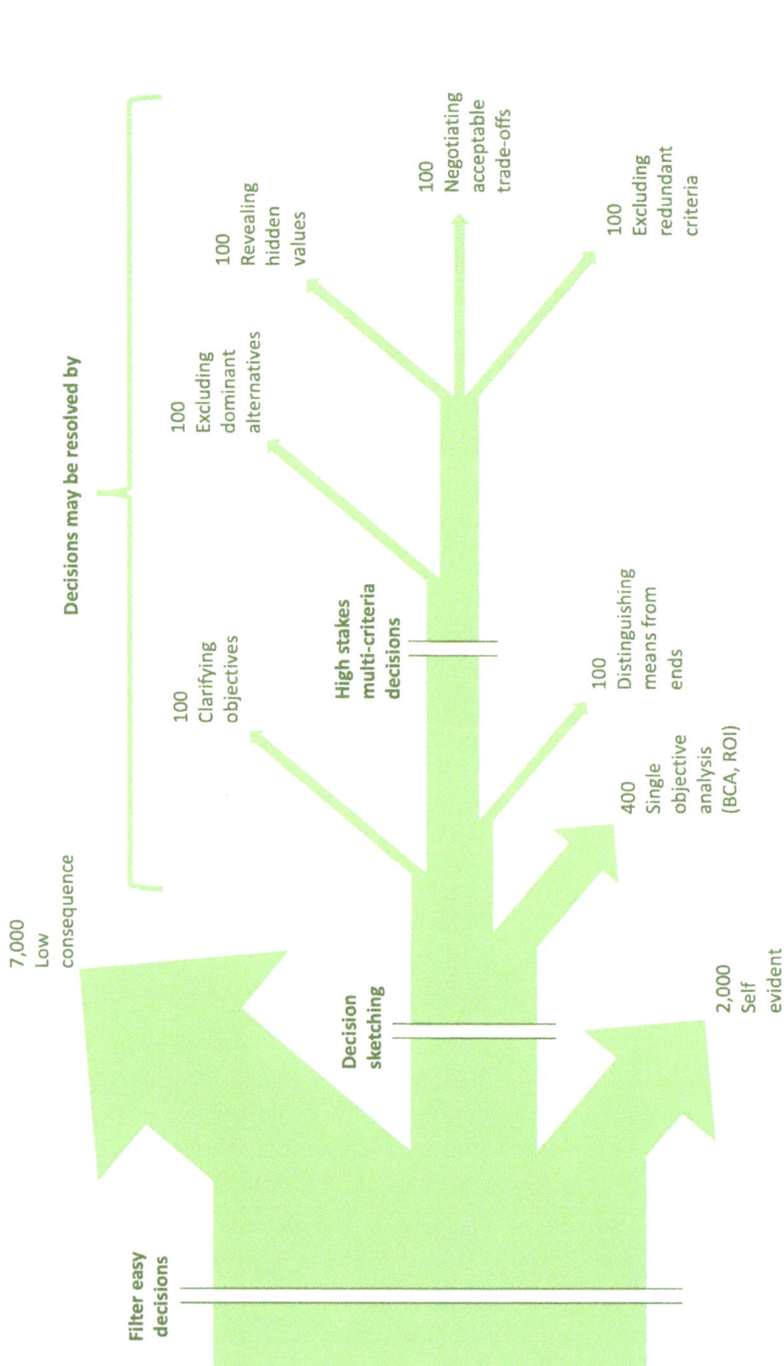

Figure 8.2 An illustration of how 10,000 decisions may be resolved (*Source*: authors, based on a figure from Keeney 2004). BCA and ROI refer to benefit-cost analysis and return-on-investment analysis, respectively.

8.2.1 Self-evident (high certainty) decisions

Many decisions are self-evident, drawing on existing, well-established knowledge. For example, a self-evident decision may include seeing a tourist on a boat pumping out their latrine into a lake and immediately asking them to stop, or discovering a harmful, invasive plant in a campsite, pulling it up and then reporting it.

Many decisions are implicit, simple continuations of the status quo. Decisions embedded in a set of procedures for the management of a protected area may include that a visitor centre opens at 9 am, a guided walk is available at 2 pm, and dogs are not allowed. Analysing every routine decision on every occasion would mean nothing else was ever done. However, it is worth occasionally reviewing the rationale and effectiveness of routine decisions, to avoid complacency (Hockings, 2003; Pullin and Knight, 2009). The process of kaizen (Section 7.4.2) may result in considerable benefits from sometimes looking at ways of improving routine practice (Sutherland, 2019).

8.2.2 Low-stakes decisions

Low-stakes decisions are those in which the outcomes matter so little that they do not justify further thought. This can be because the overall action is trivial (i.e. someone wants to sample the water from a critical bird breeding site, but after the breeding season so any disturbance will be minimal) or because the impact does not substantially extend to the project scope (i.e. a national hydrological project will affect an adjacent watershed but is expected to have minimal local impact, so local actions do not need to change).

8.2.3 High-stakes, low certainty decisions

Issues may be high-stakes for ecological, social, reputational or financial reasons (McShane et al., 2011). Some conservation decisions are high-stakes with irreversible consequences (Lindenmayer et al., 2013). An extreme example of this is species extinction. An estimated 617 vertebrate species alone have gone extinct since 1500 CE (Ceballos et al., 2015), many because of decisions that were made by humans (e.g. a bounty placed on the skins of the Tasmanian tiger, *Thylacinus cynocephalus;* Guiler and Godard, 1998), decisions that were not made (e.g. failure to ban hunting of Steller's sea cow, *Hydrodamalis gigas;* Domning, 1978) or that were delayed (e.g. monitoring the decline but not implementing management action to protect the Christmas Island pipistrelle bat, *Pipistrellus murrayi*; Martin et al. 2012).

There are also considerable risks to decision makers associated with some conservation decisions. The US Fish and Wildlife Service has been sued hundreds of times by stakeholders seeking to challenge their decisions about whether or not to list species under the US Endangered Species Act 1973 (Schwartz, 2008). In these cases, decision makers must be able to defend their rationale in court for why a species does or does not meet the required standard.

Importantly, decision makers must be cognisant that decisions can have direct implications for stakeholders (Bennett et al., 2019). Changes to policies regarding the exploitation of natural resources can lead to the loss of livelihoods, for example where marine protected areas displace local fishers or protecting native forests displace forestry activities. Even when the consequences for local communities are not financial, conflicting values can lead to contentious decisions with significant political and social implications for decision makers and the broader community. In these cases, stakeholders may have very different objectives and values, and their differing perspectives may mean that the facts of the problem are contested (Redpath et al., 2013).

8.3 Preparing to Make the Decision

Many day-to-day decisions in conservation may be considered simple or low stakes, but as noted above, most decisions are improved by briefly considering the different elements of the problem using the information currently available. Quickly working through the structure of the decision can be a cost-effective way of assessing what is already known about the context of the decision (Gregory et al., 2012; Garrard et al., 2017). This may reveal that we already know enough to make the decision or identify issues that need to be investigated further or clarified.

8.3.1 Decision sketching

Decision sketching, also known as rapid prototyping, can be used to outline the components of the decision, providing a simple way for the decision maker to structure their knowledge about the problem and possible solutions (Garrard et al., 2017). It entails asking a series of questions relating to the *who, what, why, how* and *when* of the decision (Figure 8.1, Table 8.2; Schwartz et al., 2018; Hemming et al., 2022). It is often sufficient to enable the decision to be made.

Table 8.2 A list of fundamental questions that can be used to quickly sketch a decision.

Type	Question	Chapter
Who	Who has a stake in the decision?	6
	Who should be involved in making the decision?	6
	Who will be impacted by the decision?	6
What	What is the problem the decision maker is trying to solve? E.g. seabird nests being predated by invasive rodents; or uncontrolled wildfires impacting wildlife, tourism, and water supply.	7
	What is known about the cause of the problem? E.g. non-native rats targeting seabird eggs or illegal burning of heathland.	7
	What options are available to address the problem? E.g. baited traps; or reduce area of reserve scheduled for burning	7
	What are the trade-offs among goals?	7

Type	Question	Chapter
	What resources are available to address the problem?	7
	What are the potential consequences associated with the different choices that could be made?	7
	What prompted the decisions?	7
Why	Why is this an important problem for the decision maker?	7
	Why is this an important problem for stakeholders?	6
How	How confident are they that those options will partially or fully address the problem?	4
	How much will the different options cost?	4
When	When does the decision need to be made?	7

By brainstorming the answers to these questions, decision makers can quickly identify whether they know enough to act confidently. Another advantage of decision sketching is that setting out the important elements of the decision may quickly eliminate potential choices because they are not practical given the resource constraints, or because none of the stakeholders prefer them. Likewise, considering the timeframe or urgency with which the decision needs to be made can prompt decision makers not to wait until they have perfect information (Lindenmayer et al., 2013). Where sketching the decision reveals knowledge gaps that impede a decision, the decision maker must ask how important it is to resolve that uncertainty before they act. In some cases, it may be worth investing time in reducing uncertainty before making a decision (Section 10.6); in other cases 'learning by doing' may be more appropriate (i.e. adaptive management). Because it is a rapid process, sketching the decision can be an iterative process, where issues are identified, knowledge is refined and the decision sketched again with any new information (Garrard et al., 2017).

Failing to consider a wide range of alternatives can risk decision makers defaulting to business as usual, even when that action has not been successful in the past (Gregory et al., 2012). It is likewise important that decision makers consider the consequences of delaying a decision or choosing not to act (Martin et al., 2017). Ideally, they have a strong evidence base from which to identify the range of different options and their likely consequences (Chapter 4). There are many tools that can help decision makers to organise their understanding of the system to identify leverage points that suggest potential actions (e.g. conceptual and mental models; Moon et al. 2019).

8.3.2 Benefit-cost/cost-effectiveness analyses

There are several decision-making tools to help calculate and compare cost-effectiveness of different actions, also known as return on investment (Section 2.4.4). These include economic analysis tools, such as cost-effectiveness analysis and benefit-cost analysis, which require detailed assessments of the costs and effects of actions (Cook et al., 2017). Cost-effectiveness analysis is particularly common when spending public funds but often is useful to identify how to deliver conservation effectively. It is appropriate when outcomes can be distilled into a single measure of utility, or there is a single, relevant criterion. Maximising the conservation outcomes under a given budget requires information on costs to be combined with information on the effectiveness of different actions, and for actions to be prioritised based on the evidence of their cost-effectiveness (Joseph et al., 2009). A conservation project may make claims that proposed actions are cost-effective, which can be supported or refuted by evidence.

There are many possible ways to combine information on effects and costs. For example, cost-effectiveness can be expressed as dollars per unit of conservation outcome. Cost-benefit analysis takes that process one step further, attempting to compare costs with the dollar value of the outcomes or benefits (Burgman, 2005). In this case, outcomes would be compared as the amount of conservation outcome per dollar, or in purely financial terms as the financial gain per dollar spent.

Economic analyses are widely used in healthcare, and whilst less frequently used in conservation, their use is increasing (Pienkowski et al., 2021). Other tools such as the Evidence to Decision tool (Section 9.10.3) encourages consideration of costs alongside other information in decision making (Christie et al., 2022). It is important to consider the direct (e.g. materials) and indirect costs (e.g. staff time) (Section 2.4.4; Iacona et al., 2018).

8.3.3 Clarifying objectives

If the objective is uncertain, or if individuals differ in what they see as the priorities, then there can be disagreement on the way forward. Resolving the purpose may reduce disagreement. Box 8.2 shows how conflicting ideas may be resolved by identifying core objectives.

Box 8.2 Clarifying objectives

Description: Managers of adjacent protected areas discuss their budget priorities for park management. Manager 1 rates weed control and eradication as 'low' in the list of priorities, and Manager 2 rates it as 'high'. The Managers have responsibilities for areas that are equivalent in terms of ecological conditions and the potential for harm from invasive species. The actions taken to control weeds depend on the biology of the weed species. The area manager needs to decide what proportion of their budget should be spent on weed management.

Potential actions: 1. eradicate all invasive species, 2. eradicate all harmful invasive species, 3. eradicate all large invasive species close to visitor areas and walking tracks, 4. monitor and eliminate all newly established invasive species.

Who: Stakeholders include park managers, park visitors, adjacent farmers, local community groups, NGOs focused on invasive species.

What: Values at stake include species of conservation importance, ecosystems, ecological processes that may be affected by invasive species, visitor experience, and revenue from visitors.

When: eradication and control programs over the next 12 months.

Clarify objectives: The objectives may be to 1. minimise the number of new weed species, 2. minimise the extent of all weeds, 3. minimise the ecological impact of all weeds, 4., minimize the potential for the escape of harmful agricultural species, or 5. minimise the visual impact of weeds for park visitors.

Decision: The managers agree to identify the subset of weed species that may harm the park environment, and develop a management plan to minimise the impact of potentially harmful weeds using cost effectiveness analysis. They agree to disregard weed species that have negligible environmental impacts.

Why: Manager 1 knows most weeds in her reserve have limited environmental impacts and have minor effects on visitor experience, even though some have extensive distributions and would be very expensive to eradicate. Manager 2 knows a small handful of invasive species are potentially very harmful in some habitats, but none are of concern for agriculture. The managers agree that the primary objective is to minimise the ecological impact of weeds.

8.3.4 Sketching means-ends networks

Means-ends networks can be used to separate fundamental objectives (the 'ends', the outcomes we want to achieve) from the means objectives (what we do to influence or generate those outcomes) (Gregory et al., 2012). A useful way to start is to collate statements of what is to be achieved, or avoided, in a particular decision context. Examining each statement, fundamental objectives are revealed by asking "why is that important?", and means objectives are revealed by asking "how can that be achieved?" Fundamental objectives help define the reasons for a decision and are the focus of future deliberation and analysis. Means objectives can be used to help generate management options. Means-ends diagrams can be constructed by the decision maker alone, or as part of a participatory process with stakeholders (Section 8.3.2). They can be quickly and easily sketched with the available information and can be a useful communication tool to explain the fundamental objective and point to actions that could lead to that objective being achieved (Moon et al., 2019; Burgman et al., 2021).

8.4 Making Decisions

In reality, most conservation decisions have more than one objective, even if that is to keep within existing budgets and satisfy stakeholders (Possingham et al., 2001). Sometimes these issues are easy to navigate — the action is affordable and stakeholders agree. In other cases, financial, social and political considerations may be at odds (Williams and Kendall, 2017). In these cases, unstructured brainstorming may not reveal acceptable trade-offs among objectives for different alternatives. It can be helpful to use simple tools to lay out the decision based on how the different actions might perform relative to the different considerations (Gregory et al., 2012). This can identify where alternatives perform poorly across all objectives (lose-lose consequences), are costly, ineffective and unpopular. It can hopefully find a win-win solution. At the very least, it will reveal trade-offs that allow decision makers to make explicit choices. For example, the cheapest, most effective management action to deal with invasive plants may be to treat with herbicides but this may be unacceptable to some stakeholders who are concerned about poisons in the environment. The decision maker must then decide whether to take an unpopular action to remove plants manually. Being clear about the consequences allows a trade-off to be identified and whatever the decision is, it can then be communicated (Converse, 2020).

8.4.1 All decisions have consequences

It is important to consider the full range of possible decisions. While all decisions require choices, they do not always need to result in action; choosing *not* to do something can be equally, and sometimes more consequential than choosing to *do* something (e.g. Martin et al. 2012). This is an important distinction and why considering different alternatives is a crucial part of the decision-making process (Gregory and Keeney, 2002). Chapter 7 describes how a decision maker can derive alternatives and Chapters 2, 3 and 4 describe how to interpret the evidence. Sometimes there is good information about the alternative actions available, and the probability they will be effective under different circumstances. More often, there is considerable uncertainty about the consequences of different alternatives and the conditions under which one alternative may perform better than another (Burgman, 2005; Regan et al., 2005). Weighing up the pros and cons of each action can ensure transparency and create opportunities to learn and inform future decisions (Gregory et al., 2012). Where there is a clear understanding of the consequences of doing nothing, and the only alternative has a good chance of a better outcome, then the decision may be simple. Failing to consider the range of alternatives can mean failing to identify an innovative new approach (Christie et al., 2022). Likewise failing to consider the decision context may lead to conflict over the most appropriate management alternative (e.g. the best approach to reducing the impact of feral cats in communities with pet cats). So, while some decisions are no-brainers, in many cases at least briefly considering the essential elements of the decision and justification for the choice made, can ensure decisions are transparent and create opportunities to learn and inform future decisions.

8.5 Multi-Criteria Analysis

Most decisions comprise making a choice based on a range of different criteria such as the cost, likelihood of success, or acceptability to neighbours.

8.5.1 Consequences tables

When the aims have been established, the options that might help address the problem have been shortlisted, and the evidence accessed, the next stage is to present the details of these alternative actions to stakeholders so they can be compared. Presenting this as a consequence table offers a straightforward method to compare the different options and their consequences even if only rough estimates (Gregory et al., 2012). Different actions are likely to have different consequences for the things we care about (the fundamental objectives). The nature of conservation decisions means there is almost always at least one ecological, economic and social aim (Possingham et al., 2001).

The basic structure of a consequence table is a list of objectives, each with an associated performance measure, and a list of alternative actions (Table 8.1; see Keeney and Gregory, 2005). Alternatives should always include doing nothing or continuing doing what is already being done (status quo), as a point of comparison (i.e. are the proposed alternatives better than what is being done?). Each objective needs a specific and practical performance measure to enable comparisons between different actions (Gregory et al., 2012). If possible, performance measures should be appropriate metrics for the subject of interest rather than being proxies or indirect measures (Hemming et al., 2021). Ideally, the consequence table will include evidence-based, quantitative estimates such as the financial cost of implementing the different alternatives or the expected increase in the population of a threatened species (Failing and Gregory, 2003). Social values may be expressed best by constructed scales, such as Likert scales. Where it is not possible to estimate values precisely, it is still valuable to estimate the degree to which the performance measure is likely to increase or decrease under the proposed action, to assess the relative performance of alternatives. A range of different modelling tools (Section 9.11) can generate estimates of how performance measures might respond to different actions. Each cell of the consequence table should be accompanied by a measure of the precision of the estimate, such as a confidence interval, or a subjective credible interval (Hemming et al., 2018).

In addition to helping structure the decision, consequence tables are useful to reveal when alternatives are not genuinely different from one another. It can also become clear if an alternative performs poorly across all objectives or at least performs poorly relative to the other alternatives and so it is not worth further consideration (e.g. Walshe and Hemming, 2019). Considering the consequences of different alternatives across all of the objectives can also reveal where there are unavoidable trade-offs and we will have to choose which objectives to prioritise (Converse, 2020; Moon et al., 2019).

It can be difficult to define and classify fundamental objectives unambiguously. This is best done by creating a values hierarchy (Figure 8.3) that classifies the means, objectives and measures. The specification of measures helps to clarify what is meant by them. When making tradeoffs, it is important that nothing has been omitted, leading to hidden agendas in negotiations (for example, omitting political considerations may distort discussions about other objectives). It is also important to ensure nothing is double counted (for example, the fundamental objective of conserving threatened species should be kept separate from the objective of maintaining ecosystem services). Objectives hierarchies provide a means for ordering thinking about fundamental objectives, serving to distinguish means from ends, ensuring that all relevant values at stake are considered, and that nothing is double counted.

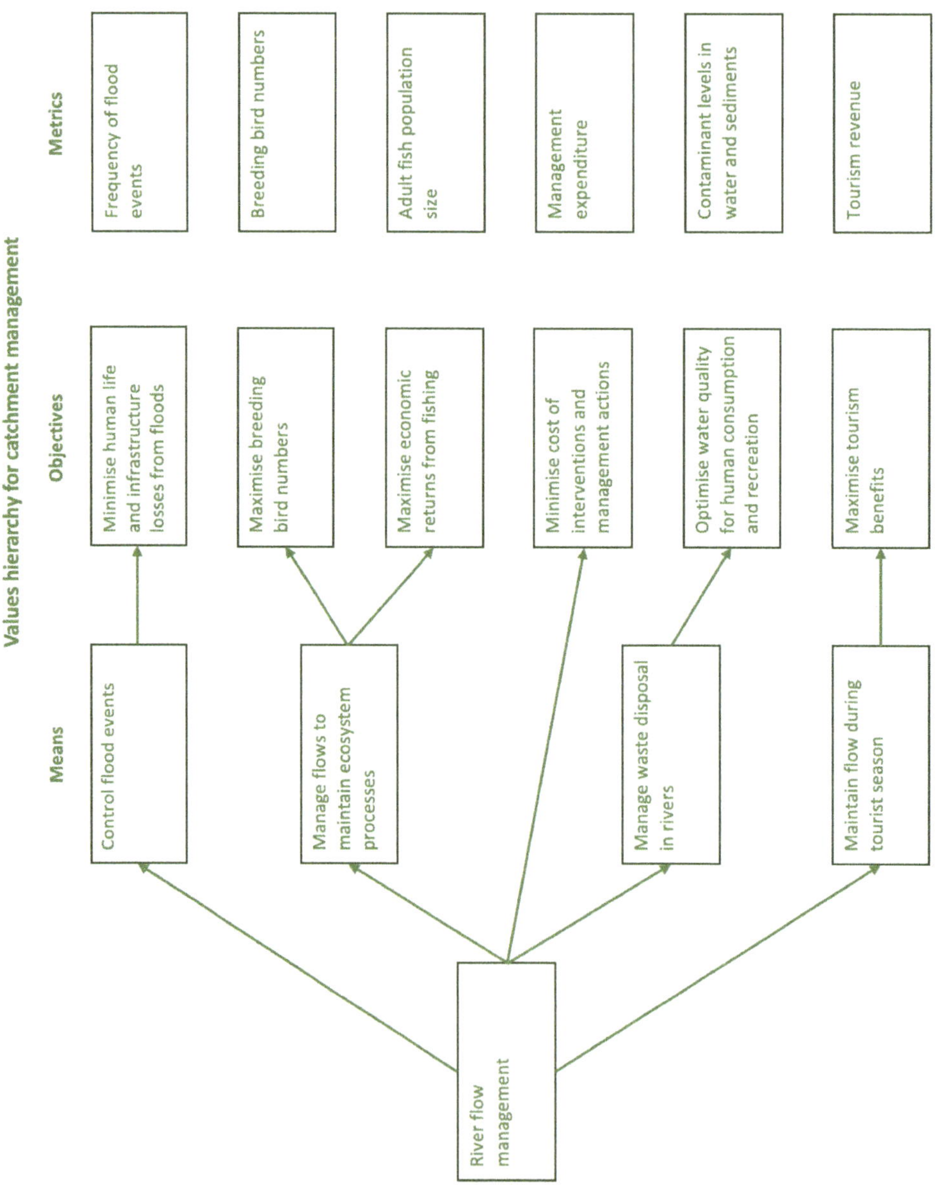

Figure 8.3 A values hierarchy for a catchment management example. (*Source:* authors)

Consequence tables are not just a valuable tool to compare alternatives and assess trade-offs among objectives, they contain the essential elements of a decision. Constructing the consequence table can be a valuable process in itself. These need to be based on evidence for outcomes (Chapter 4).

An example of a consequences table is provided in Table 8.3. How is it then possible to make a decision?

8.5.2 Checking for a dominant alternative

A dominant alternative is one that is better than all the other options being considered across all criteria. For example, option C in Table 8.3 might have been the superior option had it not been so expensive. If a dominant alternative option emerges then the decision could be considered resolved. There is no dominant alternative option in this scenario, so we need to reduce the number of alternatives based on the project aims and agreed trade-offs.

8.5.3 Revealing hidden values

All stakeholders and decision makers have unique perspectives on a decision and what is to be gained or lost from the outcomes of actions. For example, Vucetich et al. (2021) identified a wide range of motivations for conservation actions including care for future generations, present-day fairness, and utilitarian goals (Vucetich 2021). Values are amplified by cultural norms and world views (e.g., Riepe et al. 2021). Thus, for example, in managing a park, some stakeholders may oppose additional visitors because of their impacts, while others welcome visitors who pay to use facilities, some may welcome the return of stock grazing because it encourages a specific, valued bird species, while others oppose stock because fencing hinders dog walking and stock disturb wetland soils.

Typically, such values are captured and represented in values hierarchies and consequence tables. However, stakeholders may fail to disclose values for several reasons including because they were not asked about them, they did not anticipate them being affected by potential actions, or they did not know how to express them. Hidden values can derail discussions because stakeholders will negotiate for actions by weighting other factors correlated with their unstated objectives and preferred outcomes. They may be responsible when support for an action is equivocal, even though the consequence table suggests that there is a dominant alternative. Hidden values may be revealed by exploring options and objections through facilitated discussion of the consequence table, or in private discussion with a facilitator.

8.5.4 Excluding unacceptable options

Although all of the options in Table 8.3 were considered acceptable during the initial stages of the process it is apparent that some are now unacceptable for various reasons. Based on the project scope, budget and aims, it is clear from the analysis shown in Table 8.3 that, although beneficial, option C is unacceptably expensive — the cost exceeds the allocated funds. Furthermore, option F would result in an unacceptable increase in flood risk to communities downstream so is also excluded. This leaves five options.

Table 8.3 A consequences table for seven different river management options (including no change with the river continuing to deteriorate) assessed under six criteria. Evidence is shown in relation to the current status.

	Performance measure	A Do nothing	B Original DWA proposal	C Full scheme	D District suggestion	E Wildlife group proposal	F Grazers proposal	G Fisheries group proposal
Proposal		Status quo	Creating flood reservoir	Restore original water features	Connecting river with surrounding meadows	Creation of pools	Enhance river banks, plant hedges	Restore shingle, fish passes, nesting island
Ecological	Breeding bird numbers	No change	Moderate evidence of minor benefits	Moderate evidence of considerable benefits	Moderate evidence of considerable benefits	Moderate evidence of moderate benefits	Strong evidence of moderate benefits	Moderate evidence of considerable benefit
Economic	Fish numbers	No change	Weak evidence of moderate benefits	Strong evidence of considerable benefits	Moderate evidence of moderate benefits	Moderate evidence of moderate benefits	Overwhelming evidence of minor harms	Moderate evidence of considerable benefits
Water management	Frequency of downstream flood events	+5%	-60%	-80%	-60%	-10%	+200%	-20%
Water quality	Sediment in column	-10%	+10%	+40%	+10%	+10%	+5%	-10%
Cost		0	$$	$$$$$	$$	$$	$$$	$$
Tourism interest	Number non local visitors	Low	Very low	Low	No access	Low	Low	Very low

8.5.5 Excluding redundant criteria

In the example scenario in Table 8.3, the number of tourists is low and better catered for by other projects elsewhere in the valley. Tourism is not a priority for this project, so it is decided it will be excluded. The process of excluding redundant criteria can be used repeatedly as the table is simplified.

8.5.6 Removing dominated alternatives

The consequence Table 8.4 is simpler following the removal of two columns and one row. The next task is to try and simplify it further.

Dominated alternatives are options that are not superior to another in any criterion (Hemming et al., 2022). Option B is an inferior option that is equally as effective in reducing flood risk, for the same costs, as D but it has poorer ecological benefits (Table 8.4). Option E is also excluded as it has the same ecological benefits as option G but lower flood and water quality benefits. Options B and E are thus dominated alternatives and excluded. At each stage, we can exclude redundant criteria (see Section 8.5.4): water quality does not vary between the remaining options so is excluded.

As the table gets reduced more detail may be researched and added, for example, in Table 8.5 there is now a more detailed estimate of costs and the ecological consequences.

The retained options all have a unique feature (and are thus not dominated). Option A is retained as it is cheaper than all others; D is retained as it has the highest flood protection; while G is retained as it has the highest ecological benefits. It is now necessary to consider the relative importance of these benefits.

Practical dominance

Imagine in the table that there is an option with a slightly better water quality measure but is much worse in all other comparisons. The decision might be made that this small difference does not outweigh the other disadvantages. For practical reasons it is dominated — hence practical dominance — and the option is eliminated.

8.5.7 Addressing trade-offs

Trade-offs arise when there is no dominant option, and the consequences vary in important ways across the options under consideration. Decision makers need to balance two issues: how much the fundamental aims vary over the alternatives; and how much stakeholders care about what they stand to gain or lose from each option. In doing so, they aim to maximise the expected net benefit or to minimise the chances of unacceptable outcomes.

There are four main ways of addressing trade-offs:

Table 8.4 As for Table 8.3 but with options C (expensive), F (flooding) and the tourism criterion (not important) all removed

	Performance measure	A No change	B Original DWA proposal	D District suggestion	E Modified DWA proposal	G Fisheries group proposal
Ecological	Breeding birds	No change	Moderate evidence of minor benefits	Moderate evidence of considerable benefit	Moderate evidence of moderate benefits	Moderate evidence of considerable benefit
Economic	Annual fish survey	No change	Weak evidence of moderate benefits	Moderate evidence of moderate benefits	Moderate evidence of moderate benefits	Moderate evidence of considerable benefit
Water management	Flood events	+5%	-60%	-60%	-10%	-20%
Water quality	Sediment in column.	-10%	+10%	-10%	+10%	-10%
Cost		0	$$	$$	$$	$$

Table 8.5 As for Table 8.4 but with dominated options B and E removed along with water quality (as no longer differs). Details on costs, birds and fish researched and added.

	Performance measure	A No change	D District suggestion	G Fisheries group proposal
Ecological	Breeding birds	No change	An estimated 3–15 further pairs of lapwing and a very likely (90–95%) chance that cranes will start nesting	An estimated 3–15 further pairs of lapwing and a very likely (90–95%) chance that cranes will start nesting
Economic	Annual fish survey	No change	Likely to be a 5–20% increase in the main fish populations	Likely to be a 50–100% increase in the main fish population and eels about as likely as not (40–60% chance) to return
Water management	Flood events	+5%	-60%	-20%
Cost		0	$120k	$130

Apply weights to criteria

If each element of the table is given a score then one approach is to decide how important each criterion is and give each a weight reflecting their (agreed) value or utility (Gregory et al., 2012). Each score for an option is then multiplied by the relevant weight and the scores are summed to give a total score (or utility).

There are some tools for assigning weights. Analytic Hierarchic Processing is a process that entails asking a series of pairs of questions (what is the relative importance of sediment in the water against breeding birds), which then generates weights. If the uncertainties associated with the outcomes of each potential course of action are well understood then sensitivity analyses may be used to identify the options that are most likely to avoid unacceptable outcomes. This could be done by exploring 'what if' scenarios (e.g. Borsuk et al., 2003) or using models (Burgman, 2005). Weights can also be elicited from stakeholders (Walshe and Slade, 2020). In practice assigning weights is difficult (when buying lunch what is the relative importance of price, quantity, and taste?) but quickly resolves with specific examples (do you prefer this smaller tasty option or this equal price, bland but more substantial alternative?).

Common currencies

If each criterion can be converted into a common currency (usually money but can be biodiversity, time or risk) then the total net benefit of each can be calculated (e.g. cost-benefit analysis; Section 8.7). Thus the example could assess the cost of a flooding event and multiply it by the probability. The value of the water quality could be assessed just by the amount the water company might have to pay to clean the water or the value to the community of not having water with high sediment loads. Similarly, the total value of a change in fish or bird population can be assessed. With this approach, the total worth of each column can simply be added. In some cases, this is difficult (what is the value of a lapwing?) and some consider it unethical to equate environmental or social outcomes with money.

Negotiate a consensus

The consensus can be agreed upon by those involved through a process called deliberative decision making. In doing so, each stakeholder contemplates what they stand to gain or lose from each of the actions, and how much they are willing to forgo for one criterion to achieve gains on another. They also consider what others stand to gain or lose.

Even swaps

Even swaps entails considering what reduction in benefit of X would be needed to cancel out the gain in Y. For example, by asking whether people would sacrifice access along the river if there is a high probability (say 65%) of herons reoccupying the colony.

In the example, it could be asked how much people would swap an increase in flood protection for an increase in the fish population. It might be decided that flooding is really unpleasant to the community but that they would agree to a minor increase in risk for an improved fishery — perhaps 2% for moderate benefits and 5% for considerable benefits (Table 8.6).

Table 8.6 As for Table 8.5 but with D's moderate gain in fish swapped for a 2% reduction in flood risk and G's considerable benefits in fish swapped for a 5% reduction in flood risk.

	Performance measure	A No change	D District suggestion	G Fisheries group proposal
Ecological	Breeding birds	No change	An estimated 3–15 further pairs of lapwing and a very likely (90–95%) chance that cranes will start nesting	An estimated 3–15 further pairs of lapwing and a very likely (90–95%) chance that cranes will start nesting
Water management	Flood events	+5%	-58%	-15%
Cost		0	$120k	$130

The next question might be the worth of the extra flood reduction. This can be in simple financial terms such as damage caused. It can also be in terms of reputation. Suppose it is decided that each percentage reduction in flood risk is worth $5k (Table 8.7).

Table 8.7 As for Table 8.6 but with A's increase in flood risk considered equivalent $5k cost, D's reduction in flood risk considered equivalent to $290k savings and G's equivalent to $75k savings.

	Performance measure	A No change	D District suggestion	G Fisheries group proposal
Ecological	Breeding birds	No change	An estimated 3–15 further pairs of lapwing and a very likely (90–95%) chance that cranes will start nesting	An estimated 3–15 further pairs of lapwing and a very likely (90–95%) chance that cranes will start nesting
Cost		+$5k	-$170k	+$55

Option D is then a dominant alternative and selected. This approach of even swaps can be carried out for any sized table. It is better to start with the easy choices and hope that a solution appears before needing to make the more difficult choices.

8.5.8 Adding new options

Ideally, considering a wide range of alternative actions enables a decision maker to identify an option that will achieve the best outcomes for the subjects of interest. Often, there is no single option that achieves this (Gregory et al., 2012; Walshe and Hemming, 2019). The process of considering the various options, with their strengths and weaknesses, and confronting what the other stakeholders may gain or lose from each alternative option may lead to suggesting new alternative options.

It can be valuable to ask stakeholders to discuss the options and to suggest new courses of action that may satisfy all parties (Moon et al., 2019). In our scenario above, option D was preferred although it was not ideal for fish. A new alternative (H) was then suggested that provided greater benefit to fish but was slightly more expensive (Table 8.8). This was agreed as the preferred option.

Table 8.8 The consequence table with the preferred option D but with a new option (H) added, which is now considered the overall preferred option.

	Performance measure	D District suggestion	H District suggestion with meanders
Ecological	Breeding birds	3–15 further pairs of lapwing and a very likely (90–95%) chance that cranes will start nesting	6 further pairs of lapwing and a very likely (90–95%) chance that cranes will start nesting
Economic	Annual fish survey	A 5–20% increase in the main fish populations	A8 60–120% increase in the main fish population, eels likely (70–90% chance) to return and possible but unlikely (<30%) chance that salmon will return
Water manage ment	Flood events	-60%	-60%
Cost		$120k	$125k

The District suggestion (D) recommends a strategy that results in 3–15 pairs of lapwings. The best guess is that 10 pairs will establish. Suggestion H suggests that 6 additional pairs of lapwings will establish. Because appropriate ecological conditions are more certain, the prediction under suggestion H is more certain. Stakeholders or decision makers may prefer the expectation of 6 pairs over an expectation of 10 pairs, if the latter includes a more uncertain outcome that could be as low as 3 pairs. Thus, participants may trade an expectation of higher utility for an expectation of lower utility that nevertheless has a more reliable outcome. Attitude to risk is an inherent part of decision-making (Burgman, 2005; Cinner and Barnes, 2019), only possible if uncertainty is made explicit in consequence tables. In general, analyses that focus on detrimental or unacceptable outcomes are the domain of formal risk analysis.

8.5.9 Converting the table to ranks

Where possible, performance measures in a consequence table should be given realistic values. It is also possible to indicate broadly expected changes, such as a general increase in one objective and a general decrease in another. The consequence table can be confusing with a lot of information and a high number can indicate a beneficial (water clarity) or negative (costs, flood risk) outcome. It can be helpful to rank the options in each row (with 1 being the most favourable). The objectives that are most critical could also be placed at the top of the table. It is often clear as to which columns and rows can then be removed. Alternatively, it is possible to indicate the broadly expected changes, such as an increase in one objective and a general decrease in another and see if that leads to an answer without the need for more complex analysis.

8.6 Strategy Table

When there is a range of possible alternatives it can be useful to create a strategy table that organises alternatives into logical packages (see Table 8.9). This can be a useful way to see which alternatives are fundamentally different, rather than just different versions of the same thing.

Table 8.9 Strategy table for an imaginary series of programmes. The different options are placed under broad categories, such as habitat management. These options are then brought together in a set of strategies.

Strategies		**Elements**		
	Habitat management	**New habitat**	**Reintroductions**	**Monitoring**
Maintain existing status	Retain existing practice	Status quo	None	Monitor status in existing sites
Reverse decline	Reintroduce native herbivores	Status quo	None	Monitor status in existing sites
Return to former status	Reintroduce native herbivores	Identify suitable sites	Establish propagation	Monitor status in existing sites
	Recreate habitats Restore open areas	Create agreements with landowners	Programme Transplant individuals	Test success of management of new sites Monitor success of reintroductions

8.7 Classifying Decisions

8.7.1 Decisions within decisions

Problems or decisions are often spoken about as if they are a single choice, but conservation decisions are generally more like a series of linked choices. First, there is a choice about whether we need to act, often followed by a string of choices about what to do, and when and where to do it, that branch from the initial decision (Gregory et al., 2012). Mapping out the decision can help reveal the decisions within decisions and the important branch points that might lead to different considerations or trade-offs.

For example, if we decide to control an environmental weed because it is impacting something we care about, this leads to a choice about which method to use to remove the weed. Applying herbicide and physical removal can both control the weed, but physical removal is more expensive. If we choose physical removal then we can only treat half the area we could treat if we used herbicide. This leads to a decision about where to prioritise action. Herbicide needs to be applied annually, while physical removal is only needed every two or three years. We now need to decide when to act, to invest more resources now or in the future. This decision will influence whether we treat a smaller area with physical removal or treat a larger area with herbicide but reserve some of our budget to re-treat the same area. If we are not able to treat the same area of weed infestation then we may have to choose priority areas to target. Physical removal can be conducted all year round but herbicide application can only occur at certain times of the year. So the decision about how to act also creates a decision about when to act.

Decision sketching can help identify the different decisions and sub-decisions (Garrard et al., 2017). Given the nested or linked nature of many of these sub-decisions, tools like decision trees (see Figure 8.4) can be useful to identify the branch points and the different pathways and end points for the broader decision. Tools like Bayes Nets (Section 4.7.7) can also be useful to consider these linked decisions, by assigning probabilities to the different branch points.

8.8 Decision Trees

Decision trees are a means of identifying which action to adopt depending upon the circumstances by providing a series of choices. They have the merit of simplifying what separates the different options and creating a sequence of decisions that will identify the appropriate action. They are created by bringing together the range of possible actions, then removing those that are ineffective or inferior under all conditions to another. Ideally the higher choices are those that divide up the group more equally so producing a shorter tree. Figure 8.4 shows a decision tree created from the evidence for the effectiveness of different actions from Conservation Evidence for treating the invasive aquatic plant *Crassula helmsii*.

Multiple decision trees can be used to identify trade-offs between completing actions (Oliver et al., 2012). These could be taken to the stakeholders to help achieve consensus (Section 8.12).

8.9 Creating Models

Models are useful ways of representing the world. They can clarify thinking by representing the elements of a problem and revealing assumptions about how the different components of a system are connected and interact to produce outcomes. They can be used to explore possible outcomes and give indications of future outcomes. Sometimes by challenging our assumptions, models give surprising results, and provide unexpected insights as to how a system functions.

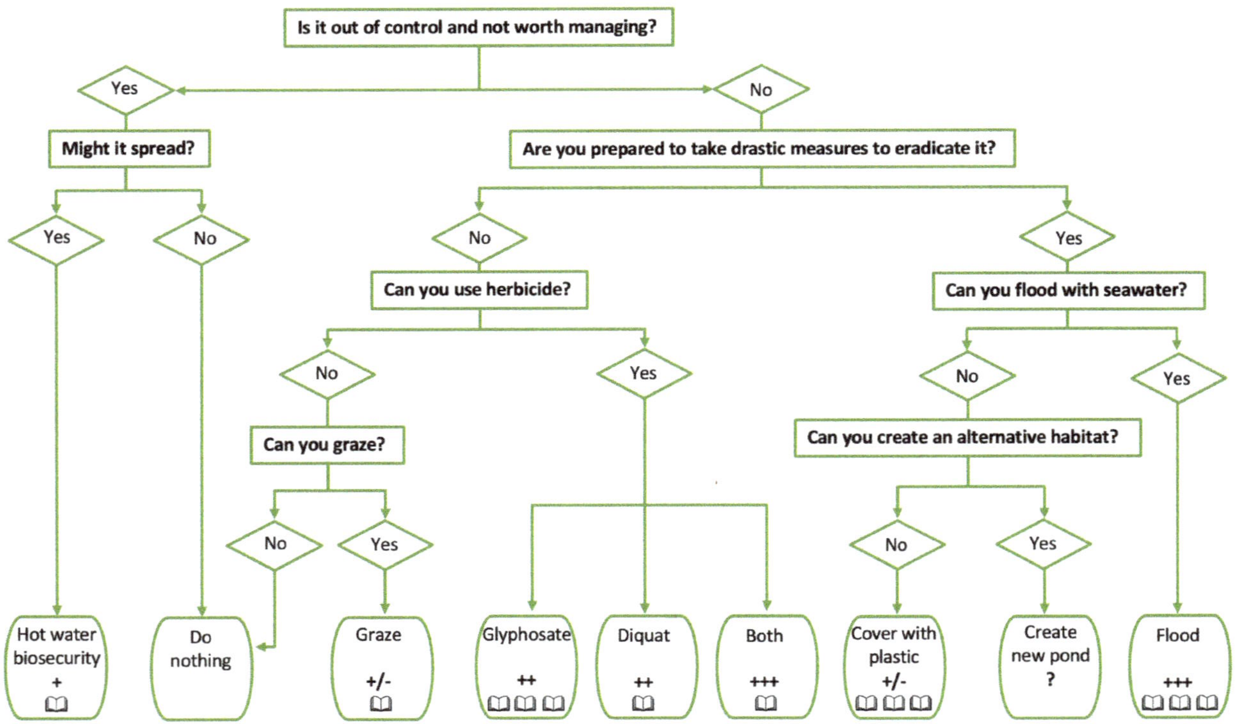

Figure 8.4 A generalised decision tree to consider options for controlling the invasive plant *Crassula helmsii*. ? = no evidence; - - - = considerable harm; - - = moderate harm; - = minor harm; 0 = no effect; +/- = mixed effect; + = little benefit; ++ = moderate benefit; +++ = considerable benefit, 🕮🕮🕮🕮🕮 = overwhelming evidence; 🕮🕮🕮🕮 = strong evidence; 🕮🕮🕮 = moderate evidence; 🕮🕮 = weak evidence; 🕮 = negligible evidence. (*Source*: authors)

However, models are only as good as the information on which they are constructed. It is therefore important that the assumptions and information on which models are built are not just accepted. This can be a particular problem for quantitative models, where numbers can be derived from a computer model and accepted without question.

Models can be simple representations of a system, that allow the complexity of a system to be reduced to an appropriate level of simplicity by focusing on the most relevant aspects and making explicit assumptions about how actions lead to outcomes. Previous sections have already presented three ways in which concepts can be linked together. Mind maps (Section 4.7.4) are an informal means of representing the links between a wide range of ideas. Means-ends networks (Section 8.3.2) provide a relatively informal means of sketching the proposed actions (means) to the desired outcomes (ends). The Theory of change (Section 7.5.2) can be a well structured means of planning a proposal by showing the links between elements. Miradi (Section 9.10.2) is a well accepted means of presenting theories of change. In making decisions, conceptual and/or explicit qualitative or quantitative models provide the links between casual observations, data, and the all-important consequences of decisions. Ideally, the underlying choices of the models are explicit and can be subject to confirmation with data, and cross-examination with theory.

Decision makers and stakeholders generally have implicit models of how a system works. Helping individuals to make those 'mental models' explicit (e.g. by creating a diagram) can have a wide range of benefits (Moon et al., 2019). Firstly, eliciting mental models can help individuals to recognise the assumptions they are making about a system (Moon et al., 2017). When done as part of a group process, eliciting mental models provide an opportunity to create a conceptual model of how a system works based on the collective knowledge of a group (e.g. Colvin et al., 2016). This group process can also help to create a shared vision for how to achieve a particular outcome. But it can also help groups with different perspectives to understand the assumptions and preferences and knowledge of other stakeholders, even if they do not share those views. By sharing mental models, it is possible to share knowledge, but also to correct misconceptions, permit solutions to be negotiated and help in conflict resolution by providing people with an opportunity to share their point of view based on their own knowledge and experiences (Moon et al., 2019).

Figure 8.5 shows the conventional process of building models to support decision making. It also shows the type of decision making adopted by the processes described in this book in which evidence use and expert elicitation play major roles. Evidence from all available sources provides the platform for developing ideas of cause and effect, transforming those ideas into equations and parameters, and validating the assumptions made in developing the model and estimating its parameters. Model building may conclude with a conceptual model that guides the creation of a consequence table. Alternatively, it may lead to a suite of different mathematical and statistical models for some or all of the elements in a consequence table. These ideas were introduced in Section 2.7.

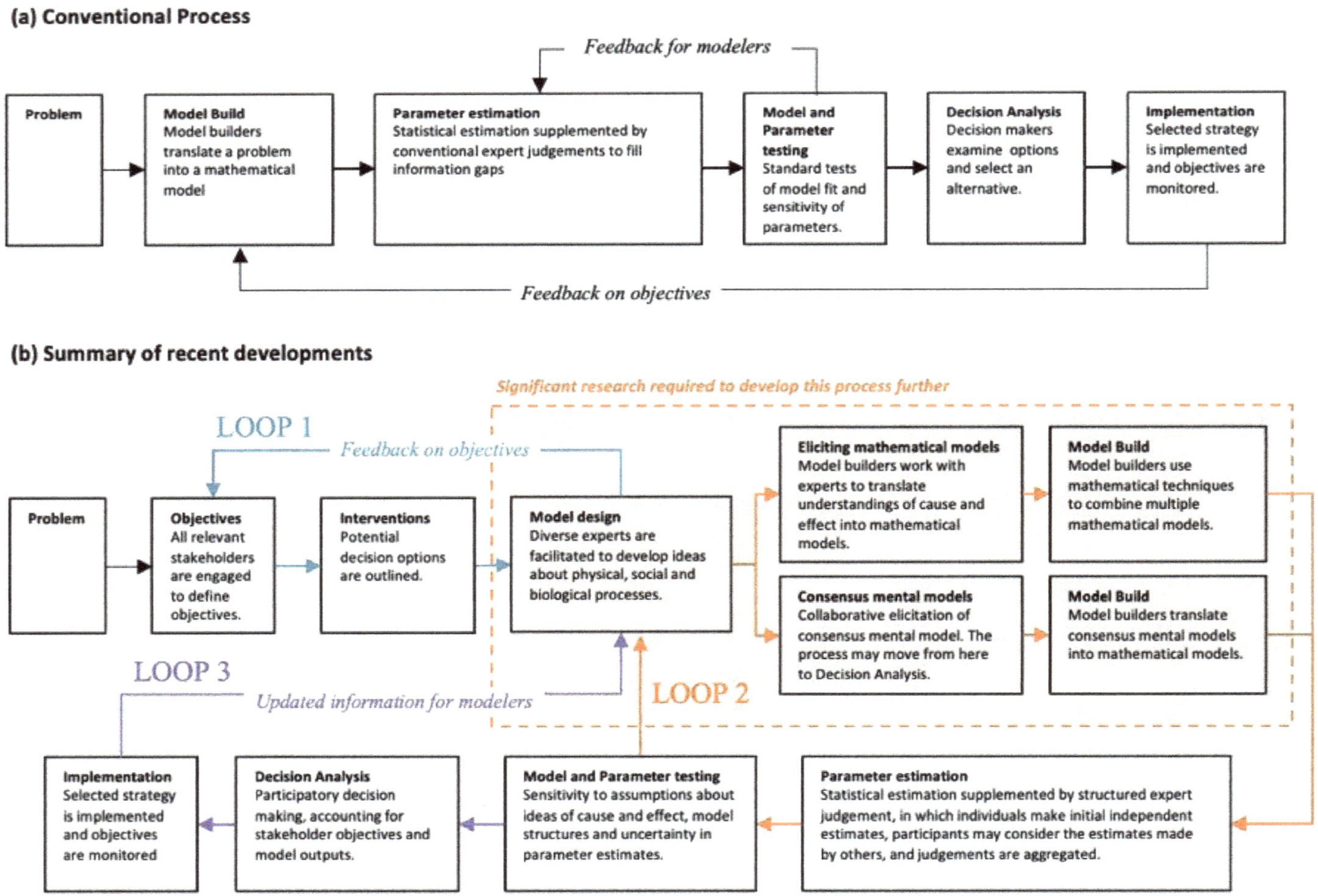

Figure 8.5 Steps in model development for decision making, summarising (A) processes typically deployed in most model developments, and (B) emerging approaches. (*Source:* modified from Burgman et al. 2021, which provides references for each step, CC-BY-4.0)

8.10 Achieving Consensus

Using the processes for stakeholder engagement presented in Chapter 6, the decision maker needs to present the different options and allow all stakeholders to evaluate the trade-offs to reach a consensus. It is most challenging when high-stakes decisions involve stakeholders with conflicting objectives. Behavioural consensus is effective when participants agree to negotiate to resolve conflict (Valverde, 2001).

The stages described previously can be approached at a range of scales from just two individuals to a range of teams. Workshops are especially useful when issues are complex and require the extraction of a wide set of evidence and views, if community involvement is key, or if there are substantially differing views, perhaps including conflict. The consequence tables described above are a key element in negotiations. They summarise what is to be expected from each management option, providing a clear picture of what each participant stands to gain or lose from each action. The fundamental objectives encapsulate what participants care about. Once dominant and dominated alternatives and redundant objectives have been removed, the remaining table focuses participants on the core issues. The measures provide unambiguous representations of these key objectives.

The facilitator's role is to look for areas in which participants may be willing to concede gains in the interests of finding a mutually acceptable, consensus solution. This requires participants to decide what they can tolerate, rather than seek an optimal outcome from their perspective. Valverde (2001) suggested a framework to achieve consensus that has its foundations in an approach developed by Kaplan (1992; see Burgman, 2005). It involves decomposing arguments (as in argument maps) into basic elements: claims, evidence, models that link claims to data and assumptions, and objections. Disagreements may be about facts, theories, data, or reasoning. Sources of disagreement may be semantic interpretations, different preferences and values, opinions about the validity of evidence, or adherence to different theories of cause and effect. Closure or resolution may be achieved through sound argument, agreeing that a particular position is 'best', or reaching a resolution that is acceptable to the participants and that is 'fair' rather than correct or optimal. Sometimes a conflict dissipates and is resolved because it turns out that the differences between actions are trivial. Participants may also trade the surety of an outcome for a particular objective for the expectation of the magnitude of the outcome. Thus, a participant may agree to a smaller expected adult fish population each year, if the variation in expected fish population size is also smaller or if the proposal includes marketing resulting in higher prices.

8.10.1 Dealing with conflict

Conflict arises when there is a difference in opinion between individuals or organisations about which options should be analysed, which objectives should be included, how options should be assessed, whether change is necessary or which trade-offs should be made (Redpath et al., 2013). The underlying principle is to discover the compromises that are most acceptable to the key stakeholders. This may include agreeing that some objectives are not essential or accepting that an option, whilst being ideal for one group, is unacceptable to others but that an alternative option is acceptable to both. In some cases the conclusion might be to acknowledge the conflict and accept that a compromise will not be found by the participants.

Conservation conflicts will likely increase (Redpath et al., 2013) and, where they do arise, it is likely due to a lack of communication and/or lack of understanding about the impacts of a particular action (Minderman et al., 2019). At a minimum, using transparent processes for decision-making can help clarify the information and rationale behind a decision.

Open and transparent processes, such as identifying the problem, gathering the evidence, and early engagement with stakeholders, can increase trust and help avoid misunderstanding. Sharing mental models can be useful to identify differences in the assumptions made by different groups. Making these assumptions clear at the start provides the opportunity to identify preferences and correct misconceptions (Moon et al., 2019).

References

Bennett, N.J., Di Franco, A. and Calò, A. 2019. Local support for conservation is associated with perceptions of good governance, social impacts, and ecological effectiveness. *Conservation Letters* 12: e12640, https://doi.org/10.1111/conl.12640.

Borsuk, M.E., Stow, C.A. and Reckhow, K.H. 2003. An integrated approach to TMDL development for the Neuse River Estuary using a Bayesian probability network model (Neu-BERN). *Journal of Water Resources Planning and Management* 129: 271–82, https://doi.org/10.1061/(ASCE)0733-9496(2003)129:4(271).

Burgman, M., Layman, H. and French, S. 2021. Eliciting model structures for multivariate probabilistic risk analysis, *Frontiers in Applied Mathematics and Statistics* 7: 1–10, https://doi.org/10.3389/fams.2021.668037.

Burgman, M. 2005. *Risks and Decisions for Conservation and Environmental Management.* (Cambridge: Cambridge University Press).

Cawson J.G., Hemming, V., Ackland, A., et al. 2020. Exploring the key drivers of forest flammability in wet eucalypt forests using expert-derived conceptual models. *Landscape Ecology* 35: 1775–98, https://doi.org/10.1007/s10980-020-01055-z.

Ceballos, G., Ehrlich, P.R., Barnosky, A.D., et al. 2015. Accelerated modern human–induced species losses: Entering the sixth mass extinction. *Science Advances* 1: e1400253, https://doi: 10.1126/sciadv.1400253

Christie, A.P., Downey, H., Frick, W.F., et al. 2022. A practical conservation tool to combine diverse types of evidence for transparent evidence-based decision-making. *Conservation Science and Practice* 4: e579, https://doi.org/10.1111/csp2.579.

Cinner, J.E. and Barnes, M.L. 2019. Social dimensions of resilience in social-ecological systems. *One Earth* 1: 51–56, https://doi.org/10.1016/j.oneear.2019.08.003.

Colvin, R.M., Witt, G. B. and Lacey, J. 2016. Approaches to identifying stakeholders in environmental management: Insights from practitioners to go beyond the 'usual suspects'. *Land Use Policy* 52: 266–76, https://doi.org/10.1016/j.landusepol.2015.12.032.

Converse, S.J. 2020. Introduction to multi-criteria decision analysis. In: *Structured Decision Making: Case Studies in Natural Resource Management*, ed. by M.C. Runge, et al. (Baltimore: Johns Hopkins University Press).

Cook, C.N., Pullin, A.S., Sutherland, W.J., et al. 2017. Considering cost alongside the effectiveness of management in evidence-based conservation: A systematic reporting protocol. *Biological Conservation* 209: 508–16, https://doi.org/10.1016/j.biocon.2017.03.022.

Domning, D.P. 1978. Sirenian evolution in the North Pacific Ocean. *University of California Publication in Geological Sciences* 118: 1–176.

Failing, L. and Gregory, R. 2003. Ten common mistakes in designing biodiversity indicators for forest policy. *Journal of Environmental Management* 68: 121– 32, https://doi.org/10.1016/S0301-4797(03)00014-8.

Garrard, G.E., Rumpff, L., Runge, M.C., et al. 2017. Rapid proto-typing for decision structuring: An efficient approach to conservation decision analysis. In: *Decision-Making in Conservation and Natural Resource Management*, ed. by N. Bunnefeld, et al. (Cambridge: Cambridge University Press).

Gregory, R.S. and Keeney, R.L. 2002. Making smarter environmental management decisions. *Journal of the American Water Resources Association* 38: 1601–12, https://doi.org/10.1111/j.1752-1688.2002.tb04367.x.

Gregory, R., Failing, L., Harstone, M., et al. 2012. *Structured Decision Making: A Practical Guide to Environmental Management Choices* (Hoboken: John Wiley & Sons).

Guiler, E. and Godard, P. 1998. *Tasmanian Tiger: A Lesson to be Learnt* (Perth: Abrolhos Publishing).

Hemming, V., Burgman, M.A., Hanea, A.M., et al. 2018. A practical guide to structured expert elicitation using the IDEA protocol. *Methods in Ecology and Evolution* 9: 169–80. https://doi.org/10.1111/2041-210X.12857.

Hemming, V., Camaclang, A.E., Adams, M.S., et al. 2022. An introduction to decision science for conservation. *Conservation Biology* 36: e13868, https://doi.org/10.1111/cobi.13868.

Hockings, M. 2003. Systems for assessing the effectiveness of management in protected areas. *BioScience* 53: 823–32, https://doi.org/10.1641/0006-3568(2003)053[0823:SFATEO]2.0.CO;2.

Iacona, G.D., Sutherland, W.J., Mappin, B., et al. 2018. Standardized reporting of the costs of management interventions for biodiversity conservation. *Conservation Biology* 32: 979–88, https://doi.org/10.1111/cobi.13195.

Joseph, L.N., Maloney, R.F. and Possingham, H.P. 2009. Optimal allocation of resources among threatened species: a project prioritization protocol. *Conservation Biology 23:* 328–38, https://doi.org/10.1111/j.1523-1739.2008.01124.x.

Kaplan, S. 1992. 'Expert information' versus 'expert opinions.' Another approach to the problem of eliciting / combining / using expert opinion in PRA. *Reliability Engineering and System Safety* 35: 61–72, https://doi.org/10.1016/0951-8320(92)90023-E.

Keeney, R.L. 2004. Making better decision makers. *Decision Analysis* 1: 193–204, https://doi.org/10.1287/deca.1040.0009.

Keeney, R.L. and Gregory, R.S. 2005. Selecting attributes to measure the achievement of objectives. *Operations Research* 53: 1–11, https://www.jstor.org/stable/25146843.

Knight, A.T., Cook, C.N., Redford, K.H., et al. 2019. Improving conservation practice with principles and tools from systems thinking and evaluation. *Sustainability Science* 14: 1531–48, https://doi.org/10.1007/s11625-019-00676-x.

Lindenmayer, D.B., Piggott, M.P. and Wintle, B.A. 2013. Counting the books while the library burns: Why conservation monitoring programs need a plan for action. *Frontiers in Ecology and the Environment* 11: 549–55, https://doi.org/10.1890/120220.

Martin, T.G., Nally, S., Burbidge, A.A., et al. 2012. Acting fast helps avoid extinction. *Conservation Letters* 5: 274–80, https://doi.org/10.1111/j.1755-263X.2012.00239.x.

Martin, T.G., Camaclang, A.E., Possingham, H.P., et al. 2017. Timing of protection of critical habitat matters. *Conservation Letters* 10: 308–16, https://doi.org/10.1111/conl.12266.

McShane, T.O., Hirsch, P.D., Trung, T.C., et al. 2011. Hard choices: Making trade-offs between biodiversity conservation and human well-being. *Biological Conservation* 144: 966–72, https://doi.org/10.1016/j.biocon.2010.04.038.

Minderman, J., Cusack, J.J., Duthie, A.B., et al. 2019. Decision trees for data publishing may exacerbate conservation conflict. *Nature Ecology & Evolution* 3: 31818, https://doi.org/10.1038/s41559-019-0804-7.

Moon, K., Blackman, D.A., Adams, V. M., et al. 2017. Perception matrices: An adaptation of repertory grid technique. *Land Use Policy* 64: 451–60, https://doi.org/10.1016/j.landusepol.2017.03.023.

Moon, K., Guerrero, A.M., Adams, V.M., et al. 2019. Mental models for conservation research and practice. *Conservation Letters* 12: e12642, https://doi.org/10.1111/conl.12642.

Oliver, T.H., Smithers, R.J., Bailey, S., et al. 2012. A decision framework for considering climate change adaptation in biodiversity conservation planning. *Journal of Applied Ecology* 49: 1247–55, https://doi.org/10.1111/1365-2664.12003.

Pienkowski, T., Cook, C., Verma, M., et al. 2021. Conservation cost-effectiveness: A review of the evidence base. *Conservation Science and Practice* 3: e357, https://doi.org/10.1111/csp2.357.

Possingham, H.P., Andelman, S.J., Noon, B.R., et al. 2001. Making smart conservation decisions. In: *Conservation Biology: Research Priorities for the Next Decade,* ed. by M.E. Soule, M.E. and Orians, G.H. (Washington: Island Press).

Pullin, A.S and Knight, T.M. 2009. Doing more good than harm — Building an evidence-base for conservation and environmental management. *Biological Conservation* 142: 931–34, https://doi.org/10.1016/j.biocon.2009.01.010.

Redpath, S.M., Young, J., Evely, A., et al. 2013. Understanding and managing conservation conflicts. *Trends in Ecology & Evolution* 28: 100–09, https://doi.org/10.1016/j.tree.2012.08.021.

Regan, H.M., Ben-Haim, Y., Langford, B., et al. 2005. Robust decision-making under severe uncertainty for conservation management. *Ecological Applications* 15: 1471–77, https://doi.org/10.1890/03-5419.

Riepe, C., Liebe, U., Fujitani, M., et al. 2021. Values, beliefs, norms, and conservation-oriented behaviors toward native fish biodiversity in rivers: evidence from four European countries. Society & Natural Resources, 34, 703–24, https://doi.org/10.1080/08941920.2021.1890865

Schwartz, M.W. 2008. The performance of the Endangered Species Act. *Annual Review of Ecology, Evolution, and Systematics* 39: 279–99, https://doi.org/10.1146/annurev.ecolsys.39.110707.173538.

Schwartz, M.W., Cook, C.N., Pressey, R.L., et al. 2018. Decision Support Frameworks and Tools for Conservation. *Conservation Letters* 11: e12385, https://doi.org/10.1111/conl.12385.

Sutherland, W.J. 2019. Kaizen Conservation? *Oryx* 53: 397–98. https://doi.org/10.1017/S0030605319000619.

Valverde, L.J. 2001. Expert judgement resolution in technically-intensive policy disputes. In: *Assessment and Management of Environmental Risks,* ed. by I. Linkov and J. Palma-Oliveira (Dordrecht: Kluwer Academic).

Vucetich, J.A., Bruskotter, J.T., van Eeden, L.M., et al. 2021. How scholars prioritize the competing values of conservation and sustainability. Biological Conservation, 257:.109126. https://doi.org/10.1016/j.biocon.2021.109126

Walshe, T.V. and Hemming, V. 2019. *Gabo Island Structured Decision-Making: Report on outcomes from a workshop held October 2011 and penguin surveys conducted 2008–2016.* (Unpublished Report, School of Biosciences, The University of Melbourne).

Walshe, T. and Slade, S. 2020. Coral reef fin fish spawning closures for coral reef fin fish. In: *Structured Decision Making: Case Studies in Natural Resource Management*, ed. by M.C. Runge, et al. (Baltimore: John Hopkins University Press).

Williams, P. J. and Kendall, W.L. 2017. A guide to multi-objective optimization for ecological problems with an application to cackling goose management. *Ecological Modelling* 343: 54–67, https://doi.org/10.1016/j.ecolmodel.2016.10.010.

9. Creating Evidence-Based Policy and Practice

Harriet Downey[1], Nicolas Boenisch[2], Des Browne[3,4], Yvonne Buckley[5], Gemma Cranston[6], Lynn V. Dicks[7], Jonathan Hughes[8], Charles Latrémouille[9], William H. Morgan[10], Matthew Muir[11], David O'Brien[12], Nancy Ockendon[13], Dilys Roe[14], Stefan Schindler[15], Tom Sheldon[16], Eleanor R. Tew[17]*

Transforming conservation depends on evidence being embedded within decision-making processes. This chapter presents general principles for embedding evidence into a wide range of approaches and processes. These include creating action plans (habitat or species), guidance documents, funding applications, policy, business environmental strategies, and management plans, as well as deciding what to fund, what to report or how to construct models. How well evidence is used in these processes can be evaluated through various processes and indices.

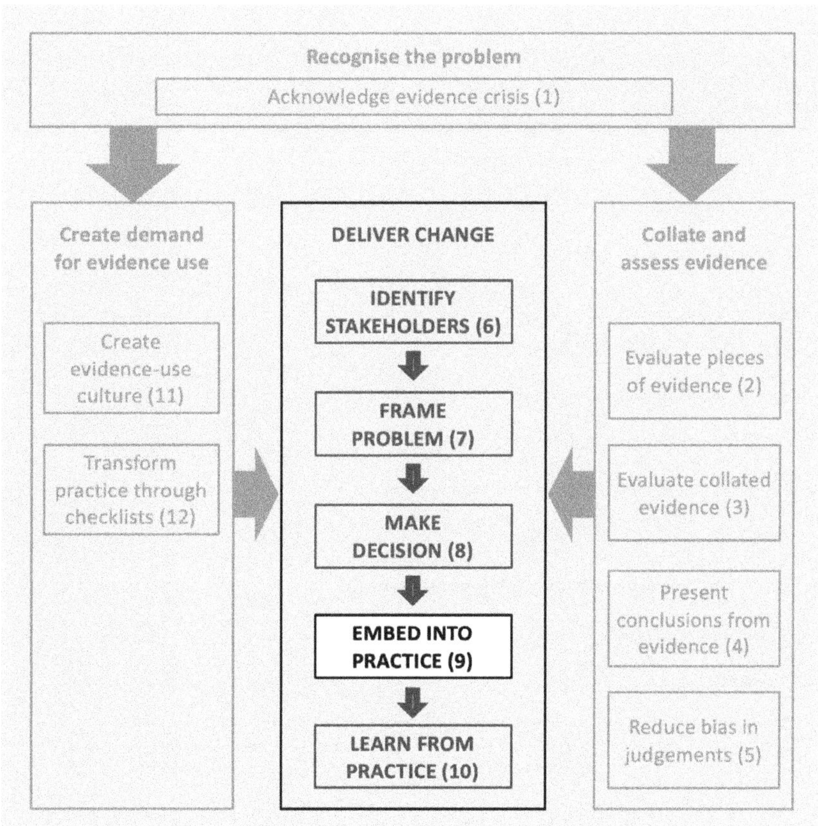

 https://doi.org/10.11647/OBP.0321.09

[1] Woodland Trust, Kempton Way, Grantham, Lincolnshire, UK
[2] Foundations of Success Europe, 7211 AA Eefde, The Netherlands
[3] Houses of Parliament, Parliament Square, London SW1A 0PW
[4] Biosecurity Research Initiative at St Catharine's College (BioRISC), St Catharine's College, Cambridge, UK
[5] School of Natural Sciences, Trinity College, Dublin, Ireland
[6] University of Cambridge Institute for Sustainability Leadership, The Entopia Building, 1 Regent Street, Cambridge CB2 1GG, UK
[7] Department of Zoology, University of Cambridge, The David Attenborough Building, Pembroke Street, Cambridge, UK
[8] UN Environment World Conservation Monitoring Centre, 219 Huntingdon Rd, Cambridge, UK
[9] Partner, Conservation Measures Partnership (CMP), Montreal, Canada
[10] Conservation Science Group, Department of Zoology, University of Cambridge, The David Attenborough Building, Pembroke Street, Cambridge, UK.
[11] United States Fish and Wildlife Service International Affairs, 5275 Leesburg Pike, Falls Church, VA 22041, USA
[12] NatureScot, Great Glen House, Leachkin Road, Inverness, UK
[13] Endangered Landscapes Programme, Cambridge Conservation Initiative, The David Attenborough Building, Pembroke Street, Cambridge
[14] International Institute for Environment and Development, 235 High Holborn, London UK
[15] Science Media Centre, 183 Euston Road, London, UK
[16] Forestry England, 620 Bristol Business Park, Coldharbour Lane, Bristol, UK

** The findings and conclusions in this chapter are those of the author and do not necessarily represent the views of the US Fish and Wildlife Service.*

Contents

9.1 How Embedding Evidence Improves Processes

A common barrier to effective, evidence-based decision making is that many processes and systems do not appropriately integrate the available relevant evidence. This includes drafting policy recommendations, producing written guidance, writing management plans or devising organisational policies.

There are many benefits of adopting evidence-based processes, including:

- More effective decisions and actions. Transparently reviewing the evidence enables an assessment of the likely efficacy of the proposal.
- Retention of institutional knowledge. When staff leave an organisation, information and institutional memory may also be lost. Embedding processes that capture the information that is used to guide decision making minimises the risk of knowledge erosion or loss.
- Self reflection and self challenge. Creating time to pause and reflect on approaches can help break the cycle of 'doing things the way they have always been done', instead identifying where new approaches might be more effective.
- Filling knowledge gaps. Identifying key gaps in knowledge that underpins evidence may encourage further evidence creation.
- Skills building. Including the consideration of evidence in decision-making processes will often provide staff with new skills. As staff build these skills, evidence-based processes will naturally become part of everyday thinking and help effect cultural change (Chapter 11).
- Improved consistency and accountability. This is often useful when decisions need to be justified, for example to funders. Having consistent transparent processes makes this much clearer and provides a means of improving future decisions.
- Efficient actions and savings. By enabling more effective actions to be put in place, evidence use can reduce the resources required to reach given goals, thus saving money.
- Risk Reduction. Reducing the unsatisfactory outcomes, which may also lead to bad publicity or reputational damage.
- Improved morale. Staff and volunteers can be motivated by having the knowledge that similar interventions have been successful in the past.

There are, however, major barriers faced by conservation practitioners and policy makers when using evidence. Walsh et al. (2019) highlighted the particular importance of organisational structure, decision-making processes and culture, together with practitioner attitudes and relationships between scientists and practitioners. Additional specific barriers include assuming guidelines and advice are based on science; poor databases or dysfunctional information management systems; limited access to relevant evidence; confusing decision-making processes; deficient or absent planning processes that ensure use of evidence and no process to publish or document outcomes. These can all result in poor decision making, missing windows of opportunity from being insufficiently prepared, poor investments, and information and institutional memory being lost when staff leave (Possingham et al., 2012; Wilson et al., 2009). An effective way of ensuring that evidence is at least considered in decision making is to embed evidence review within organisational and governance processes. Fortunately, many such processes exist, as outlined in this chapter. Adopting evidence in these processes can help deliver a shift towards organisational-level evidence-based practice and create a culture of evidence use (see Chapter 11).

Processes in practice

Processes that automatically embed evidence into decisions need to be simple and operate smoothly, to be affordable and sustained in the long term. Systems that add significant workload without obvious benefit are likely to detract motivation for evidence-based practice. Of course, learning and adopting these processes will take some initial investment but the long-term result will be more robust, reliable, and effective end-products that enable quicker and more efficient updates. This in turn will motivate staff and volunteers, and increase the confidence of funders and project partners

9.2 General Principles for Embedding Evidence into Processes

9.2.1 When, what and how much evidence?

In any process embedding evidence there are some general questions that need answering.

When is evidence needed?

Not all the decisions need to be evidence-checked (Keeney, 2004; Hemming et al., 2022, Sutherland et al., 2021). As explained in Section 8.2, it may not be worthwhile to seek evidence for minor interventions, where the evidence is well known, or where the outcome is obvious.

Evidence is worth checking further where there is uncertainty of the outcome and the decision is important because of the consequences or the scale of investment or other sensitivities (e.g. perhaps there are public concerns around erecting fences or increasing populations of carnivores).

As discussed in Chapter 7, this can be thought about in terms of the assumptions or claims that are made when designing a project (e.g. in a Theory of Change approach). Some of these claims will be very critical to the success of the project, others may be inconsequential. Evidence should be focused on the claims being made where we are unsure of the evidence base and which are critical for success.

What evidence is needed?

As described in Chapter 2, the evidence base for a conservation decision may come from published literature (books or journals), manuals and reports (grey literature), previous personal experience, or be received through word of mouth from colleagues, peers, expert advisors, or local knowledge. Evidence may be on a mixture of subjects including the status of biodiversity, threats, stakeholder values, costs, and effectiveness of action. The two most important characteristics of each piece of evidence are its reliability and its relevance; both these components need consideration when assessing how much weight to put on a particular piece of evidence in the decision-making process (see Chapter 2). The type of audience will also determine the type and relevance of evidence; decision makers in environmental agencies may be interested in robust biological data but this may be meaningless to decision makers in, for example, ministries of finance or planning who will want to understand issues such as numbers of jobs created, returns on investment or contributions to GDP. It is critical, thus, to understand the political economy of decision making in order to understand the type of evidence that may help influence policy and practice (Bass et al., 2021).

How much evidence?

The level of detail of evidence needed for each decision will also depend on reliability and relevance. For example, a planned action to create nesting platforms at a lake to encourage breeding ospreys may require a few sentences to describe the relevant evidence, summarising previous studies on platform usage and breeding success. In comparison, a costly proposal to reintroduce a large-herbivore population to a new site is likely to warrant a detailed description of the status and population trends of the species, the suitability of the habitat, and the probability of success of the proposed translocation (including the source and number of animals, the method of translocation and the release approach). In such examples, it may be pertinent to include evidence about the context, importance, and feasibility of a proposed action. This could include information about acceptability, cost, logistical practicality, and the availability of necessary equipment and expertise.

How to present evidence

Box 9.1 outlines general principles for collating and presenting evidence, based upon Downey et al., (2022). Following these principles should become easier as information becomes more accessible, evidence synthesis becomes more common, and the culture changes described in Chapter 11 result in increased training in evidence use. These principles underpin all the approaches described in subsequent sessions.

When presenting evidence, referencing should be clear and consistent throughout the document, including in the methodology. If a recommendation is based upon expert opinion this should be stated. Chapters 2–4 describe how to assess and summarise evidence.

In some cases, it is inappropriate to give references, for example in leaflets or popular documents. In these cases, the evidence base can be alluded to by an appropriate choice of phrasing (Table 9.1). However, even if references are not included in a document, it is important that they are available should stakeholders wish to see them. Doing so increases trust and helps reduce the research needed for subsequent projects. With an increasing move to online information, these references could be included on project websites, with a related QR code or link on the leaflet.

Box 9.1 General principles for presenting evidence

Collating and assessing evidence to support decision-making

1. The evidence base underlying key claims should be reviewed and where available incorporated when formulating recommendations on specific actions.
2. The types of evidence assessed should be made clear.
3. Presentation and interpretation of the evidence should be neutral.
4. Any bias and limitations of the reviewed evidence base should be stated explicitly. This should include any language biases where appropriate and what sources of evidence were consulted.
5. Where possible, assess and report on the cost (financial and other), cost-effectiveness, side effects and acceptability of potential conservation actions.

Deciding upon actions, and making recommendations

1. Specify the type and source of evidence used to make recommendations.
2. The strength of the evidence behind recommendations should be transparent.
3. Make explicit where recommendations have been made in the absence of effectiveness information.
4. Communication of evidence should be tailored to the audience for whom it is intended.
5. Make explicit where recommendations are based on factors besides the evidence of effectiveness (e.g. costs, social acceptability).

Table 9.1 Examples of wording to describe different evidence support when omitting evidence sources (Modified from Downey et al. 2022)

Evidence base	Examples of wording
Evidence assessed and straightforward	Evidence shows that .... Scientific studies indicate that .... The accumulated evidence to date demonstrates ....
Evidence assessed but results mixed	Most evidence supports the idea that .... The evidence is equivocal but overall seems .... The evidence suggests it is likely that ....
Evidence shows no effect	The evidence shows no benefit from .... The science shows no support for ....
Evidence assessed but recommendations based on other factors	Evidence shows that ... However, due to high financial costs the action is not recommended.
No or weak evidence	There are no studies testing .... There is little research about .... There are no studies testing ... But because of low risk and cost, practitioners should consider .... There are studies but they have flaws or exhibit serious bias that makes findings dubious ....
No evidence but experience used	Although no scientific evidence exists, practitioner experience suggests ... Our experience is .... It seems likely that...
No evidence considered but uncontroversial	Do not treat shrubs in bird nesting season. Attach boxes by bands rather than nails if the tree is grown for timber.

9.3 Evaluating Evidence Use

9.3.1 Evidence-based capability maturity models

Organisations can gradually increase their capacity and their commitment toward evidence-based decision making. Capability Maturity Models (CMM) can help organisations initiate discussions on where they are now, where they want to be, and how to get there. CMMs usually depict 4–5 levels of maturity for a given discipline, outlining the characteristics applying at each level. The levels describe an evolutionary improvement from ad-hoc immature processes to disciplined high-quality effective processes (Stewart, 2016). Table 9.2 shows a CMM developed to help organisations improve their evidence-based practices. This can be used at various scales, but is most effective for organisations, programmes, or operational regions where it can encourage discussion about means of improving practice (Stewart, 2016).

Table 9.2 Evidence Use Capability Maturity Model (Adapted from Stewart, 2016)

#	Level	Characteristics
4	Embedded	Projects and programmes routinely adapted based on new evidence. Information systems enable analysis, reporting on the use of evidence, and learning. Workflows embedding evidence are well defined and efficient.
3	Adopted	Standard evidence-based decision-making practices are used widely and consistently, with strong leadership support. Evidence-based decision-making practices are appropriately resourced within projects.
2	Developing	Organisation adopts evidence-based decision making best practices, defined processes and information management systems, but use is voluntary. Evidence use is not fundamental to the organisation.
1	Initial	Organisation has no consistent way of using evidence or documenting its use. Processes are often reinvented for different projects. Evidence is not readily accessible. Culture of 'do the best you can'. Success depends on individual efforts.

Currently, many organisations would sit at level one. Moving to the higher levels may require an increase in capacity, training or investments in processes alongside commitments to routinely use the sorts of processes described in this chapter.

9.3.2 Learning agendas

Learning agendas, also known as evidence-building plans, are a tool for organisations to coordinate and elevate the use of evidence in policy, budget preparation and decision making. In developing a learning agenda, organisations identify and prioritise the questions that, once answered, are likely to have the biggest positive impact on organisation functioning, performance, value creation and impact generation. Learning agendas can thus be defined as a set of prioritised research questions and activities that guide an organisation's evidence-building and decision-making practices (Nightingale et al., 2018).

Learning agendas in government

In the United States, government agencies are increasingly expected to develop and use learning agendas to identify priority questions and engage with the public and other key stakeholders on evidence. Guidance and requirements for learning agendas were developed following the 2018 law, *The Foundations for Evidence-Based Policymaking Act* (abbreviated as the Evidence Act), which advances data and evidence-building functions in the US government and builds on more than a decade of legislative actions to strengthen federal evidence-building (GAO, 2019). Learning agendas are described by the US Office of Management and Budget (OMB) as the 'driving force for several of the activities required by and resulting from the Evidence Act,' including the development of annual evaluation plans and assessments of an agency's ability and infrastructure to carry out evidence-building activities (OMB, 2019). Box 9.2 describes a process for agenda development, intended as a cycle for continuous learning. Additional resources and US agency guidance include the *Evidence Act Toolkit: A Guide to Developing Your Agency's Learning Agenda* (GSA OES, 2020), the OMB memo *Evidence-Based Policymaking: Learning Agendas and Annual Evaluation Plans* (OMB, 2020), and www.evaluation.gov.

Box 9.2 Creating a learning agenda

Adapted from the GSA Office of Evaluation Sciences (2020) *Learning Agenda Toolkit.*

Engage with key stakeholders. Receive input on priority questions, document any learning activities, and understand available data, tools, methods, and analytic approaches.

Identify priority questions. Select the questions that, when answered, will have the biggest impact on organisational functioning and performance.

Write the learning agenda. Incorporate feedback and input gathered through stakeholder engagement activities, align with organisational strategic goals and objectives, and solicit leadership input and buy-in.

Implement and update. Carry out the activities specified in the learning agenda and update accordingly.

9.3.3 Evidence-use index

To meet the expectations and requirements established under the Evidence Act and related policies, two US agencies are routinely incorporating evidence within programme implementation. The international conservation programmes of the US Fish and Wildlife Service have proposed annually reporting an index of the percentage of financial awards implementing at least one action for which there is evidence of effectiveness. In this context, an action with evidence of effectiveness will be defined as a conservation intervention that has been categorised as Effective/Beneficial or Likely to be Effective/Beneficial by an independent public repository of evidence (e.g. Conservation Evidence www.conservationevidence.com), or similarly assessed by systematic review (e.g. CEEDER database environmentalevidence.org/ceeder/about-ceeder, or the Evidence for Nature and People Data Portal https://www.natureandpeopleevidence.org/#/).

Another US agency, AmeriCorps, has been tracking a similar metric since 2017. Through a programme that provides strategic grants to organisations that engage in service to address local and national challenges, AmeriCorps has increasingly invested in what works by supporting programmes with moderate to strong levels of evidence. Since establishing a baseline percentage of 29% in 2017, AmeriCorps programmes have annually improved and, by 2021, 68% of competitive grant funding was invested in programmes where interventions were underpinned by evidence (Figure 9.1). AmeriCorps uses evidence to allocate programme resources effectively, such that over time, 'more grant dollars were awarded to applicants with strong and moderate levels of evidence... for proposed interventions, and fewer grant dollars were awarded to applicants with little to no evidence of effectiveness' (CNCS, 2020).

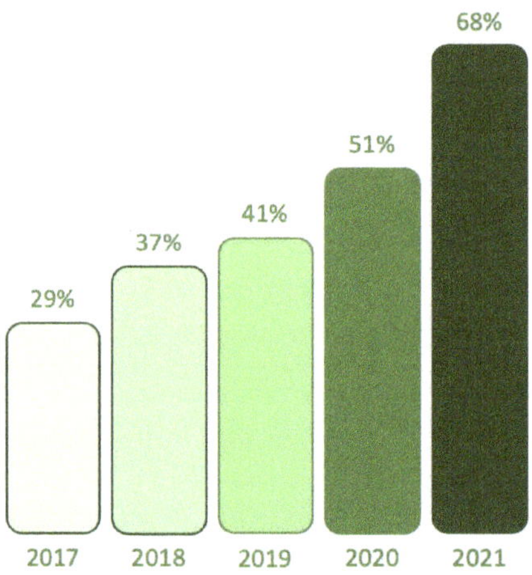

Figure 9.1 The percentage of ASN competitive funding awarded to projects with moderate or strong levels of evidence of effectiveness. (*Source*: authors using data from CNCS, 2020)

These are innovative approaches using indices for measuring evidence use within organisations. Such metrics of evidence use can help drive improvements across an organisation and are likely to encourage further experiments and collation of evidence.

9.4 Evidence-Based Species and Habitat Management Plans

A management plan is a fundamental tool to translate objectives into practice. The process of writing a management plan allows management options and targets to be explored and defined, and the resultant plan guides practical interventions. A management plan is also an important record of intentions, which can be referred back to in the future and retained through staff successions to ensure continuity in management. It may also be a requirement of funding. Using evidence appropriately throughout the creation of the management plan helps substantiate the rationale for management decisions and ensures that the plan is robust, and well informed with the best chance of delivering its objectives.

The following sections describe key components in the process of creating an evidence-based management plan.

9.5 Evidence-Based Guidance

Guidance provides a practical means of obtaining synthesised advice on which interventions are effective, and how to implement them, often without the need to undertake time-consuming literature searches (Brancalion et al., 2020). Studies have shown that practitioners often assume that guidance is based on the scientific literature (Walsh et al. 2019), but this may not be the case.

A recent review of UK and Irish conservation guidance documents revealed a range of problems. These included a lack of referencing making explicit what evidence had been used, insufficient evidence to support the rationale for the recommendation; outdated guidance (often over ten years old); lack of clear methodology and explanation on how recommendations were derived; and an absence of consideration of uncertainty in the key evidence (Downey et al., 2022).

To improve the process, Downey et al. (2022) provide a detailed explanation of how guidance creation can become more evidence-based. The main idea is to adopt the principles shown in Box 9.1 so that the justification of all main claims is transparent and uses the types of phrasing suggested in Table 9.1. This would allow the reader to know what underpins the statement and where the advice is more speculative. One approach is to provide a user-friendly version, alongside a link to a more in-depth report that details the methodology and provides the underlying evidence (e.g. Cruickshanks, 2018).

Box 9.3 Preparing an evidence-based plan

Defining the system

Defining and describing the scope of the system is a fundamental first step. This includes describing the system extent (location, site and land tenure), physical conditions (hydrology, geology and soils, topography), biological conditions (habitat types, flora and fauna) and socio-cultural conditions (archaeology, human history, human use, legal considerations and socio-cultural considerations). This description is vital as it underpins decisions about the objectives and determines which management interventions are possible. Both current and future environmental conditions need considering, for example by integrating local climate change projections and potential land developments. A range of tools and evidence sources can assist this, such as publicly available geospatial datasets, but, in most cases, these freely available data need to be enhanced with targeted surveys and analysis to add further detail to key features and conditions of the system.

Defining objectives

Setting the vision and objectives for the management of a system helps to define the rationale for the plan and vice versa. The scope of the objectives can be narrow (such as promoting the conservation of one target species) or broad (such as maximising the natural capital value of the site). More often than not, a site will have multiple objectives and uses. Ideally, these should be considered together in the same plan, so that the wider effects of actions to enhance different objectives can be evaluated. The relative priorities of different objectives should also be specified.

The evidence underpinning the objectives should be stated. For example, the reason for focusing conservation efforts on a target species may be the importance of the site for the wider population or the prioritisation of specific ecosystem services may be based on a strategic assessment of the potential natural capital value of the site. The inclusion of evidence must be impartial and, where possible, include an unbiased assessment of other potential objectives for the site and reasoning for their exclusion.

Planning interventions

Describing the practical interventions that will ultimately realise the plan's objectives will often make up the bulk of the management plan. These detailed actions should follow logically from the definition of the system and objectives. A range of interventions should be considered and their efficacy and inclusion evaluated based on evidence (Chapter 4). Conservation Evidence is a crucial resource here, alongside other sources such as the wider scientific literature, evidence-based guidance, and expert knowledge. Ideally, the management plan should state the sources of evidence for each intervention and include a confidence rating of its likely effectiveness (see processes described in Chapter 4).

Monitoring and evaluation

As described in Chapter 9, the management plan should describe the desired results from the practical interventions (based on evidence), and include indicators of success that can be monitored throughout, and ideally after, the timespan of the plan. These can be used to evaluate the effectiveness of the plan in delivering the overall objectives. Review points may also be specified to prompt a re-assessment of the plan and to revise and update its contents if necessary. Sharing and publishing results from testing interventions will contribute to the evidence base, so can be incorporated into future plans.

Using templates

The quality of management plans, and whether they are used at all, greatly varies. For example, a recent survey of British woodland owners found that a third did not have a forest management plan at all, and 69% did not have a management plan compliant with the UK Forestry Standard (Hemery et al, 2020). Creating a management plan can be simplified by using templates, such as the Forestry Commission Woodland Management Templates https://www.gov.uk/guidance/create-a-woodland-management-plan, which can be created for sectors, organisations, or habitats.

As an example of how evidence can be collated and converted into guidance, The European Commission sought to support EU Member States in conserving farmland birds and wider biodiversity. Conservation of farmland birds is a Birds Directive obligation that is essential to the success of the EU Green Deal and the EU Biodiversity and Farm to Fork Strategies. The Birds@Farmland Initiative was launched in 2020 to co-develop evidence-based conservation schemes for farmland birds and their habitat with experts, agricultural and environmental authorities, farmers, NGOs and other relevant stakeholders. Development of the schemes followed a standardized approach and Member States were invited to consider these schemes for their National Strategic Plans of the EU's Common Agricultural Policy (CAP), instruments that allow ambitious Member States to improve the environmental performance of the CAP (Pe'er et al., 2019).

Twenty-two conservation schemes were developed in ten Member States (Austria, Bulgaria, Czech Republic, Finland, France, Germany, Hungary, Italy, Portugal, and Spain), with each scheme focusing on one of ten specific agricultural systems or one of fifteen specific flagship farmland bird species (https://bit.ly/farmlandbirds). The schemes outlined the measures that could be adopted, their technical and financial implications, and their likely ecological and social consequences. The availability of collated evidence was essential for an effective co-development of this guidance. As shown in Table 9.3, if evidence was not readily available, then literature searches and expert opinion helped fill the gaps.

Table 9.3 Evidence used during different stages of creating agricultural schemes to benefit biodiversity including whether the necessary information is already collated and easily available (YES, PARTLY, NO) and how evidence gaps were filled.

Task	European scale	National scale (EU Member States)	Scale of particular agricultural systems in Member States
Occurrence and predominance of agricultural systems.	NO. Data were finally derived by combining European datasets on land cover, land use intensity and biogeography.		
Occurrence and population trends of bird species.	YES. Data from European Bird Atlas, Reporting for Article 12 of the Birds Directive, IUCN Red List of threatened species.		NO.
Most relevant pressures and threats for birds.	YES. Reporting for Article 12 of the Birds Directive.	PARTLY. Expert opinion and literature (including evidence from single case studies, other areas, and other scales).	
Most relevant conservation measures for birds.	YES. Conservation Evidence.	PARTLY. Expert opinion and literature (including evidence from single case studies, other areas, and other scales).	
Social and economic factors, making schemes for biodiversity/birds successful.	PARTLY. Literature.	PARTLY. Expert opinions of national authorities, farmers, NGOs, etc.	

Task	European scale	National scale (EU Member States)	Scale of particular agricultural systems in Member States
Performance of existing bird conservation schemes under the CAP.	YES. Overall CAP evaluations.	NO. In most Member States few schemes exist and existing ones are not sufficiently evaluated.	
Co-benefits for other species.	PARTLY. Conservation Evidence.	NO. Expert opinion (and literature) used.	
Co-benefits for farmers and for wider society.	NO. Expert opinion (and literature) used.		

9.6 Evidence-Based Policy

The advantages to basing policy on evidence may seem obvious, but despite this, evidence is not always used and evidence use can be mixed. Evidence that supports a popular policy may be stretched beyond its scope, while contrary evidence can be ignored or even suppressed (e.g. Hutchings, 2022). This is because policy-making is inherently political rather than scientific or technical (Bass et al., 2021). There is, however, increasing pressure from society through traditional and social media, citizen activism and from politicians themselves for policies to be grounded in sound science (reviewed in Parkhurst, 2017). Whilst scientific evidence usually informs policy, it does not necessarily determine policy development (Gluckman et al., 2021).

One of the major advantages of using evidence, from a policy point of view, is that any intervention that evidence suggests has worked in the past will be more appealing to risk-averse policy makers than untried approaches. Furthermore, solid evidence for a problem or intervention provides justification should decisions be questioned or if the policy is ineffective.

Chapter 1 describes the policy hexagon in which evidence is refined as it is incorporated into the decision making.

Suggestions for enhancing the uptake of evidence in policy creation include the following:

- Identify the broad change sought (e.g. conserving a species or site).
- Identify the specific change sought (e.g. species listed as protected, planning decision changed, modification of agri-environment scheme).
- Take advantage of any policy windows (Rose et al., 2020) resulting from an identified problem, public concern, a pledge to act or legislative horizon scanning to identify forthcoming legislation (Sutherland et al., 2022).
- Legislation and regulation typically follow a consistent series of papers and decisions. Understand the process, including which reports or papers underpin each stage, and how to embed evidence in these. For example, in the UK parliament, 'Green Papers' outline thinking and alternatives used for seeking views from interested parties,

'White Papers' present proposals for legislative changes, and a 'Bill' is the proposed legislation.

- Determine the key opinion formers, both inside and outside the policy-making institution, as well as who makes decisions.
- Understand stakeholder motivations, relationships and power dynamics, the 'political economy' of decision making, in order to improve understanding of when, how, and by whom information is used (Hou-Jones et al., 2021). For example, policy makers need to show change within the timescale of their mandate.
- Get into the decision maker's mindset. What is politically advantageous? What will be seen to be popular? What is an appropriate response to current concerns? What will be seen to appeal to a higher self and not appear selfish? What is likely to be financially beneficial to society?
- Identify where evidence can make a difference such as identifying problems, elaborating on the consequences of issues or showing the effectiveness, or not, of possible actions.
- Change typically requires a mix of publicity to drive a concern, lobbying through credible organisations and attracting political support.
- Ensure language used fits local cultural norms (Barnes and Parkhurst, 2014; Rose, 2014).
- Build on successes. Evidence of successful interventions may also reinforce policy makers' appetite for continued investment.

At a global level, key drivers of conservation policy include the Convention on Biological Diversity (CBD), the Intergovernmental Science-Policy Panel on Biodiversity and Ecosystem Services (IPBES) and the International Council for the Exploration of the Seas (ICES). Each of these produces reports that are heavily grounded in evidence and are transparent, with methods published in multiple languages and sources of data clearly referenced. The assessments tend to be associated with texts often thousands of pages long. Whilst aiming to provide authority, Sutherland (2013) pointed out that they can undermine the venture if they contain inaccuracies, or are unclear about the reliability and relevance of underlying evidence sources. One approach is to break down the assessment into more focused reports on the key issues and carefully assess the evidence for each, using some of the processes described in earlier chapters on evidence assessment, use of experts and decision making.

9.7 Evidence-Based Business Decisions

The success of any business is determined by the effectiveness of the strategy it follows. The strategy sets out the business's vision and objectives. Management plans and policies can be set to support the delivery of these objectives and inform business decision making. An appropriate evidence base can inform these plans and guide them in delivering upon the prescribed objective. This is the case in standard business operations, but also when considering environmental and climate-related impacts and actions.

Biodiversity is rising on the business agenda (WEF, 2022). Although many companies are still not fully addressing biodiversity impacts of their operations, it is becoming increasingly common for businesses to have high-level sustainability strategies that incorporate biodiversity impacts and actions (Addison et al., 2019; de Silva et al., 2019). These strategies may lay out what a company's impacts are on biodiversity, and what actions a company is taking to mitigate that impact and restore biodiversity values. This can include impacts through direct operations, supply chains, and in their investments. Through its guidance, methods, and tools, the Science Based Targets Network (SBTN) aims to help companies determine what they can do today to begin aligning with science (or evidence) to ensure they are doing their part for an equitable, net-zero, nature-positive future.

A company may also have specific projects or activities that require a detailed assessment of impacts (e.g. through the environmental impact assessment process), and the creation of management plans for biodiversity. For example, a company may have a management plan to enhance biodiversity in an urban office site, a programme focused on taking action to reduce the impact of upstream supply chains on biodiversity, or a management plan to minimise and mitigate the impact of new infrastructure on biodiversity. Such strategies and plans will vary depending on the sector. Companies can also integrate biodiversity considerations into wider environmental programmes such as nature-based solutions for climate change, or developing new investment opportunities that have positive impacts on biodiversity and climate. With growing ambitions and commitments around net zero and nature restoration, businesses are looking for approaches that will support them in delivering on their objectives.

However, there are some barriers to evidence use in this sector (CISL, 2022): much of the relevant science is still developing, as are the methodologies and metrics. Executive management often feels there is insufficient data, information and evidence to make decisions. There is not always sufficient internal knowledge of biodiversity, which can slow acceptance. This is amplified by the fact that alternative, environmentally-beneficial approaches can be perceived as complex, costly, uncertain, or unable to deliver compared to traditional solutions.

Drivers for evidence use

There are a number of possible drivers for businesses to embed evidence in their environmental strategies and programmes. However, the drivers for evidence are not common across all sectors, all value chains within a sector, or all functional areas and leaders within a business.

- **Effective action** — A clear evidence base can support a business in determining the appropriate steps to meet environmental goals (for example, creating reductions in carbon emissions or creating resilient supply chains) by ensuring future supplies of raw materials. It may be that a company wants to reduce its impact on biodiversity, and an evidence base can be used to identify what the most effective and efficient practice or approach would be to achieve this.

- **Lowered risk and realised opportunities** — Action to mitigate biodiversity impacts in some sectors is driven by the operational, financial and reputational risks associated with negative impacts on nature. By increasing the effectiveness of proposed mitigation actions, evidence use can help lower these risks. Other sectors may be driven by operational opportunities associated with biodiversity restoration (e.g. improving pollination of agricultural crops, improving water quality), and using evidence can help realise these opportunities.
- **Providing leadership, and raising the bar** — Some companies will want to lead and raise the bar within their sectors. This could be something as ambitious as wanting to transform a sector such that it delivers sustainability opportunities and biodiversity benefits across a landscape. Evidence, decision tools, and sensible use of experts are needed to undertake this transformation. For example, in agribusiness it is often not clear what interventions can be undertaken at the farm and landscape level to successfully reduce the risks associated with nature degradation in order to deliver a more sustainable supply chain.

Using evidence in business decisions

White et al., (2022a) look at principles for incorporating evidence into business biodiversity strategies, which we briefly summarise below. These principles can help ensure that business-biodiversity actions to mitigate impacts and restore biodiversity are based on evidence. They can also be used by consultants working with businesses to ensure that recommendations are based on the best available information.

- **Collating evidence of status, impacts, and actions** — Relevant evidence, from a range of sources (see Chapter 2), should be collated and reviewed to inform proposed actions. This should include information on the status of biodiversity, the impacts of business activities (including through direct operations, upstream and downstream supply chains, and investments), and the effectiveness, costs, acceptability, and feasibility of the proposed action.
- **Prioritising action based on evidence** — The collated evidence should be used to decide upon and prioritise actions as part of strategies and environmental programmes, to ensure they are likely to be effective at delivering action to avoid and minimise impacts, and restore biodiversity. The rationale behind taking (or not taking) an action should be documented.
- **Transparency** — Information compiled on the status of biodiversity, negative impacts of business activities, the evidence base underlying mitigation and restoration actions, and the observed impact of positive actions should be transparently reported.
- **Monitoring** — Where feasible, actions taken should be monitored to assess their impact, and changes made where unexpected or ineffective outcomes are occurring.

This is particularly important where the evidence base behind an action may be limited or if a business may carry out similar projects in the future

- **Embedding evidence** — Evidence use can be embedded across the organisation. This should involve engaging colleagues from across the business in evidence use, where a consistent evidence base can offer clarity and help streamline decision-making processes for example in new investments, or new products. Doing so can help remove barriers to action, such as scarcity of data and knowledge around particular subjects, challenges around identifying concrete benefits and the financial implications of biodiversity actions, and complexity of working in partnership with internal and external partners.

Businesses move at a fast pace and often want to make decisions swiftly. There is a challenge around wanting to make the right decision versus wanting to make a decision in a given timeframe. The 80–20 rule can be applied whereby the information needed to inform a decision is 80% there: it is not perfect but it is near enough to enable a business to move forward. Expansion of evidence bases from which to draw would be advantageous and could provide businesses with the confidence that they are making decisions based upon scientifically rigorous information.

9.8 Evidence-Based Writing and Journalism

Science is a rich source of stories on a wide variety of public-interest subjects, from climate change and energy to electronic cigarettes and vaccines. Over time, a body of evidence is built up to establish the facts. What journalism often does is report these pieces of evidence in isolation and without clarity around the uncertainty of any results. This is especially true for controversial subjects. These days, much journalism is written to shock and keep the interest of readers as long as possible. Whilst this can lead to entertaining headlines, the reality is that many articles are published daily that are not using and communicating evidence correctly. This misinformation can often have gross consequences.

There is an important distinction between evidence-based writing and conventional writing. Accuracy is important in all kinds of journalism, but in many science stories, it can be particularly misleading — and even harmful — if reporting is based on what people said or published, without considering the quality and the weight of the evidence. Ignorance of the quality of the evidence of the infamous Wakefield 'MMR and autism' study, and ignorance of the weight of evidence around climate change, did immense harm to public understanding of two of the defining issues of our times.

A key element of an evidence-based journalist is to make sources as clear as possible. If material is from a journal paper or report then provide sufficient clues to make it easy to find (such as author and journal name or doi or a hotlink if online).

There are a number of things journalists should do when reporting the science behind a story.

When researching a study:

- Consider the strength of the evidence, as described in Chapters 2 and 3. Is this a peer-reviewed paper in a reputable journal, a report from a recognised scientific body or a conference abstract with no data available for scrutiny? Not all science has equal weight.
- Take care with preprints. Preprints are scientific papers that have been posted online without any external peer review, meaning they have not yet been scrutinised by the wider scientific community. Whilst many of these may be excellent, others will never make it into the published body of scientific literature. Journalists should take even more care with conference abstracts, which usually do not even have data available. If used as a source for a story, it should be made clear when a piece of work has not been peer-reviewed or published.
- Look at the author list, especially if the claims strike you as hyperbolic. A group of authors with a track record of solid science may carry more weight than one or two individuals with a history of campaigning on the subject.
- Try to seek the opinions of other scientists actively researching the same field. If new findings attract serious scientific concerns, these need exploring; they will also be able to give you clues about the trustworthiness of an author or journal and help establish where the weight of evidence lies.
- Pay attention to the study design and consider whether the statistical analysis is appropriate, whether the sample size was sufficient, whether the trial was properly blinded/randomised, if there are any major limitations, and whether there is any extrapolation in their discussion. Not all of these will be relevant to every study but they are the kinds of things other scientists should be able to spot. Report these in the article where possible. Sutherland et al. (2013) provide a list of twenty tips for interpreting scientific claims and cover many common misinterpretations.
- Correlation does not equal causation. Be especially careful of this when reporting observational studies, as they are almost always completely incapable of establishing cause. These are very common, they are useful for generating scientific hypotheses or suggesting where future research should be directed, but they often prove nothing and can be misleading.

When reporting a study:

- Try to frame them in the context of other evidence. It is important to investigate the current scientific consensus on a subject; results that challenge or overturn the received wisdom should be handled with care — especially if the findings change or reinforce public beliefs or behaviour on an important subject. Extraordinary claims need extraordinary evidence!
- Always try to give a sense of the stage of the research (are these early provisional findings or is this the final conclusions of a long, major trial?)

- In all cases ensure you should represent the overall findings of a study, rather than focussing on extreme values, using 'up to', or cherry-picking data. It is important to distinguish clearly between the actual findings of a study and interpretation or speculation.
- Headlines should not mislead the reader about a story's contents and quotation marks should not be used unless for a direct quote.
- Be clear about any uncertainties, because no scientific paper can ever answer all the questions. Take note of whether the authors themselves have been candid about the uncertainties and limitations in their own paper. Make clear the distinction between findings, interpretation and extrapolation.

9.9 Evidence-Based Funding

Our ability to meet conservation targets is highly dependent on how biodiversity conservation is funded. Because of this, funders have the potential to drive major decisions in conservation. However, it is surprisingly difficult to gather information on how much is spent on conservation projects and where it comes from. The data are scarce, and those which are available are often incomplete (Waldron et al. 2013; White et al. 2022b). In addition, it can be difficult to assess whether funding has been given primarily for a biodiversity-related project or for a project that also benefits biodiversity in addition to its primary goal, such as development (Miller, 2014). It was recently estimated that, globally, we spend $100–180bn a year on nature conservation, with previous studies suggesting that funding needs to be an order of magnitude greater (McCarthy, 2012) and better distributed (Waldron et al., 2013) to meet targets. Understanding the success of investing in biodiversity conservation has been limited by the lack of evidence to show how investments in interventions measurably affect biodiversity (Possingham and Gerber, 2017). If we are lacking in funds, then ensuring the funding that is available is going to the right places is vital. For this to be possible, we need to look at how funding is awarded and what approaches may help to improve this process.

Evidence-based grant-giving could be the gateway to effective, evidence-based conservation decision making. Under an evidence-based system, funds would be awarded to projects and programmes judged on the evidence of their effectiveness, potential effectiveness, or ability to fill a knowledge gap. This approach would be cost-effective, reduce risk, and promote a positive societal change, therefore, it is in the interest of the giver, the grantees, and society as a whole.

The ability to move to an evidence-based funding approach may be limited by the capacity of the funding organisation. Whilst some organisations may be able to spend money and time investigating the evidence behind projects from their end, many will not. Having applicants complete this process, and having free and easy-to-use resources (e.g. Conservation Evidence, Applied Ecology Resources) and guidance (provided in this chapter), significantly reduces this challenge. This simply requires that grant givers ensure that their grantees show the use of evidence, either by including a section on their application forms or demonstrating how they will monitor and include evidence in all aspects of their project, from planning to the dissemination of results. As well as helping the funders, this process should encourage grantees to scrutinise their own work and challenge assumptions more. In addition, if funders require that grantees report back their findings, whether results are successful or not, a better record of where funding has been effective can be kept and prevent the waste of future resources on projects that are unlikely to work (Catalano et al., 2019).

How could this be done?

Whilst the solution seems simple and obvious, the implementation may not be straightforward. There are now many platforms available to freely access evidence on a variety of topics (conservationevidence.com/content/page/127). This makes it easier for grantees to search for and find evidence, if it exists, in their subject area.

Box 9.3 shows different ways that funders can ask grantees to provide evidence during their projects at various different stages. These were developed by a group of conservation funders (Parks et al. 2022).

9.9.1 Assessing evidence-based funding applications

When assessing the evidence base provided by applicants, it is important to remain aware that reviewing evidence may be a new process for applicants and it is likely to be most useful if the process is supportive and collaborative rather than a strict judgement of the evidence base. It is particularly important to ensure the process is fair to projects working on species, ecosystems or regions where evidence is scarce.

The aim of asking applicants to describe the evidence base for their proposal is to aid the applicant, encourage transparent decision making, and improve the effectiveness of the final project. Therefore, it may, or may not, be appropriate for evidence use to be a criterion on the assessors' score sheet; the main aspiration is to ensure that applicants have gone through a rigorous process as part of their decision-making, even if the formal scientific evidence is sparse.

Some key questions to consider when reviewing the application are as follows:

1. If the project proposal is delivered as described, would it be a good investment?

This question is asking about the main objectives of a proposal and whether they are a priority for funding.

- Is the target species/habitat/ecosystem a conservation priority?
- Are the proposed gains in the target species and/or habitats sufficient to warrant the funding requested (i.e. is it cost-effective)?
- Is there suitable monitoring in place to assess and demonstrate whether the project is meeting its proposed objectives?
- What is the legacy of the project? Will the benefits be lost as soon as funding ceases? Is the local community involved in a way that is more likely to ensure longevity?

Box 9.4 Means by which funders ask applicants about the evidence underpinning the proposed actions

1. Application form has a question asking about the evidence

A section in the application form states something like 'If undertaking an action, reflect on any evidence relating to its effectiveness'. This is a straightforward approach and funders can help applicants by providing a list of recommended places to search for evidence. It may not be appropriate if most applications do not involve actions or if decisions about actions are considered at a different stage.

2. Applicants are asked to describe their use of evidence somewhere within their proposal

This is a less formal version of approach (1), but without a specific section to complete. Perhaps more appropriate for smaller grants.

3. Second application stage asks about evidence

Some funders have an initial short application form, followed by a second stage of proposal development in which a subset of applicants provide further details of their project. Funders can ask for evidence of effectiveness of the proposal for projects that reach the second stage where this is relevant. This may have the advantage of reducing the work for the first stage. It also allows for a general application process for a wide range of grants, with only practical conservation grants requested to provide evidence in the second stage.

4. Asked to justify assumptions underpinning the Theory of Change

Many applications request a Theory of Change to elucidate how the proposed project is likely to result in the desired outcomes. Here funders ask for an assessment of the evidence base for the assumptions underlying the Theory of Change.

5. Grantees asked to report on evidence use as part of reporting

The grant application and contract state that it is expected that the decision-making process will be evidence-based. The grantees' reports to the funder then describe how the evidence was used in decision-making and why key actions were chosen. This approach is appropriate for projects where key decisions are made during the project rather than before the application is submitted.

6. Funders check the evidence themselves

If assessing the evidence is considered too onerous or off-putting for applicants, the funder can decide to check the evidence for the actions proposed by applicants as part of the selection process.

7. The evidence is considered during project co-development

Some funders may not use a straightforward application process but instead collaborate and codesign a project with potential grantees. The funder may then identify the evidence and discuss how to use this in project planning with the potential grantee.

8. The process for using evidence is described

The applicant is asked to describe the process by which relevant evidence will be identified and considered during their project. This may be appropriate for complicated projects where numerous decisions have to be made.

9. Applications for a programme comprising numerous projects describe how they will ensure that evidence is considered

The grant may be to fund an organisation or a programme, who then run or fund a set of projects. Applicants are asked to describe how evidence use will be embedded in the project selection, probably using one of the approaches described above.

10. Evidence use is included in the selection criteria with the criteria made transparent to applicants before applying

Funder makes it clear that only proposals examining evidence will be considered for funding.

Sometimes the objective may be easily justified, such as the protection of a globally threatened species in one of its few remaining sites. In other cases, evidence to support the local, regional or national importance of the target needs to be provided. Such evidence may include issues such as connectivity of habitats or populations or the delivery of public benefits, such as clean water or access to nature for local communities.

2. Is the proposal likely to succeed in achieving the described outcome?

This question relates to the likelihood that the proposal, and the actions described within it, will lead to the desired outcomes.

- Are the intended outcomes well defined?
- What actions have been proposed in order to achieve those outcomes?
- After reflecting on the available evidence, and its relevance, are the suggested actions likely to be effective for this project?
- Are there the resources needed available, for example, are there water sources, saplings, animals for grazing, or individuals for reintroducing?

3. Is there capacity to deliver the project?

This question relates to the likelihood that the organisations, institutions, structures, and people involved in the proposal are in place to enable successful delivery.

- Are people with the skills needed available and willing to contribute to delivery of the project?
- Is there agreement, acceptance and involvement within the local community/across relevant stakeholders?
- Is the funding requested adequate for successful delivery of the proposal?

4. Is there capacity to learn?

- Does the proposal suggest processes and capacity are in place to monitor progress and to learn and adjust plans accordingly?

If applicants are able to demonstrate that they have considered the evidence base behind key decisions in planning the objectives and implementation of their proposal this will provide assurance of transparent and carefully thought out planning processes, increasing the chances of delivering a successful conservation project.

Table 9.4 gives examples of potential text for describing how the evidence has been checked.

Table 9.4 Example text describing range of approaches for evidence checking.

Context	Description	Example text
Evidence assessed	Appropriate if straightforward, key for deciding, a major investment or already completed as part of evidence-based practice.	*A systematic review in SR in 2020 showed, in field studies, Francis hydroelectric turbines resulted in higher immediate mortality risk for fish than Kaplan turbines. Every six months we check the literature for other tests of turbine design on fish populations.*
Evidence available but review and assessment delayed until grants is awarded.	Appropriate where the assessment of the evidence for a decision will not affect the overall feasibility of the proposal, but the details of how it is to be implemented will be important in determining success.	*There is substantial evidence on meadow creation, including 36 actions reviewed on Conservation Evidence and considerable expertise at the local agricultural college. Prior to carrying out the grassland restoration, we will review this literature and consult widely at the local and regional level to determine the most effective methodology.*
Providing assurance that future decisions will be evidence based	Applicable in situations where substantial decision making takes place after the grant is awarded. Here, the aim is to demonstrate that evidence review is already embedded into organisational processes.	*The precise conservation actions to be implemented are not yet known, as they will depend on the land acquired and the results of stakeholder consultation. Since 2020 we have become an evidence-led organisation, employing an Evidence and Implementation Manager who oversees the use and creation of evidence. This includes establishing a new test of an action each year and organising an annual meeting of regional wetland reserve managers to share experiences and lessons learned. Before undertaking any actions we will undertake an evidence review and expert consultation to inform decision making.*

9.9.2 Using Evidence to Learn from Funding Portfolios

Funders can play a vital role in collecting and providing evidence, and often have a unique opportunity through the sheer volume of evidence that is available across their portfolios. The evidence may exist in the form of primary data from field projects of grantee organisations, published documents and literature from funded projects and programmes, or project impact and effectiveness reporting. Collecting and sharing this evidence can improve the evidence base, including many topics, especially organisational, for which evidence is thin.

The MAVA Foundation has adopted a process to collect, assess, and share available evidence from its portfolio. The foundation has been funding conservation efforts for over 25 years and closes in 2022. With the closing date approaching, MAVA was keen to leave behind tangible conservation outcomes. In addition, the foundation also wanted to promote evidence-based conservation and take responsibility for making its insights and learnings accessible to the broader conservation community. While collecting the evidence for selected learning topics, the foundation sought to bring together key actors in evidence-based conservation to compare and align their approaches and concepts and eventually ensure a broader uptake of a common evidence-based conservation practice in the discipline.

Building on a recent approach for defining and using evidence in conservation practice (Salafsky et al. 2019, 2022), MAVA and its partners Foundations of Success and Conservation Evidence adopted a five-step process to solicit learnings on a range of topics.

Step 1. Define the learning topic

To collect evidence from a considerable portfolio of grants in retrospect requires finding focus. The full portfolio of MAVA grants was grouped by conservation action, using the IUCN standard classification for conservation actions (IUCN 2022). A shortlist of actions with many grants was selected, with the final list based on the perceived relevance for the wider conservation community, and the interest of the project team in exploring those topics. This resulted in a final list of four topics: 1) flexible conservation funding; 2) partnerships and alliances; 3) capacity building; and 4) research and monitoring.

Step 2. Design learning questions

A theory of change (see sections 5, 4.7.5 and 7.6.5) for each topic can capture the pathway from conservation intervention to desired outcome. This helped formulate specific learning questions and associated assumptions that could be tested with evidence.

Step 3. Collect evidence

Evidence to test assumptions came from a range of sources, including MAVA grant proposals and reports, questionnaires distributed to grantees, systematic searches of the Conservation Evidence database, and exploratory searches of the wider literature.

Step 4. Assess evidence

Each evidence piece entailed two issues: 1) whether the evidence supported or refuted the assumption, and 2) the weight of the evidence. The assessment of weight was based on the reliability of the information contained within the evidence, and the relevance of that information to the assumption in question.

Figure 9.2 shows all the combined evidence pieces for the assumption 'Organisations use flexible funding to invest in organisational development and/or maturity (that is not affordable otherwise)', displayed using a ziggurat plot (see Section 4.6.1).

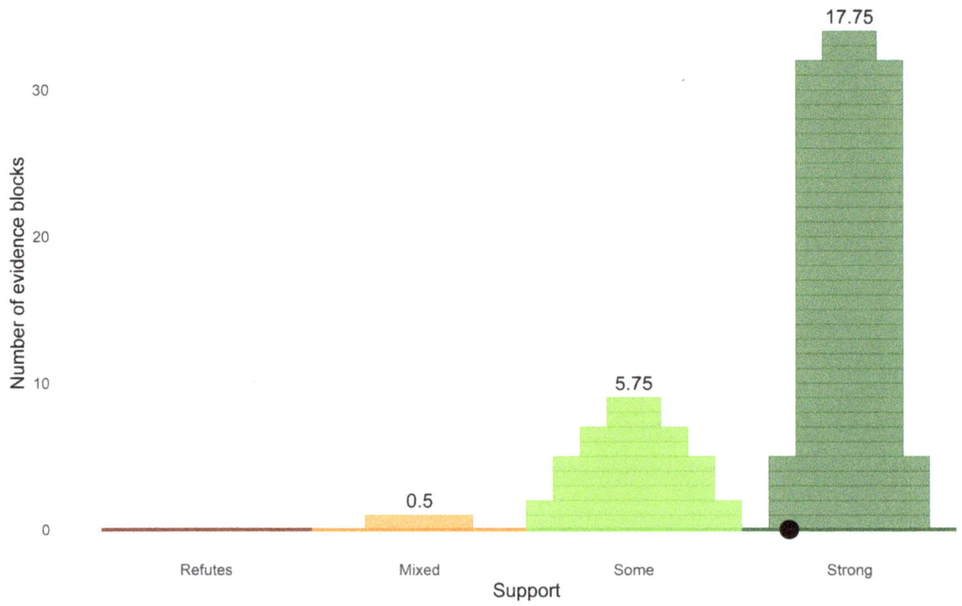

Figure 9.2 A ziggurat plot showing strong support for the assumption that organisations use flexible funding to invest in organisational development and/or maturity. Each piece of evidence is a horizontal block whose width represents its weight. The maximum potential weight of a single block is one. The number above each pile of evidence blocks shows the total evidence score for that pile. The filled black point is a weighted mean, and represents the balance of evidence for the assumption. (*Source*: authors)

Step 5. Draw conclusions

The last step was to use the summary plots, along with a detailed consideration of all evidence pieces, to draw conclusions. This highlighted important knowledge gaps, where the available evidence was not sufficient for drawing conclusions.

The findings of the evidence assessment, key learnings, and further details on the five-step approach is provided on https://conservation-learning.org.

9.10 Evidence-Based Decision-Support Tools

Decision-support tools can help decision makers by leading them through logical decision steps. A number of decision-support tools are available to support evidence-based decisions in conservation, ranging from complex models to simple software specifically designed to be used by non-experts (Dicks et al., 2014, Christie et al., 2021). Extending the range of such tools is key to delivering evidence-based practice at scale for a range of communities.

Some of these tools embed evidence behind the scenes, drawing on a specific data set (e.g. the Cool Farm Tool Biodiversity metric, 9.10.1), whilst others allow the user to search for the relevant data themselves (e.g. Evidence-to-Decision Tool, 9.10.3). It is important that these tools also incorporate other information such as values, resource availability, and stakeholder views and are transparent and auditable to ensure stakeholder confidence. Here, we provide examples, to illustrate different ways to embed evidence in decision support tools.

9.10.1 Cool Farm Tool

The 'Cool Farm Tool', owned and managed by the Cool Farm Alliance, is a suite of online software tools that provide multi-metric sustainability assessments for farms. The tools provide a quantitative or semi-quantitative assessment of water footprint, greenhouse gas emissions and biodiversity management at the farm scale, based on the best available evidence. They are designed to be easy to use by growers, agronomists and suppliers of agricultural products, but also freely accessible for individual growers, and globally applicable. The Cool Farm Tool is widely used in global supply chains, enabling hundreds of thousands of growers globally to make more informed on-farm decisions that reduce their environmental impact.

The greenhouse gas and water footprint calculators in the Cool Farm Tool embed evidence by using models, emissions factors and approaches from published literature, where possible those recommended by the Intergovernmental Panel on Climate Change (IPCC), or approved for international standards (Hillier et al., 2011; Kayatz et al., 2019). Such evidence is not always entirely relevant (see section 9.11), and biases in evidence can lead to inaccurate calculations if tools are used in contexts not related to the evidence. For example, two greenhouse gas calculators, including the Cool Farm Tool, were shown by Richards et al. (2016) to frequently over-estimate emissions from farms in the tropics and to incorrectly predict the direction of change in response to changes in management in 41% of cases. Richards et al. argued that this was because the majority of data incorporated into the tools (e.g. over 90% of studies analysed for N_20 fluxes following fertiliser application) came from temperate, not tropical agricultural systems (Richards et al., 2016).

The Cool Farm Biodiversity Metric (https://coolfarmtool.org/coolfarmtool/biodiversity/) takes a different approach than used for greenhouse gas measures. It embeds evidence for the effects of specific management interventions on biodiversity by scoring actions for their expected benefits to 'general biodiversity' (reflecting species — or habitat richness) and to a set of defined 'species groups', or biodiversity targets. The scores are partly drawn from the Conservation Evidence (https://www.conservationevidence.com) database of evidence assessments, and thus embed the rigorous subject-wide evidence synthesis method (section 3.3). The set of actions and species groups are selected in partnership with stakeholders, following participatory methods described by Macleod et al. (2021, 2022) for a particular biome, or set of biomes. The tool has different versions for temperate forest systems, Mediterranean and semi-arid systems, and (still in development) tropical forest systems. A similar tool is available for New Zealand farms (Macleod et al., 2021).

Expert judgement (Chapter 5) also contributes to the scores for each action in the Cool Farm Biodiversity Metric, for the simple reason that the Conservation Evidence database does not have complete coverage of farm management actions; for example, it does not consistently cover agronomic practices intended to reduce impacts of agrochemical use, or enhance agrobiodiversity (crop and livestock genetic diversity) while many other actions have insufficient evidence considering how widely they are practised. This mismatch between evidence generated by research and the evidence desired by practitioners has been called 'evidence disparity' and is a persistent problem in evidence-based conservation (Macleod et al., 2022).

The approach to embedding evidence in the Cool Farm Biodiversity Metric differs from the other metrics in the Cool Farm Tool largely because of the complex nature of biodiversity. Biodiversity outcomes are more varied and context-dependent than greenhouse gases, or water, both of which can be measured, predicted and reported in relatively simple units (tonnes of CO2-equivalent, or litres of water used, for example). Predictive models of biodiversity impacts or responses to land management, or even simplified numerical relationships between land management and outcome, are hard to build and even less likely to be reliable for biodiversity than for greenhouse gases or water. Such models are being actively developed to deal with certain aspects of biodiversity (e.g. Newbold et al., 2015; Duran et al., 2020; Schipper et al., 2020). To be embedded in an easy-to-use decision support tool, such as Cool Farm Biodiversity Metric, these models will need to be fully accessible and robust to any context, without vast data and processing demands.

9.10.2 Miradi

Miradi, https://www.miradishare.org (named after a Swahili word meaning 'project' or 'goal') has been developed over the past two decades to support the design, implementation, monitoring, and adaptive management of conservation and natural resource management projects at all scales. Miradi supports the assessment of the status of conservation targets, contributing factors and threats, setting conservation goals, and clarifying and tracking how actions are leading to desired outcomes and impacts. Miradi helps teams create diagram-based situation models and Theories of Change that can be used to model assumptions about a current situation and how an organisation's strategies will lead to measurable impacts. Miradi helps document the information used throughout the planning cycle (preventing the loss of knowledge due to staff turnover) and identify knowledge gaps where more information is needed. Evidence can be documented in Miradi using open text for most factors that make up situation models and theories of change (e.g. target, threats, strategies, results), as well as using standard evidence typology, that is more easily used for data analysis (e.g. indicator viability ratings and measurements, rating of threats, selection of strategies, goals, and objectives). An upcoming release of the software will allow users to document analytical questions and assess the degree of support, weight, and confidence in assumptions based on the available evidence (Salafsky et al., 2019). Miradi also allows users to query the back-end relational database, analyse data across projects and information systems, and produce dashboards to inform evidence-based decision making.

The Conservation Standards (https://conservationstandards.org/) can be used as a framework to use evidence for decision-making and the supporting Miradi software (https://www.miradishare.org) can be used to manage information and to define processes (Salafsky et al., 2019). Business Process Management can be used to discover, model, and improve business processes (Dumas et al., 2013). Simple models can also be used to support decisions on what actions to implement depending on the available evidence (see Figure 9.3).

9.10.3 Evidence-to-Decision Tool

This tool www.evidence2decisiontool.com is a template designed to make clear the reasoning and evidence behind conservation management decisions (Christie et al 2022). The tool has three major steps: (1). Define the decision context; (2). Gather evidence; and (3). Make an evidence-based decision. In each step, practitioners enter information (e.g., from the scientific literature, practitioner knowledge and experience, and costs) to inform their decision-making and document their reasoning. The tool packages this information into a customized downloadable report. This report can be embedded into other material, simply stored to document decisions by the group (such as a reserve) and an organisation or exchanged so others can see the logic of decision made by others for shared problems.

The experience is that this is useful in bringing together those making the decision to complete the tool. The intention of the group who created this tool was that by enabling practitioners to revisit how, and why, decisions were made can help increase the transparency and quality of decision-making in conservation.

9.11 Evidence-Based Models

Chapter 2 describes how models can be evaluated as sources of evidence and Chapter 8 describes their use in decision making. This section considers how evidence can be embedded into models.

Models are critical sources of evidence and can be highly influential in decision-making. Despite this, they can be highly variable in the manner in which they use evidence. The outputs of models depend critically on the structure of the model, parameter values, and assumptions used in their formulation. Because of this, transparency around the sources of model structures, parameters, and assumptions used, any uncertainties around these, and how uncertainty affects conclusions and decisions is essential.

Models can vary from very general, with conclusions that can be broadly described, to the very specific, where conclusions are highly dependent on model structure and parameter values. General models can be useful in identifying broadly applicable principles. However,

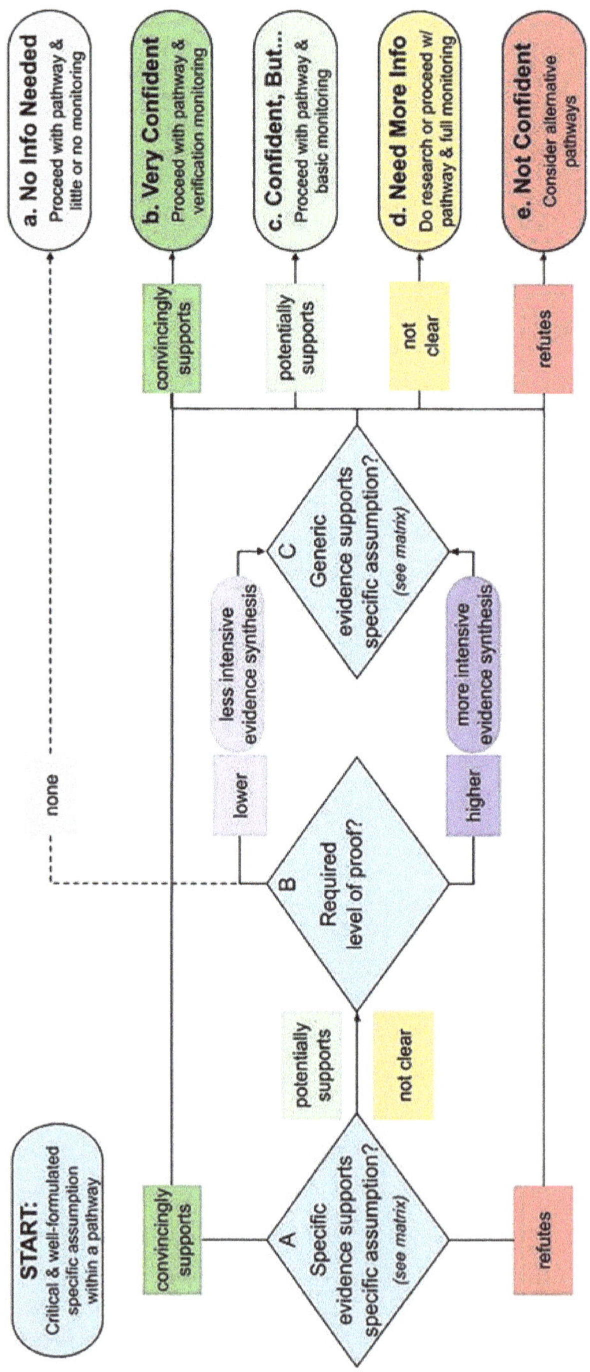

Figure 9.3 Decision tree for using evidence in assessing a potential conservation action. (*Source:* Salafsky et al, 2022, CC-BY-4.0)

context dependence requires more specific models to be constructed, which rely strongly on particular parameter values derived from the available evidence. The transfer of model conclusions from one context to another can be problematic, making it particularly important that the boundaries and applicability of the model be clearly specified.

Models can be evidence based by adopting the same principles described in this book in which evidence is assessed and expert judgement carried out in ways that reduce bias. One model may require numerous sub-models, functions, and parameter values and some structures and values may be critical to the conclusions drawn from the modelling process. Identification of structures and parameters that strongly influence the conclusions of the modelling process can greatly facilitate the process. Indeed, a major advantage of models is the ability to explore plausible scenarios where evidence is lacking or even unobtainable. What then is responsible behaviour for creating evidence-based models?

- Determine how models will be used to inform decisions and where in the process models will be used. For example, models may be used to explore the effects of costs and/or benefits of different management actions or be used to contrast the effects of different ecological contexts.
- Explore which model structures, sub-models, functions, and parameter values are critical for decision making. Value of Information analysis and sensitivity analysis can show which assumptions and parameter values are critical for decision making. These analyses can inform on whether evidence collation has been sufficient and identify critical gaps in available evidence.
- Assess what evidence is available. Strong a priori evidence for a particular ecological process may reduce the need to explore multiple structures, whereas weak or missing evidence may necessitate a wider exploration of model structures and parameter space.
- Establish principles for extracting data and associated uncertainties to include in models. This is critical (garbage in — garbage out). As in Chapter 2, collate studies and extract values. Use analysis such as meta-analysis but also Delphi Technique to consider values.
- Be explicit about models for uncertainty, there can be big differences in outcomes from different probability distributions of a particular parameter (e.g. the difference between uniform and triangular distributions of a parameter value).
- When reporting the results of modelling, describe the effects of relaxing or changing assumptions, values of parameters, and uncertainties.
- Modelling is an imperfect and iterative procedure; be honest about the limits of the conclusions and where improvements need to be made.

References

Addison, P.F., Bull, J.W. and Milner-Gulland, E.J. 2019. Using conservation science to advance corporate biodiversity accountability. Conservation Biology 33: 307–18, https://doi.org/10.1111/cobi.13190.

Barnes, A. and Parkhurst, J., 2014. Can global health policy be depoliticized? A critique of global calls for evidence-based policy. In *Handbook of Global Health Policy* ed. by G.W. Brown, et al. (Chichester, UK: John Wiley & Sons, Ltd), https://doi.org/10.1002/9781118509623.ch8.

Bass, S., Roe, D., Hou-Jones, X. et al. 2021. Mainstreaming nature in development: A brief guide to political economy analysis for non-specialists, UNEP-WCMC, Cambridge, UK, https://pubs.iied.org/sites/default/files/pdfs/2021-10/20566G_0.pdf.

Brancalion, P.H.S., Holl, K.D. 2020. Guidance for successful tree planting initiatives. *Journal of Applied Ecology* 57: 2349–61, https://doi.org/10.1111/1365-2664.13725.

The University of Cambridge Institute for Sustainability Leadership (CISL). (2022). Decision-making in a nature positive world: a corporate diagnostic tool to advance organisational understanding of nature-based solutions projects and accelerate their adoption. Cambridge: The University of Cambridge Institute for Sustainability Leadership, https://www.cisl.cam.ac.uk/resources/publications/decision-making-nature-positive-world.

Catalano, A.S., Lyons-White, J., Mills, M.M. et al. 2019. Learning from published project failures in conservation. *Biological Conservation* 238: 108223, https://doi.org/10.1016/j.biocon.2019.108223.

Christie, A.P., Amano, T., Martin, P.A. et al. 2021. The challenge of biased evidence in conservation. *Conservation Biology*, 35: 249–62, https://doi.org/10.1111/cobi.13577.

Christie, A.P., Downey, H., Frick, W.F., et al. 2022. *A practical conservation tool to combine diverse types of evidence for transparent evidence-based decision-making*. Conservation Science and Practice 4: e579, https://doi.org/10.1111/csp2.579.

Christie, A.P., Downey, H., Frick, W.F. et al. 2022. A practical conservation tool to combine diverse types of evidence for transparent evidence-based decision-making. Conservation Science and Practice 4: e579, https://doi.org/10.1111/csp2.579.

Corporation for National and Community Service (CNCS). 2020. Strategic Evidence Plan. Available for download: https://americorps.gov/sites/default/files/documents/CNCS%20Strategic%20Evidence%20Plan%20FY2020_508.pdf.

Cruickshanks, K. 2018. Building biodiversity through land management: An evidence-based assessment of the needs of butterflies and moths and the opportunities for a countryside rich in insects. Unpublished report: Butterfly Conservation Report Number S18–02.

de Silva, G.C., Regan, E.C., Pollard, E.H.B. et al. 2019. The evolution of corporate no net loss and net positive impact biodiversity commitments: understanding appetite and addressing challenges. *Business Strategy and the Environment* 28: 1481–95, https://doi.org/10.1002/bse.2379.

Dalbeck, L., Hachtel, M. and Campbell-Palmer, R. 2020. A review of the influence of beaver *Castor fiber* on amphibian assemblages in the floodplains of European temperate streams and rivers. *The Herpetological Journal* 30: 135–46, https://doi.org/10.33256/hj30.3.135146.

Dicks, L.V., Walsh, J.C. and Sutherland, W.J., 2014. Organising evidence for environmental management decisions: a '4S' hierarchy. *Trends in Ecology & Evolution* 29: 607–13, https://doi.org/10.1016/j.tree.2014.09.004

Downey, H., Bretagnolle, V., Brick, C. et al. 2022. Principles for the production of evidence-based guidance for conservation actions. *Conservation Science and Practice,* e12663. https://doi.org/10.1111/csp2.12663.

Dumas, M., La Rosa, M., Mendling, J. et al. 2013. F*undamentals of Business Process Management.* (Heidelberg: Springer).

Durán, A.P., Green, J.M.H., West, C.D. et al. 2020. A practical approach to measuring the biodiversity impacts of land conversion. *Methods in Ecology and Evolution* 11: 910–21, https://doi.org/10.1111/2041-210X.13427.

General Services Administration Office of Evaluation Sciences (GSA OES). 2020. Evidence Act Toolkit: A Guide to Developing Your Agency's Learning Agenda. Available for download: https://oes.gsa.gov/toolkits/.

Government Accountability Office (GAO). 2019. Evidence-based policymaking: selected agencies coordinate activities but could enhance collaboration. (GAO Publication No. 20–119). Washington, D.C.: U.S. Government Printing Office. Available for download: https://www.gao.gov/assets/gao-20-119.pdf.

Gluckman, P.D., Bardsley, A. and Kaiser, M. 2021. Brokerage at the science–policy interface: from conceptual framework to practical guidance. *Humanities and Social Sciences Communications* 8: 1–10, https://www.nature.com/articles/s41599-021-00756-3.

Henderson, M. 2012. *The Geek Manifest: Why Science Matters,* Transworld Digital.

Hemery, G., Petrokofsky, G., Ambrose-Oji, B., et al. 2020. Awareness, action, and aspirations in the forestry sector in responding to environmental change: Report of the British Woodlands Survey 2020. 33pp. www.sylva.org.uk/bws.

Hemming, V., Camaclang, A. E., Adams, M. S., et al. 2022. An introduction to decision science for conservation. *Conservation Biology,* 36, e13868. https://doi.org/10.1111/cobi.13868.

Hillier, J., Walter, C., Malin, D. et al. 2011. A farm-focused calculator for emissions from crop and livestock production. *Environmental Modelling & Software* 26, 1070–78, https://doi.org/10.1016/j.envsoft.2011.03.014.

Hou-Jones, X., Bass, S., Roe, D. 2021. Mainstreaming biodiversity into government decision-making. (Cambridge: UNEP-WCMC). Available from: https://pubs.iied.org/20426g.

Hunter, S.B., zu Ermgassen, S.O.S.E., Downey, H. et al. 2021. Evidence shortfalls in the recommendations and guidance underpinning ecological mitigation for infrastructure developments. *Ecological Solutions and Evidence* 2: e12089, https://doi.org/10.1002/2688-8319.12089.

Hutchings, J.A., 2022. Tensions in the communication of science advice on fish and fisheries: northern cod, species at risk, sustainable seafood. *ICES Journal of Marine Science* 79: 308–318, https://doi.org/10.1093/icesjms/fsab271.

Hillier, J., Walter, C., Malin, D. et al. 2011. A farm-focused calculator for emissions from crop and livestock production. *Environmental Modelling & Software* 26: 1070–1078, https://doi.org/10.1016/j.envsoft.2011.03.014.

IUCN. 2022. Conservation Actions Classification Scheme (Version 2.0). https://www.iucnredlist.org/resources/conservation-actions-classification-scheme.

Kayatz, B., Baroni, G., Hillier, J. et al. 2019. Cool Farm Tool Water: A global on-line tool to assess water use in crop production. *Journal of Cleaner Production* 207: 1163–1179, https://doi.org/10.1016/j.jclepro.2018.09.160.

Keeney, R.L. 2004. Making better decision makers. *Decision Analysis* 1: 193–204, https://doi.org/10.1287/deca.1040.0009.

MacLeod, C.J., Brandt, A.J., Collins, K. et al. 2021. Giving stakeholders a voice in governance: Biodiversity priorities for New Zealand's agriculture. *People and Nature* 4: 330–350, https://doi.org/10.1002/pan3.10285.

MacLeod, C.J., Brandt, A.J., Dicks, L.V., 2022. Facilitating the wise use of experts and evidence to inform local environmental decisions. *People and Nature* 4: 904–917, ttps://doi.org/10.1002/pan3.10328.

McCarthy, D.P., Donald, P.F., Scharlemann, J.P. et al. 2012. Financial costs of meeting global biodiversity conservation targets: Current spending and unmet needs. *Science* 338: 946–949. https://www.science.org/doi/10.1126/science.1229803.

Miller, D. C. 2014. Explaining global patterns of international aid for linked biodiversity conservation and development. *World Development* 59: 341–359. https://doi.org/10.1016/j.worlddev.2014.01.004.

Newbold, T., Hudson, L.N., Hill, S.L.L. et al. 2015. Global effects of land use on local terrestrial biodiversity. *Nature* 520: 45–50, https://www.nature.com/articles/nature14324.

Nightingale, D.S., Fudge, K., and Schupmann, W., 2018. Learning Agendas. Evidence-Based Policymaking Collaborative. Available for download: https://www.urban.org/sites/default/files/publication/97406/evidence_toolkit_learning_agendas_2.pdf.

Office of Management and Budget 2019. Phase 1 Implementation of the Foundations for Evidence-Based Policymaking Act of 2018: Learning Agendas, Personnel, and Planning Guidance. (OMB Memo M-19-23). Available for download: https://www.whitehouse.gov/wp-content/uploads/2019/07/M-19-23.pdf.

Office of Management and Budget 2021. Evidence-Based Policymaking: Learning Agendas and Annual Evaluation Plans. (OMB Memo M-21-27). Available for download: https://www.whitehouse.gov/wp-content/uploads/2021/06/M-21-27.pdf.

O'Brien, D., Hall, J.E., Miró, A. et al. 2021. A co-development approach to conservation leads to informed habitat design and rapid establishment of amphibian communities. *Ecological Solutions and Evidence*, 2: e12038, https://doi.org/10.1002/2688-8319.12038.

Parkhurst, J. 2017. *The Politics of Evidence: From Evidence-Based Policy to the Good Governance of Evidence* (London: Taylor & Francis), https://doi.org/10.4324/9781315675008.

Parks, D., Al-Fulaij, N., Brook, C. et al. 2022. Funding evidence-based conservation. *Conservation Biology*, 36, e13991. https://doi.org/10.1111/cobi.13991:

Pe'er, G., Zinngrebe, Y., Moreira, F., et al. 2019. A greener path for the EU Common Agricultural Policy. Science 365: 449–51, https://doi.org/10.1126/science.aax3146.

Possingham, H. P., Wintle, B. A., Fuller, R. A. et al. 2012. The conservation return on investment from ecological monitoring. In: *Biodiversity Monitoring in Australia*, ed. by D.B. Lindenmayer and P. Gibbons (CSIRO Publishing).

Possingham, H.P. and Gerber, L.R. 2017. The effect of conservation spending. *Nature* 551: 309–310, https://doi.org/10.1038/nature24158.

Richards, M., Metzel, R., Chirinda, N. et al. 2016. Limits of agricultural greenhouse gas calculators to predict soil N2O and CH4 fluxes in tropical agriculture. *Scientific Reports* 6: 26279, https://doi.org/10.1038/srep26279.

Rose, D.C. 2014. Five ways to enhance the impact of climate science. *Nature Climate Change* 4: 522–524, https://doi.org/10.1038/nclimate2270.

Rose, D.C., Mukherjee, N., Simmons, B.I. et al. 2020. Policy windows for the environment: Tips for improving the uptake of scientific knowledge. *Environmental Science & Policy* 113: 47–54, https://doi.org/10.1016/j.biocon.2021.109126

Salafsky, N., Irvine, R., Boshoven, J. et al. 2022. A practical approach to assessing existing evidence for specific conservation strategies. *Conservation Science and Practice* 4: e12654. https://doi.org/10.1111/csp2.12654.

Salafsky, N., Boshoven, J., Burivalova, Z. et al. 2019. Defining and using evidence in conservation practice. *Conservation Science and Practice,* 1: e27, https://doi.org/10.1111/csp2.27.

Schipper, A.M., Hilbers, J.P., Meijer, J.R. et al. 2020. Projecting terrestrial biodiversity intactness with GLOBIO 4. *Global Change Biology* 26: 760–771, https://doi.org/10.1111/gcb.14848.

Stewart, A. 2016 Conservation Capability Maturity Model. A model for assessing organisational performance and identifying potential improvements. Conservation Standards, https://conservationstandards.org/wp-content/uploads/sites/3/2020/10/Conservation-Capability-Maturity-Model.pdf.

Sutherland, W.J., Downey, H., Frick, W.F. et al. 2021. Planning practical evidence-based decision making in conservation within time constraints: the Strategic Evidence Assessment Framework. *Journal for Nature Conservation* 60: 125975, https://doi.org/10.1016/j.jnc.2021.125975.

Sutherland, W.J. 2013. Review by quality not quantity for better policy. *Nature* 503: 167, https://www.nature.com/articles/503167a.

Sutherland, W., Spiegelhalter, D. and Burgman, M. 2013. Policy: Twenty tips for interpreting scientific claims. *Nature* 503: 335–337, https://doi.org/10.1038/503335a.

Sutherland, W.J., Burke, E., Chamberlain, B. et al. 2022. What are the forthcoming legislative issues of interest to ecologists and conservationists in 2022? *The Niche, British Ecological Society* 53: 28–35.

Tuttle, L.J., Donahue, M.J. 2022. Effects of sediment exposure on corals: a systematic review of experimental studies. *Environmental Evidence* 11: 4, https://doi.org/10.1186/s13750-022-00256-0.

Waldron, A., Moeers, A., Miller, D. C. et al. 2013. Targeting global conservation funding to limit immediate global biodiversity declines. *Proceedings of the National Academy of Sciences* 110: 12144–12148, https://doi.org/10.5061/dryad.p69t1.

Walsh, J.C., Dicks, L.V., Raymond, C.M. et al. 2019. A typology of barriers and enablers of scientific evidence use in conservation practice. *Journal of Environmental Management* 250: 109481, https://doi.org/10.1016/j.jenvman.2019.109481.

WEF (2022) Global Risks Report 2022. World Economic Forum, https://www.weforum.org/reports/global-risks-report-2022.

White, T.B., Petrovan, S.O., Bennun, L. et al. 2022a. Principles for using evidence to improve biodiversity impact mitigation by business. *OSF Preprints,* https://doi.org/10.31219/osf.io/427tc.

White, T.B., Petrovan, S.O., Christie, A.P. et al. 2022b What is the Price of Conservation? A Review of the Status Quo and Recommendations for Improving Cost Reporting, *BioScience* 72, biac007, https://doi.org/10.1093/biosci/biac007.

Wilson, K.A., Carwardine, J. and Possingham, H.P. 2009. Setting conservation priorities. *Annals of the New York Academy of Sciences* 1162: 237–264, https://doi.org/10.1111/j.1749-6632.2009.04149.x.

10. How Conservation Practice Can Generate Evidence

Nancy Ockendon[1], Marc W. Cadotte[2], Johanna Eklund[3], Paul J. Ferraro[4], Martin Fisher[5], Mark Hancock[6,7], Thomas A. Worthington[8]

Conservation practice provides a considerable opportunity to generate new evidence to inform future decision-making. Substantial resources are currently invested in data collection and monitoring, yet too often these are ineffectively designed, meaning the data gathered contributes little to building an evidence base. However, thinking in advance about how actions are implemented, data are collected, and results are shared can greatly increase the usefulness of the results. Controls, comparisons, replication, randomisation, and preregistration can all improve the value of the data collected.

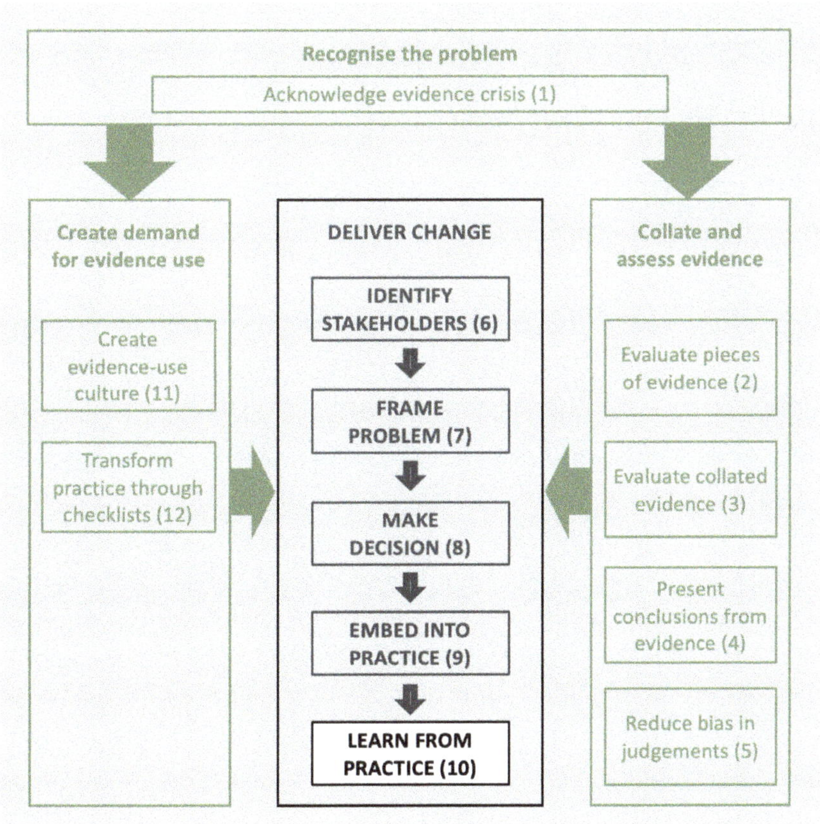

 https://doi.org/10.11647/OBP.0321.10

[1] Endangered Landscapes Programme, Cambridge Conservation Initiative, The David Attenborough Building, Cambridge, UK
[2] Department of Biological Sciences, University of Toronto-Scarborough, 1265 Military Trail, Toronto, ON, M1C 1A4, Canada
[3] Digital Geography Lab, Department of Geosciences and Geography, Faculty of Sciences, University of Helsinki, Helsinki, Finland
[4] The Center for Behavioral & Experimental Agri-Environmental Research, Johns Hopkins University, Baltimore, Maryland, USA
[5] Fauna & Flora International, The David Attenborough Building, Pembroke Street, Cambridge, UK
[6] RSPB Centre for Conservation Science, RSPB Scotland, North Scotland Office, Etive House, Beechwood Park, Inverness, UK
[7] Cairngorms Connect, Achantoul, Aviemore, UK
[8] Conservation Science Group, Department of Zoology, University of Cambridge, The David Attenborough Building, Pembroke Street, Cambridge, UK

Contents

10.1 Ensuring Data Collection is Useful

Conservationists need better evidence about the effectiveness of their actions. There are substantial gaps in the evidence (Christie et al., 2020) and many straightforward questions are unanswered (Sutherland et al., 2022). Filling such gaps would greatly enhance the effectiveness of conservation practice.

Such gaps persist despite the fact that conservation practitioners routinely collect large amounts of data that describe the interventions they undertake and the subsequent ecological and socio-economic conditions. These monitoring efforts generally aim to measure the progress of conservation projects but, despite considerable effort and resources dedicated to data collection, conservation projects are rarely designed to demonstrate the link between action and effect (Legg and Nagy, 2006). Embedding experimental designs in conservation projects is the most effective means of demonstrating a link between an action and its effect, yet is seldom used. On the occasions when projects do link action and effect, the results may not be disseminated sufficiently to ensure they contribute to the wider evidence base.

The beauty of planned experimental designs is that conservation projects can focus their attention on collecting data from a specified set of sites for a focused set of informative indicators. This allows them to move away from long lists of indicators collected across large areas that may have used in the past, with little idea of how the data collected could be applied. With a little forward planning of intervention design, data collection and dissemination, conservation projects could yield much more useful information. The resulting improved evidence base may lead to result in better-informed, and hence more effective, conservation decision-making.

This chapter has three main sections. Firstly, we discuss which different approaches to data collection, including measuring outputs, measuring outcomes, testing actions, or evaluating impact, are best applied to selected circumstances. Secondly, we describe the principles of experimental design. Finally, we describe how the evidence generated from well-designed experiments can be effectively shared with the wider community.

There are two main messages that we wish to convey. Firstly, embedding experimental tests into conservation practice is often less challenging than it may initially appear and should become routine. Secondly, the details of the experimental design really matter — better designed experiments are much more likely to produce useful and accurate results.

10.1.1 Standardising methods and outcomes

One major challenge in assessing the effectiveness of conservation actions is a lack of standardisation in how projects are implemented and evaluated. For instance, Cadier et al. (2020) found that 238 different indicators had been used to measure coastal wetland restoration across just 133 projects. The complexity of natural systems means that a range of methodological approaches and indicators is warranted, but such an extreme lack of consistency renders it difficult to synthesise, compare, and draw inferences across different projects.

One possible solution is that expert working groups develop and agree on standards for measuring and reporting conservation outcomes. For example, Sutherland et al. (2010)

developed a set of minimum standards for documenting and monitoring bird translocation and reintroduction projects. These aimed to facilitate the collection of comparable data that could be more effectively combined to detect patterns in the causes of successes and failures.

Another approach to ensuring consistency in reporting across projects is demonstrated by the Mangrove Restoration Tracker Tool (Leal and Spalding, 2022). This tool has been co-designed by mangrove scientists, NGO staff and restoration practitioners from around the world, and provides a framework to collect data on all aspects of a mangrove restoration project, from design, through implementation to monitoring. Each project that uses the tool will record a comprehensive and consistent, yet easily collected, set of ecological and social metrics, alongside baseline information describing the site and actions undertaken. As this tool accumulates comparable data from a wide range of projects, it will permit rapid synthesis, making it easier to identify the most successful and cost-effective approaches and allowing a better assessment of progress toward national and global restoration targets.

Standardised methods also allow harmonised experiments to be carried out (Ferraro and Agrawal, 2021). Here, multiple teams carry out parallel experiments, using agreed standards and data collection to examine the generalisability of results and determine whether results are condition-dependent.

In the absence of an agreed method for monitoring a conservation intervention, a sensible rule is to adopt an already widely used technique where possible. The aim is to avoid seemingly trivial changes in methodology and data collection that render direct comparisons between studies impossible.

10.1.2 Adaptive management and learning

An advantage of embedding experimental designs into conservation projects is the potential to improve future management. Adaptive management entails information describing the progress and effectiveness of project actions being fed back to inform future management decisions (Figure 10.1, Walters and Holling, 1990). Furthermore, generating evidence within a project provides project managers with confidence about its relevance.

Evidence will have much greater benefits if it is also shared with the wider conservation community in addition to being used in adaptive management (Figure 10.1, Section 10.6). If the majority of conservation projects committed to some routine testing and documentation, then the massively enlarged evidence base could revolutionise effectiveness, fully justifying the costs required by each organisation (Rey Benayas et al., 2009).

10.2 Collecting Data Along the Causal Chain

Project planning often involves developing a theory of change (Figure 10.2) showing how actions are expected to lead to the desired change in outcomes and the target. Data can be collected at each stage along this causal chain (Table 10.1): this section considers the relative merits of each stage.

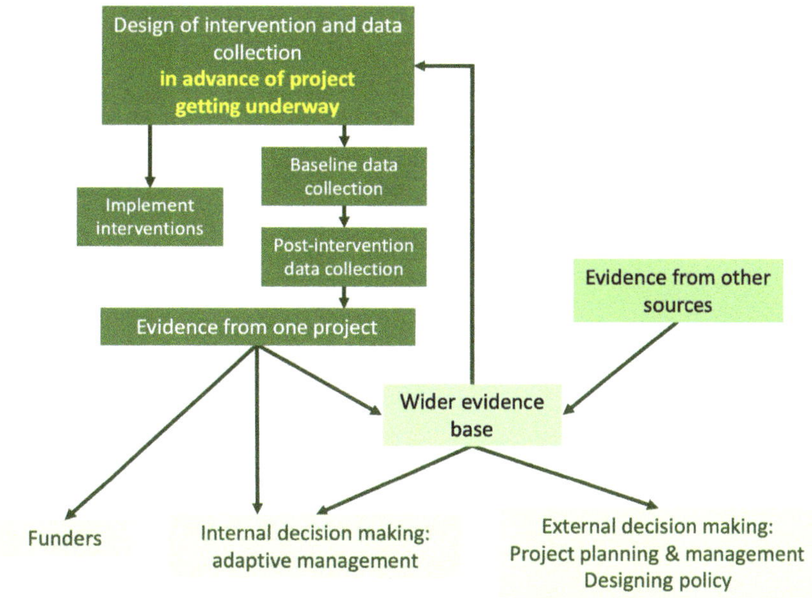

Figure 10.1 The flow of evidence, from appropriately designed data collection to a wider evidence base to inform internal and external decision making. (*Source*: authors)

There is often a demand, from funders or others, to demonstrate the ultimate impact of a project — the changes in the target as a consequence of the actions undertaken (right-hand box of Figure 10.2). However, in moving through the causal change from left to right, the link to each action becomes weaker due to the effects of other actions and known and unknown external factors. Therefore, unless a project is deliberately designed to measure or control for these effects, practitioners cannot separate the effect of their actions from that of other external, uncontrolled factors that also affect the project outcomes or the ultimate target variable. To isolate the contribution of the conservation actions, practitioners need a way to estimate the counterfactual value of the target that would be seen in identical conditions except for the absence of the conservation actions. The effect of the project actions can then be attributed to the difference between the project site and the counterfactual. Estimating counterfactual outcomes requires data from control or comparison sites where no project actions take place. Because of the challenges of finding such identical conditions in the natural ecosystems where conservation takes place, collecting data at multiple sites, both with and without the project action, will greatly increase our confidence that any observed difference is due to the actions implemented.

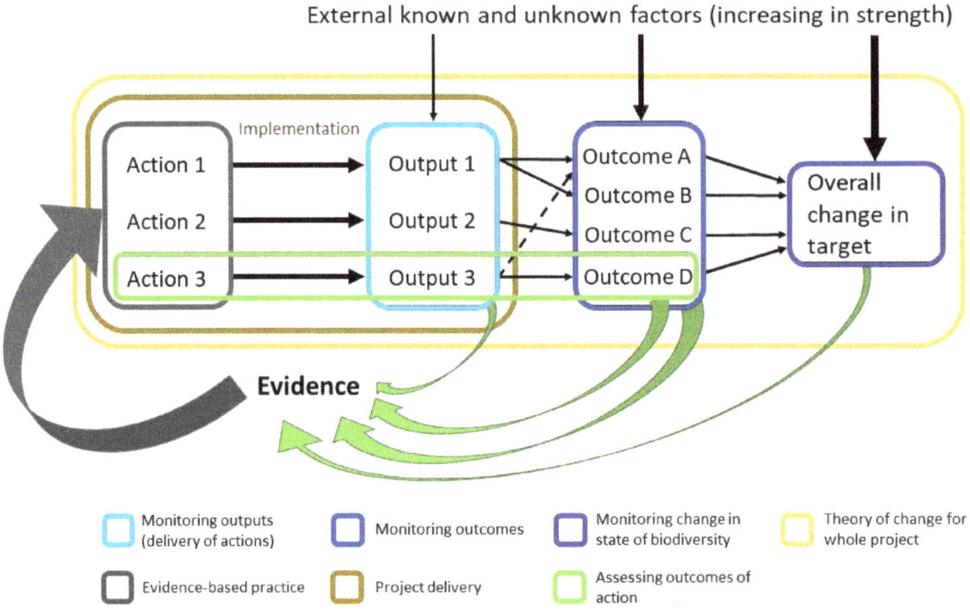

Figure 10.2 Options for collecting data along the causal chain. A conservation project consists of a series of actions. Each action (organise workshops, fund creation of islands for nesting birds) is implemented at a particular scale, context and efficiency, which determines the outputs (fishers attend workshops, islands created). These outputs may then have a direct outcome (bycatch of seabirds reduced, birds nest on islands), which may be additive or interactive (dotted line). The outcomes can result in changes in the ultimate target (seabird populations). Each stage may be influenced by known (climate, another project) and unknown (changes in fish abundance, an undocumented predator) external factors. The relative importance of external factors will usually increase along the chain. The figure also shows a range of processes: evidence-based practice (adopting measures shown to be effective), project delivery (ensuring actions are delivered), assessing the outcomes of actions (testing effectiveness) and testing the theory of change. (*Source*: authors)

Table 10.1 The different stages of a project life cycle (see Figure 10.1) at which data can be collected, with examples of the type of data at each stage. Not all projects will collect data at all stages. Each stage is described in the relevant section in the text below.

Stage	Description	Examples of monitoring data	Section
Output	The direct deliverables produced by the project	Number of ditches blocked in a drained peatland Area of reed replanted	10.2.1
Outcome	Changes in direct target of actions	Abundance of invertebrates in rewetted peatland Area of reed the following spring	10.2.2

Stage	Description	Examples of monitoring data	Section
Test of the effectiveness of an action	Change in the outcome of an action compared to an untreated control or an alternative treatment	Abundance of dragonfly larvae in ditches with weed cutting compared to those with no cutting Survival of an invasive plant treated with herbicide at two different times of year	10.2.3
Causal impact	Change in ultimate target of a project, compared to a control without action (thus usually larger scale than test of an action and sometimes involving programmes with multiple actions)	Change in the bird community in areas with restoration programmes compared to control areas Difference in fish populations in areas with community fisheries programmes compared to control areas without programmes	10.2.5
Descriptive impact	Change in ultimate target of the project, without a comparison to demonstrate causality	Change in number of waterfowl, plants and insects in a site after ditch blocking, flooding and changing grazing Change in water quality at the site after a programme of reed planting, regenerative farming and beaver reintroduction	10.2.6
Ambient monitoring	Ongoing monitoring not directly related to actions	Number of eagle pairs nesting within a protected area each year	10.2.7
Case study	Narrative account of actions/ project, including context, action taken, data collected and lessons learned. Rarely includes tests.	Description of the site, its history and characteristics, the actions taken, any challenges met, results observed, and lessons learned	10.6.1

In a review of monitoring in conservation, Mascia et al. (2014) identified that conservation practice tends to be good at ambient monitoring (measuring status and change in ambient social and ecological conditions, independent of any conservation intervention), measures for assessing how management is progressing (inputs, actions, and outputs) and measures for assessing performance (assessing progress toward desired levels of specific actions, outputs, and outcomes). However, it is much weaker at determining the intended and unintended causal impacts of conservation interventions or synthesising findings to improve practice. They concluded that this is a serious gap as ensuring that such learning is generated, and builds on existing knowledge, is key to delivering improved conservation outcomes (as shown in Figure 10.1).

In practice, a monitoring and evaluation programme for a project is likely to comprise a number of the approaches described in Table 10.1. For example, the landscape restoration projects funded by the Endangered Landscapes Programme, based in the Cambridge Conservation Initiative, collect information to monitor their outputs (generally used internally by organisations to assess project progress), and their outcomes (to report to funders and others). Each project also has to include a documented test of an action (to improve the project and society's understanding of what does and does not work).

10.2.1 Monitoring project progress — outputs

A routine element of project management is to keep a record of the delivery of project outputs — the activities or deliverables produced by the project. For example, a project that aimed to reduce mangrove loss by raising awareness and changing the behaviour of shrimp farmers may include the output 'Hold a series of workshops for shrimp farmers'. The output could then be measured via the number of workshops held and the number of farmers that participated. This is useful information that can be used internally, or reported to funders, to ensure that the project is on track with its planned programme of work.

However in conservation, outputs and outcomes may not be directly linked (the workshops may be well attended but not result in behaviour change); unless this link is proven, outputs cannot be used to reliably measure the success of an action or project. This has been a major criticism of many recent large-scale tree-planting initiatives, carried out as part of efforts to mitigate climate change, where success has often been measured by the number of trees planted, rather than by any long-term ecological or carbon sequestration benefit (Lee et al., 2019). In many cases the survival rate of trees is very low, meaning simple monitoring of outputs is not an accurate assessment of success in the medium or long term.

However, in cases where the action is already known to be effective (i.e. the link between output and outcome is proven), measuring outputs may be sufficient to assess project success. Thus in health, a clinic delivering vaccines that are known to be effective at improving health outcomes may report success as the number of people vaccinated, rather than looking at subsequent disease rates in the community. This is the 'project delivery' box in Figure 10.2. An ecological equivalent might be to provide barn owl nest boxes, which are well known to be effective (Johnson 1994, Bank et al. 2019), to farms with suitable habitat along with instructions for their positioning, but not ask for any monitoring.

10.2.2 Monitoring project results — outcomes

In most circumstances, monitoring outcomes rather than outputs will provide a much better measure of a project's success (Figure 10.2). Outcomes tend to refer to the medium-term effects of actions taken, and often relate to the objectives of a project. Examples could include the number of trees that are alive after 5 years (rather than the number of trees planted), the rate of adoption of the new fishing technique (rather than the number of fishers attending workshops), or the number of tourists visiting a national park (rather than the creation of a new visitor centre).

The challenge when monitoring outcomes is to know how much of any observed change can be attributed to a project's actions. For actions known to be effective, where the casual chain is simple and understood, and where there are few other potentially confounding factors, it may be reasonable to attribute the change in the outcome to the actions undertaken. For example, creating some small islands and counting the number of birds nesting on them is likely to be a reasonable approach to assessing effectiveness. But in many cases, with more complex causal chains and the potential for considerable impacts from external factors, it is difficult to confidently attribute any observed changes in an outcome to the project's actions.

10.2.3 Testing effectiveness of actions

An excellent option for improving our understanding of the consequences of conservation actions is to identify opportunities to experimentally test single actions as part of conservation practice (Section 10.3). Although such tests are rarely currently included in conservation projects, there is an appetite to routinely include an element of testing in funding proposals for conservation projects amongst funders and practitioners alike. When asked what proportion of conservation grants should be allocated to testing intervention effectiveness, there was considerable overlap in their responses, with practitioners tending to prefer slightly larger percentages (median 3–6%) than funders (median 1–3%) (Tinsley-Marshall et al., 2022).

10.2.4 Various definitions of 'impact'

Project managers may be asked to demonstrate the difference that a project or programme of work has made to its ultimate target(s) (Figure 10.2). Funders in particular are often interested in such measures of overall impact, to show that the interventions that they have supported have brought about change. However, across different parts of the conservation community, there is variation in the use and meaning of the terms *impact* and *impact evaluation*. In much of the conservation community, impact describes the changes in the ultimate target that take place between the beginning and end of a project, without necessarily rigorously demonstrating attribution. We refer here to this general meaning of impact as *descriptive impact* (Table 10.1). In contrast, in the impact evaluation community *impact* has a different and more precise meaning, where it refers to the change in target status during a project in comparison to a control or counterfactual where no action is taken (White, 2010). Here, we refer to this experimental form of impact as *causal impact*. As an example, the descriptive impact of a project could describe the changes observed in its target at the end compared to the start (e.g. the change in fish populations and local livelihoods at the end of a reef restoration programme), while the causal impact would require a comparison of these changes with a counterfactual (e.g. the difference in the changes in fish populations and livelihoods at the project site compared to another similar site where no actions were taken, 10.3.5).

Prior to applying any specific impact evaluation method or study design, it is important to clearly define the scope of the project and the outcome variables of interest. The following five steps outlined by Glewwe and Todd (2022) are useful when starting to design a causal impact evaluation:

1. Clarify the project and outcome variables of interest.
2. Formulate a theory of change to define and refine the evaluation questions.
3. Depict the theory of change in a results chain i.e. inputs => action=> outputs => outcomes => target (impact)
4. Formulate specific hypotheses for the impact evaluation.
5. Select performance indicators for monitoring and evaluation.
6. Design a sampling program and the details of analyses.

10.2.5 Monitoring final target of actions: descriptive impact

Descriptive impact is the change seen in the final target when a project finishes, without any attribution of the change to the actions undertaken. This approach is very commonly used across conservation projects for a number of reasons: a package of actions is often implemented together, it is challenging and resource intensive to find and monitor a comparable control site, the mechanisms linking actions and outcomes are poorly understood, and a culture of experimental testing is not widespread among conservation organisations or funders. Although simple monitoring of changes in the targets of a project is widely used and potentially useful, it is important to recognise the limitations and to ensure that appropriate caveats are attached.

As with the discussion of outcomes (section 10.2.2), the adequacy of using descriptive impact to assess the success of a project depends on the strength of the causal change and the likely impact of external factors. Consider a project removing a barrier in a river to allow the movement of migratory fish. If there were no such fish present upstream when the barrier existed, but after removal both fish and spawn are recorded, it is probably unnecessary to monitor another river, where no such barrier removal took place, in order to attribute the improved status of the fish population to the project. In this case, simply monitoring a before-and-after change along the length of the river is sufficient to demonstrate project impact. In another project, deer numbers are reduced to allow tree regeneration, and after two years the change in the number of tree seedlings is monitored. Without comparing these results to an area without reduced deer numbers, it is difficult to assess the success of the project: any regeneration may actually be due to the unusually suitable weather or a collapse in the rabbit population. Descriptive impacts often comprise a significant component of the results reported in project case studies (10.6.1).

10.2.6 Monitoring final target of actions: causal impact

To confidently assess the difference a conservation project has made, and hence quantify the causal impact, in most cases comparative data describing the target state either before or without the intervention (or both) are needed, in addition to an assessment of the final state of the target of the project. Section 10.3 describes how measures of causal impact can be achieved.

10.2.7 Ambient monitoring

Ambient monitoring involves collecting data describing ecological or socio-economic states but which is not directly aiming to understand the effects of particular actions or projects. For example, it may provide surveillance of the spread of an invasive species, the shift in a coastline or the national or regional trends in a species or ecosystem. This can provide a wider understanding, identify issues and indicate whether overall conditions are improving or deteriorating.

10.3 Incorporating Tests into Conservation Practice

Implementing agencies, funders and policy makers want to know the effectiveness of actions. However, as previously described, simply monitoring changes in the target variable rarely discriminates whether any changes seen were due to the actions undertaken, rather than other factors. Is the provision of predator-proof nest boxes responsible for an increase in a bird population, or was it a particularly good year for insects? Is a reduction in snaring related to a project's support for alternative livelihoods or because young people are leaving the villages for the cities? Furthermore, the actions may have had beneficial effects on the target population, but these could be masked by a wider overall decline caused by other factors.

In order to attribute observed changes to an action, tests in which the changes in the target variable are compared in the presence and absence of the action are needed. This helps eliminate rival explanations, such as external factors, for the patterns observed in the monitoring data. The most robust tests, such as randomised controlled trials, include replication of treatments, comparisons or controls, and random allocation of treatments and controls. Studies that contain only one or two of these components of experimental designs are less robust, but still more informative than those that contain none. Quasi-experimental designs may be appropriate for interventions in which the treatment is already determined, such as an assessment of the effectiveness of marine reserves in which their location has already been decided (Section 10.4.3).

10.3.1 Why include a test?

Experimental tests are usually the best way of generating rigorous evidence for the effectiveness of particular actions (Ockendon et al., 2021). Routinely including tests in conservation projects will both increase the number of actions tested and the range of contexts in which they are tested. An increased number of studies on any particular action increases confidence in the results and enables consideration of heterogeneity and the variation in outcome between studies, meaning the evidence will be relevant for a wider range of users.

Of course the results of tests can also be used immediately by conservation organisations and projects to enable adaptive management (Section 10.1.3) and improve practice. This requires a deliberate process to embed learning and evidence generation into project implementation (Wardropper et al., 2022).

10.3.2 When to test actions?

A key skill is to be able to recognise where and when trials can be included most easily and effectively in conservation practice. The best opportunities occur when it is relatively straightforward to integrate a test into ongoing work (for example the same action is being carried out on multiple independent occasions); the necessary skills and capacities are available; and the results from the test are likely to be of interest to the organisation or others (Figure 10.3). Collaboration and codesign between stakeholders, the practitioner and academic partners can be an effective way to identify and capitalise on these opportunities (Kurle et al., 2022).

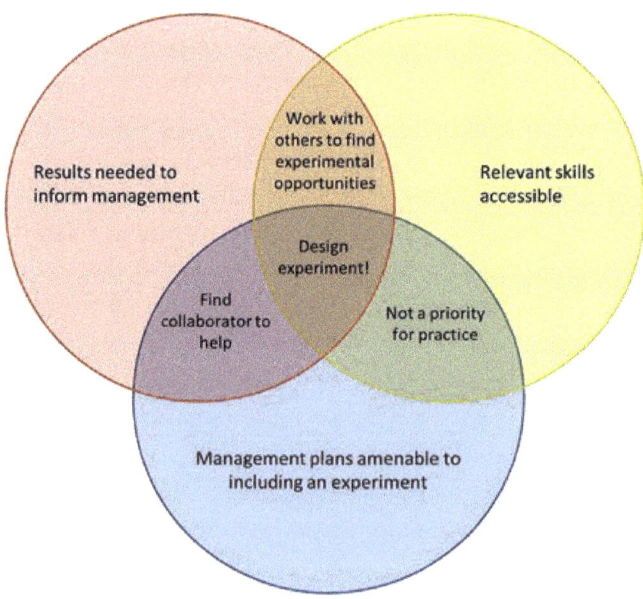

Figure 10.3 How a combination of skills, the need for results, and the existence of opportunities determines whether an experiment can usefully be included in conservation management. The optimal conditions for carrying out an experiment in practice arise when all three overlap. (*Source*: Ockendon et al., 2021, CC-BY-4.0)

Below we briefly consider the key principles of experimental and test design in the context of conservation practice.

10.4 Design of Experiments and Tests

There are five key elements (Crawley et al., 2015) that can improve the design of experiments and tests: (1) Randomisation; (2) Controls or comparisons; (3) Data sampled before and after an intervention or impact has occurred; (4) Temporal and spatial replication; (5) Preregistration. These elements can be combined in different ways, to give the six most commonly used test designs: Randomised controlled trials (RTC), Randomised before-after control-intervention (R-BACI), Before-after control-intervention (BACI), Before-after; Control-intervention; and After designs. These different designs vary considerably in their likelihood of producing an accurate answer, as tabulated in Table 2.5 (Christie, 2019, 2020b). Thus it is important to note that the components included in the design of a study make a significant difference to the likelihood of it giving useful results.

10.4.1 Randomised controlled trials

Randomised controlled trials (RCTs) and Randomised Before-After Control-Impact (R-BACI) experiments are the most rigorous study design for reliably and accurately estimating the magnitude and direction of an effect (Table 2.5, Christie, 2019, 2020b); therefore, where they are practically feasible, they should be the first choice for testing conservation actions. Although RCTs have a reputation for being complicated and difficult to implement in conservation, in practice they are carried out routinely, particularly when testing simple actions that can easily be replicated. This is reflected by the presence of over 1400 RCTs of conservation in the Conservation Evidence database — 16.7% of the total number (Christie et al. 2020b). For example, a simple replicated controlled trial demonstrated that the addition of an artificial, moulded 'form' into nest boxes for swifts *Apus apus* increased occupancy rates compared to boxes without such a 'form'. This study, which monitored multiple swift nest boxes across four sites, demonstrated how replication, stratification and controls can be applied to generate evidence to inform future design using a straightforward and easy-to-implement trial design (Newell, 2019). There are innumerable opportunities to design such simple yet robust experiments, as shown by a group of conservation researchers and practitioners identifying a hundred possible experiments that could test the effectiveness of actions using an RTC and would also produce useful results for practice (Sutherland et al., 2022).

Carrying out RCTs at the level of conservation programmes or projects, as used in formal impact evaluation, may be more complicated in conservation, although the approach has been widely adopted in the medical and international development sectors (Ferraro and Pattanayak, 2006; White, 2010). As with any aspect of project design, the decision of whether to use an RTC will depend on the value of the data produced versus the time and resources required to set up, implement and monitor the trial, within any practical constraints that exist. Time and financial resources are frequently cited as limiting factors in the more widespread adoption of randomised controlled trials as a project evaluation approach in conservation (Curzon and Kontoleon, 2016).

10.4.2 Randomisation

If the controls and replicated plots are spatially clumped, with treatments in one area and controls in another, it is possible that confounding factors may bias the results. For example, if a study testing the effect of nest box height on breeding success was designed such that the boxes placed at 2m height were all on one side of the site, those at 3m in the middle section, and those at 4m at the other end of the site, then any gradient across the site in an underlying environmental factor that influences nesting success could override any effect of nestbox height. In such cases, controls and replication are insufficient; experimental units also need to be interspersed in space so that spatial confounding factors are random with respect to the treatments.

The most commonly used method to allocate interventions to plots or experimental units is to randomly assign them as controls or treatments. The simplest way to achieve this is to number each unit and use a random number generator (there are many online options) to assign each unit to treatment or control. This method works well if the study area is homogeneous and has a relatively large number of replicates. However, if the study includes a small number of larger plots, then randomisation can result in suboptimal plot distributions, if, for example, the three replicate control plots all happen to be assigned to one side of the study area. This is a particularly important issue if there is a known underlying gradient at the site. In these cases, blocking can be used prior to randomisation (either by creating contiguous experimental blocks, each containing a number of plots or sites or by matching plots into blocks by some other criterion, e.g., slope in peatlands, which links to hydrology and peat depth) to reduce the potential differences in environmental conditions between treatment and control sites.

10.4.3 Quasi-experimental designs or natural experiments

Sometimes, experimental tests with randomised allocations of treatments and controls are not feasible in conservation for ethical, logistical or political reasons. Alternatively, in some circumstances, for example, where an intervention is being carried out numerous times albeit not under the control of the data analyst (e.g. agri-environment prescriptions) statistical approaches can be used to estimate the impact of an interventions (Schleicher et al. 2020). Identifying opportunities to take advantage of such 'natural experiments' can greatly improve understanding of the effectiveness of national or regional policy decisions. In such cases, methodological advances in statistical approaches have allowed causal inferences to be drawn from non-experimental data (Ferraro and Pattanayak, 2006). These quasi-experimental designs allow us to maximise the potential learning from large observational datasets by using statistical approaches to overcome the potential biases that are likely where randomization is not possible (Schleicher et al., 2020). Such quasi-experimental statistical methods, including matching, instrumental variables, or difference in difference, have been widely used in other disciplines, particularly economics. In combination with remotely sensed data, they have huge potential for building the evidence base in conservation.

These methods attempt to identify treatment and control groups that are similar in their observed characteristics. An example could be a comparison of rates of forest loss inside a protected area with rates in areas that are not protected but are otherwise similar in terms of remoteness and accessibility (e.g. Eklund et al., 2016). In this example, distance from urban centres is a likely so-called confounding variable, as sites far from human settlements tend to be more likely to be assigned for protection but are also less prone to suffer from deforestation as they are harder to reach. Using quasi-experimental designs makes it possible to control for the effect of such confounding variables in order to obtain a true estimate of the impact of the treatment in question (here site protection).

Matching is one of the more commonly used impact evaluation methods in conservation because it lends itself so well to evaluating an intervention post hoc. With matching methods, it is possible to reduce differences between treatment and control units in terms of confounding variables, thus aiming to isolate the intervention effect. The idea is simply to pair each treatment unit with an observably similar control unit and then interpret any difference in their outcomes as the effect/impact of the project intervention. A major benefit of matching is that it has relatively few data requirements, but on the other hand it does assume that there are no unobserved confounders (Schleicher et al., 2020). In conservation, another possible difficulty is if control units that are similar enough to the treatment units simply no longer exist. For example, if forest only remains at high altitudes and these areas of forest are all protected, then there are no control units of unprotected forest at high altitudes available for the comparison. The best designs often combine approaches (Ferraro and Miranda, 2017).

It is important to acknowledge that not all conservation projects or programmes are amenable to experimental or quasi-experimental study designs and for many projects it will not be possible to collect data in a way that can reliably distinguish a treatment effect from the most plausible hidden biases (Ferraro, 2009). However, an understanding of counterfactual thinking and confounding factors is beneficial for any conservation project. As a practical step, most conservation projects, when preparing theories of change, could include a consideration of the possible confounding factors that might also affect the outcome of interest. Informed guesses or judgements can then be used to adapt project implementation to minimise or account for these biases. Key here is that project implementation could be adapted in response to both evidence generated from the project in question and evidence from analyses elsewhere with better internal and external validity (Ferraro, 2009).

10.4.4 Controls and comparisons

Some form of comparison is vital for assessing the magnitude of the impact of an intervention, although monitoring areas where no action has taken place has traditionally not been a priority for conservation. Comparing a species, habitat or community before and after an action has been carried out can provide some information about the change that has taken place, but results are vulnerable to other external factors and sources of bias that change through time, such as weather conditions, disease or climate change. The inclusion of an untreated control, with such a before-and-after comparison (a BACI design), significantly improves the value of the results (Christie et al., 2019).

Practical challenges are likely to have contributed to the rarity of the inclusion of controls or comparisons in conservation projects. These include that the immediate benefits of comparative treatments are not obvious to landholders or funders or that control areas are not covered by the same funding as areas of conservation activity. Related to this, management decisions at these control sites made over the course of the study may sometimes be made independently of the data collection needs of the experimental design. However, in many cases these challenges can be overcome once the value of including a comparison site is realised and communicated.

Although an untreated control is often the default comparator, it may be more informative to compare the results of a new intervention with ongoing or traditional management practice to assess their relative effectiveness and see if the new approach works better (Smith et al., 2014). For example, if a site manager is wondering whether changing the timing of herbicide application improves control of an invasive plant, then it would be much more useful to compare the results with an area where herbicide is applied at the conventional time, rather than an untreated area where no herbicide is applied at all.

There is no guarantee that another site selected as a control will not differ from the treatment site in some key aspects (such as soil moisture, land-use history or elevation), limiting its ability to serve as a robust comparison. To check this, vegetation, soil and environmental sampling should be conducted at the outset to determine if sites differ in any important aspects. Where feasible, subdividing each management site and retain areas within each that are untreated (blocking) will help minimise these biases. Replication (10.3.3) of treatment and control sites is also important in helping to demonstrate a causal link between interventions and changes in the target. For example, a study looking at the effectiveness of creating deadwood in forests in the Scottish highlands is testing three possible approaches (winching, cutting and ring-barking, plus an untreated control) on trees at five blocks of replicates in different areas of the forest. If the three treatments result in the same changes in the invertebrate community across each of these blocks, this will provide good evidence that it is the treatments, rather than another factor, that are causing these changes.

10.4.5 Replication

Replicating, the independent repeating of the action and control, increases the precision of the estimated effect and so will increase the confidence that observed results are due to the action taken rather than any other factor in the wider environment. Replication can take place either in space (e.g. multiple plots, community groups, protected areas) and/or through time (e.g. repeating treatments on multiple dates or across several years, Stewart-Oaten et al., 1986).

Formally, the number of replicates needed to detect a certain effect size with a specified probability can (if sufficient appropriate information is available) be calculated using a power analysis (Crawley, 2015). However, within the constraints of conservation, a general rule is 'the more replicates the better'. More practically, a minimum of three replicates is required to estimate mean and variance. Larger numbers of replicates are more powerful and also allow for unforeseen and uncontrolled circumstances, such as part of the study site being flooded.

However, for many large-scale conservation interventions, such as reintroductions of animals or alterations to whole ecosystems (e.g. wetland creation or dam removal), there may be substantial or even insurmountable challenges to replication. In these cases, where the options for monitoring the effect of an action is to carry out an unreplicated trial or not to do it at all, there may well be benefits to carrying out unreplicated actions (e.g. Davies and Gray, 2015; Ockendon et al., 2021), especially if before- and after-treatment sampling is possible. If an unreplicated trial is undertaken, it is especially important that the necessary caveats are included in any interpretation of the results, in particular in relation to ascribing causality between the action and any changes seen in the target. Developing an understanding of the mechanism by which actions cause any observed changes in the target can aid the drawing of inferences from unreplicated studies.

Another justification for carrying out tests with very small numbers of replicates (or even none) is that, if multiple studies of the same or similar actions exist, these can be combined in meta-analyses to investigate the generality of the results (Gurevitch et al., 2001). For example, an analysis of the effects of eradicating invasive mammals on seabirds across 61 oceanic islands (where each individual study had n = 1) found that most seabird populations had increased, with a mean annual recovery rate of 1.12 (Brooke et al., 2018). Here, the results of an intervention that would be very difficult for a single project to replicate were combined across a large number of individual studies to produce an estimate of the average impact of the action.

10.4.6 Preregistration and project declarations

Conservation actions take place within highly complex and dynamic settings (Catalano et al., 2019). Therefore, the failure of conservation actions should be accepted as unsurprising. However, publication bias, whereby organisations and researchers only communicate positive outcomes and where journals preferentially publish positive results, is a pervasive problem in conservation (Wood 2020), with successful projects reported at four times the rate of unsuccessful ones (Catalano et al., 2019). From an individual or organisational standpoint, this may be driven by the need to maintain reputation, particularly when reporting to funders or applying for more funding to work on similar conservation actions. However, a bias towards reporting positive outcomes will inflate the apparent effectiveness of an action and may lead to other individuals or organisations applying it, in the mistaken belief that the action is generally effective. The failure to report negative outcomes means we lose a valuable opportunity to learn from past mistakes or to understand factors affecting success, and will likely result in wasted effort using less effective actions (Wood, 2020).

There is therefore a need to develop a model whereby the reporting of well-designed actions is the norm, and rewarded even when results are neutral or negative (Burivalova et al., 2019). One proposed solution is 'pre-registration', whereby the project design, data collection and analysis are defined and published prior to collecting data to measure the outcomes; this is the norm for medical trials (Nosek et al., 2019). One can always deviate from the pre-registered plan, but one should report that deviation. One can also do exploratory analyses that were not anticipated, but again, such analyses ought to be reported as such. The idea is to make it clearer what kind of evidence confirms theories of change and what kind of evidence is more speculative or indicative of a theory of change because the evidence was only discovered after an ex post exploration of the data.

Given the time and effort required for detailed, refereed pre-registration, such as is seen for medical trials, this is unlikely to become widespread in conservation. This is likely to be a particular issue for small organisations with limited resources, where there are a large number of small projects that apply a similar approach, or where actions are time sensitive and delays caused by pre-registering would have negative impacts. A less onerous alternative pathway is for simple preregistration reporting the intent to do a study and including the hypothesis (adding X will increase Y) and the primary and secondary outcome variables. The website AsPredicted shows how the main aspects of a trial can be covered by answering a series of questions: Have any data been collected for this study already? What's the main question being asked or the hypothesis being tested in this study? Describe the key dependent variable(s) specifying how they will be measured. Specify exactly which analyses you will conduct to examine the main question/hypothesis. How many and which conditions will be used? Describe exactly how outliers will be defined and handled, and your precise rule(s) for excluding observations. How many observations will be collected or what will determine the sample size? AsPredicted then creates a time-stamped URL that can be shared with referees and linked back to when the final results are written up. Such an approach could easily be adopted in conservation.

Another related option is project declaration, whereby collaborations are established to create repositories in which conservation projects on certain topics can be efficiently recorded pre-intervention. Providing project data to such repositories should be free, simple and quick, and should describe whether the outcomes will be monitored. By publicly sharing their plans before implementation, projects can be encouraged (for example by funders) to follow this up by adding their findings as they become available; this should help reduce the bias associated with only recording successful actions.

10.5 Value of Information: When Do We Know Enough?

Decisions often need to be made during the planning of conservation projects and ongoing management about the balance of resources and effort that are allocated to monitoring and data collection versus action on the ground (McDonald-Madden et al., 2010). Given the scale of the biodiversity crisis and the limited funding available to tackle it, it can often feel that any diversion of limited resources and funds from practical conservation action is hard to justify. However, as we have discussed in this chapter, there are many reasons why generating new knowledge is an important component of any nature conservation project that is likely to improve the ultimate outcome.

The value of information describes the benefit a decision maker would gain for additional information prior to making a decision. This is important in deciding when you should delay a decision until more information about its likely effectiveness is known and when you should act on the information you already have. For example, a project is working on the recovery of a highly endangered bird species where chick survival is low; one proposed option is to take a number of chicks into captivity to attempt a captive rearing programme before releasing them back into the wild. However, there is uncertainty around the likely success of both the rearing programme and the release of sub-adults back into the dwindling population. How important is it to reduce these uncertainties? Is it better to collect more data to understand the likely effectiveness versus the risk of delaying, during which the population is likely to decline still further (Bolam et al., 2019)? In general, collecting further information is most worthwhile when uncertainty is high, further data collection is likely to substantially reduce the uncertainty, and the gain in effectiveness is important (Canessa et al., 2015).

10.6 Writing Up and Sharing Results

10.6.1 Generating case studies

Case studies are often written by conservation practitioners at the end of a project or programme of work, to document their experiences and share lessons learned. They generally contain descriptions of the site and context, and the actions and monitoring activities carried out, followed by the results observed and lessons learned, in a narrative form. Case studies often contain substantial practical details, but rarely include any experimental component to the design. Examples of case studies could range from descriptions of how a community engagement programme has been implemented to reduce unsustainable harvesting, to describing the process of restoring a grassland site.

Case studies are often written and shared between practitioners and may be particularly useful for those working in a similar context. Drawing together a large number of case studies can also begin to allow generalities and patterns to be drawn out (e.g. a database of dam removals in Europe https://damremoval.eu/case-studies; Panorama provides an extensive database of case studies in conservation https://panorama.solutions/en). However, challenges that are frequently associated with drawing of generalities from case studies include a failure to report final outcomes, particularly for projects or components that are deemed unsuccessful, that information may be scattered across various media and formats making synthesis extremely time consuming, and variability in the consistency and scope of the data recorded (Gatt et al., 2022). Overall, case studies tend to be weak at providing evidence for the effectiveness of actions, as they lack controls or replication, and often include multiple actions carried out simultaneously or adapted over time. This means that effects cannot be clearly disentangled, and it is likely to be difficult to extrapolate results to other contexts. Case studies describing the story of a project are usually more readable than scientific papers, easier to relate to and can provide inspiration for others. They provide a useful means of describing general approaches and how problems were overcome and may also include tips for practical application. They are useful in providing the bigger picture used alongside other methods that are more effective at assessing effectiveness.

The creation of common reporting frameworks could improve consistency, reduce reporting bias and ease the tracking of the progress of multiple projects toward national and global targets (Eger et al. 2022).

10.6.2 Publishing peer-reviewed articles

Publication in a peer-reviewed journal is another option for disseminating the results of a test of a conservation action, with the benefits of verification and authentication.

Publication in this form has the additional advantage that the journal curates the report of the research, securing it for the future and guaranteeing its availability. Some journals actively seek papers written by practitioners in formats more appropriate to them (e.g. *Conservation Evidence Journal, Oryx, Conservation Science and Practice,* and *Ecological Solutions and Evidence*).

Communicating ideas and findings in a way that is both informative and interesting may require some practice. Some journals provide help with writing (e.g. Fisher, 2019, *The British Ecological Society Short Guide to Scientific Writing*). An additional resource is AuthorAID, a global network that provides support, mentoring, resources and training for researchers.

Box 10.1 summarises the details to include in a published article to facilitate the extraction of the results for evidence collation (e.g. Conservation Evidence, systematic review, meta-analysis).

One consideration is whether you (or your funder) would like to make the article open access, ensuring it is available for all to read online, free of charge. There is variation between journals in this regard: *Conservation Evidence Journal* is open access and there are no publication charges. Most other open access journals levy an article-processing charge, but in practice this may be waived or reduced, in particular for residents of countries on the Research4life eligibility list.

Here is a recommended pipeline for the writing, submission and sharing of results in a peer-reviewed journal:

1. Speak to colleagues and/or search online to identify journals that could potentially be suitable for the publication of your findings. Pertinent considerations include whether a journal supports the publication of case studies, results of interventions and of negative or non-significant results. Is the journal open access (if this is your preference) and does it have an article processing charge? If an article processing charge is levied and you do not have access to funds for this purpose, are waivers/discounts of the charge available? If yes, does the process of applying for a waiver/discount appear straightforward?

2. To ensure that you have selected a trusted journal, use the tools and resources at thinkchecksubmit.org.

3. Use an article template for your writing. If the journal you have chosen does not provide a template, repurpose a template from one that does.

Box 10.1 Details to include in publications to enable data to be included in evidence collations

- Study design, e.g. controlled, replicated, randomised (Table 4.5)
- Year(s)/season(s)/month(s) when action(s) were implemented and when monitoring was undertaken
- Number of independent sites/individuals/replications
- Habitat type(s)
- Location(s) of study including specific site(s) where possible, country and latitude and longitude
- Target species/habitat
- Description of the action(s) tested including methods of implementation and of the control/comparison, size of sites (i.e. area over which action was applied), distance between sites, duration of action, number of times carried out per site
- Costs of action (see Chapter 4, Section 4.3.8)
- Monitoring methods for each outcome measured, number of replicates, size of plots/sampling units, distance between replicate plots/sampling units
- Description of statistical tests used
- Summary data describing the treatment and control (i.e. no treatment), or before and after each action, or for comparing actions. Provide numbers if possible (with units) rather than qualitative measures, e.g. how many individuals were treated, or how many species were present in each treatment and control. Provide the mean value for both the treatment and control group, plus the standard deviation or standard error and n (sample size) and a way of quantifying the uncertainty about the impact estimate that comes from sampling variability (e.g. p-values, confidence intervals, posterior distributions, etc.).

4. Study the content of a few articles in your chosen journal for format and style, and for inspiration. All trusted journals provide guidelines/instructions for authors that detail their preferred format and style.
5. Use a bibliography manager, such as Zotero or Mendeley, to manage citations and references.
6. Include the details covered in Box 10.1 to enable results to be used in a collation of evidence (e.g. meta-analysis).

7. Ensure that all figures convey their message or purpose clearly and unambiguously (see Fisher 2019, for guidance) and the data can be extracted easily. Provide detailed, self-explanatory captions for all figures and tables.
8. Obtain a free ORCID iD and include it with address and affiliation details. Using this persistent unique identifier will ensures recognition for all publications.
9. Full datasets can be shared in the form of supplementary material to the article, or in a data repository. Follow the 10 simple rules of Contaxis et al. (2022) to ensure your research data are discoverable.
10. Ensure any co-authors have read and approved the manuscript (or better, involve them in the writing).
11. Aim to optimise your text for search engines (see Fisher, 2019, for guidance), as this will improve the discoverability of your article following publication.
12. Following peer review, but before acceptance, the journal is likely to call for a revision of your article: provide the revision promptly and supply a list of the changes you have made.
13. Finally, publication is not the end of the story, but rather the beginning: promote the findings using additional means such as a blog, Twitter or other social media feeds, and a press release (your institution or the journal's publisher will normally be able to help with this). A public Google Scholar profile will help in the discoverability of your article.

Not all data collected by conservation projects and organisations will be written up and published as a peer-reviewed article. Peer-review publication can be a lengthy and time-consuming process that conservationists may not have the time or enthusiasm to undertake. Instead, the priority might be to use results to produce a report that communicates project outcomes to stakeholders or other practitioners, or to inform adaptive management. These reports and case studies can still be published on organisational websites or in a practitioner-focused information clearing house like Applied Ecology Resources (https://www.britishecologicalsociety.org/applied-ecology-resources/about-aer/), a repository of conservation reports, articles, case studies, fact sheets. Applied Ecology Resources contains thousands of documents and is fully searchable, thus ensuring that reports are sharable, discoverable and permanent.

Whatever the format, data collected describing the outcomes and impact of conservation actions is likely to have immense value. Evidence describing many conservation actions across a wide range of contexts and stages in the causal chain (Figure 10.2) is still far too scarce (Christie et al., 2020a), and the value of this information will grow as more evidence is accumulated, allowing patterns of success and failure to be understood. This evidence can result in better-informed decision making and prioritisation across conservation practice and policy.

References

Bank, L., Haraszthy, L. Horváth, A et al. 2019 Nesting success and productivity of the common barn-owl: results from a nest box installation and long-term breeding monitoring program in outhern Hungary, *ornis Hungarica* 27, 1–31, doi:10.2478/orhu-2019-0001

Bolam, F.C., Grainger, M.J., Mengersen, K.L., et al. 2019. Using the value of information to improve conservation decision making. *Biological Reviews* 94: 629–647, https://doi.org/10.1111/brv.12471.

Brooke, M. D. L., Bonnaud, E., Dilley, B. J., et al. 2018. Seabird population changes following mammal eradications on islands. *Animal Conservation* 21: 3–12, https://doi.org/10.1111/acv.12344.

Burivalova, Z., Miteva, D., Salafsky, N., et al. 2019. Evidence types and trends in tropical forest conservation literature. *Trends in Ecology & Evolution* 34, 669–679, https://doi.org/10.1016/j.tree.2019.03.002.

Cadier, C., Bayraktarov, E., Piccolo, R. et al. 2020. Indicators of coastal wetlands restoration success: A systematic review. *Frontiers in Marine Science* 7: 600220, https://doi.org/10.3389/fmars.2020.600220.

Canessa, S., Guillera-Arroita, G., Lahoz-Monfort, J.J. et al. 2015. When do we need more data? A primer on calculating the value of information for applied ecologists. *Methods in Ecology and Evolution* 6: 1219–1228, https://doi.org/10.1111/2041-210X.12423.

Catalano A.S., Lyons-White J., Mills, M.M. et al. 2019. Learning from published project failures in conservation. *Biological Conservation* 238: 108223, https://doi.org/10.1016/j.biocon.2019.108223.

Christie, A. P., Amano, T., Martin, P. A., et al. 2019. Simple study designs in ecology produce inaccurate estimates of biodiversity responses. *Journal of Applied Ecology* 56: 2742–2754, https://doi.org/10.1111/1365-2664.13499.

Christie, A.P., Amano, T., Martin, P.A., et al. 2020a. Poor availability of context-specific evidence hampers decision-making in conservation. *Biological Conservation* 248: 108666, https://doi.org/10.1016/j.biocon.2020.108666.

Christie, A.P., Abecasis, D., Adjeroud, M., et al. 2020b. Quantifying and addressing the prevalence and bias of study designs in the environmental and social sciences. *Nature Communications 11*: 1–11. https://doi.org/10.1038/s41467-020-20142-y

Coetzee, B.W. and Gaston, K. J. 2021. An appeal for more rigorous use of counterfactual thinking in biological conservation. *Conservation Science and Practice* 3: e409, https://conbio.onlinelibrary.wiley.com/doi/pdf/10.1111/csp2.409.

Contaxis N, Clark J, Dellureficio A, et al. 2022. Ten simple rules for improving research data discovery. *PLoS Computational Biology* 18: e1009768. https://doi.org/10.1371/journal.pcbi.1009768

Crawley, M. J. 2015. *Statistics: An introduction using R* (2nd ed.) (Chicester: John Wiley & Sons).

Curzon, H.F. and Kontoleon, A., 2016. From ignorance to evidence? The use of programme evaluation in conservation: Evidence from a Delphi survey of conservation experts. *Journal of Environmental Management* 180: 466–475, https://doi.org/10.1016/j.jenvman.2016.05.062.

Davies, G. M. and Gray, A. 2015. Don't let spurious accusations of pseudoreplication limit our ability to learn from natural experiments (and other messy kinds of ecological monitoring). *Ecology and Evolution* 5: 5295–5304, https://doi.org/10.1002/ece3.1782.

Eger A., Earp H., Friedman K., et al. 2022. The need, opportunities, and challenges for creating a standardized framework for marine restoration monitoring and reporting. *Biological Conservation* 266: 109429, https://doi.org/10.1016/j.biocon.2021.109429.

Eklund, J., Blanchet, F. G., Nyman, J., et al. 2016. Contrasting spatial and temporal trends of protected area effectiveness in mitigating deforestation in Madagascar. *Biological Conservation* 203: 290–97, https://doi.org/10.1016/j.biocon.2016.09.033.

European Commission (EU). 2015. The Habitats Directive. Brussels, Belgium, EU.

Ferraro, P. J. 2009. Counterfactual thinking and impact evaluation in environmental policy. In: *Environmental Program and Policy Evaluation: Addressing Methodological Challenges. New Directions for Evaluation*, ed. by M. Birnbaum and P. Mickwitz, 122: 75–84, https://doi.org/10.1002/ev.297.

Ferraro, P. and Agrawal, A. 2021. Synthesizing evidence in sustainability science through harmonized experiments: Community monitoring in common pool resources. *Proceedings of the National Academy of Sciences* 118: e2106489118. https://doi.org/10.1073/pnas.2106489118.

Ferraro P. J. and Hanauer M. M. 2014. Advances in measuring the environmental and social impacts of environmental programs. *Annual Review of Environment and Resources* 39: 495–517, https://www.annualreviews.org/doi/pdf/10.1146/annurev-environ-101813-013230.

Ferraro, P.J. and Miranda, J.J. 2017. Panel data designs and estimators as substitutes for randomized controlled trials in the evaluation of public programs. *Journal of the Association of Environmental and Resource Economists* 4: 281–317, https://www.journals.uchicago.edu/doi/10.1086/689868.

Ferraro, P. J., and Pattanayak, S. K. 2006. Money for nothing? A call for empirical evaluation of biodiversity conservation investments. *PLoS Biology* 4: e105, https://doi.org/10.1371/journal.pbio.0040105.

Fisher, M. 2019. Writing for Conservation. Fauna & Flora International, Cambridge, UK. https://www.oryxthejournal.org/writing-for-conservation-guide.

Fraser E.D., Dougill A.J., Mabee W.E. et al. 2006. Bottom up and top down: Analysis of participatory processes for sustainability indicator identification as a pathway to community empowerment and sustainable environmental management. *Journal of Environmental Management* 78: 114–127, https://doi.org/10.1016/j.jenvman.2005.04.009.

Gatt Y.M., Andradi-Brown D.A., Ahmadia G.N., et al. 2022. Quantifying the reporting, coverage and consistency of key indicators in mangrove restoration projects. *Frontiers in Forests and Global Change* 5: 720394, https://doi.org/10.3389/ffgc.2022.720394.

Glewwe, P. and Todd, P. 2022. Impact Evaluation in International Development: Theory, Methods and Practice. Washington, DC: World Bank. https://openknowledge.worldbank.org/handle/10986/37152 License: CC BY 3.0 IGO

Gurevitch, J., P. S. Curtis, and M. H. Jones. 2001. Meta-analysis in ecology. *Advances in Ecological Research* 32: 199–247, https://doi.org/10.1016/S0065-2504(01)32013-5.

Hughes, F. M. R., Adams W.M., Butchart S.H.M., et al. 2016. The challenges of integrating biodiversity and ecosystem services monitoring and evaluation at a landscape-scale wetland restoration project in the UK. *Ecology and Society* 21: 10, http://dx.doi.org/10.5751/ES-08616-210310.

Jellesmark, S., Ausden, M., Blackburn, T.M., et al. 2021. A counterfactual approach to measure the impact of wet grassland conservation on UK breeding bird populations. *Conservation Biology* 35: 1575–1585, https://doi.org/10.1111/cobi.13692.

Johnson P.N. (1994) Selection and use of nest sites by barn owls in Norfolk, England. *Journal of Raptor Research*, 28, 149–153.

Kurle, C. M., M. W. Cadotte, H. P., Jones, et al. 2022. Co-designed ecological research for more effective management and conservation. *Ecological Solutions and Evidence* 3: e12130, https://doi.org/10.1002/2688-8319.12130.

Lahoz-Monfort, J. J., and Magrath, M. J. 2021. A comprehensive overview of technologies for species and habitat monitoring and conservation. *BioScience* 71: 1038–1062, https://doi.org/10.1093/biosci/biab073

Leal, M. and Spalding, M. D. (Eds.), 2022 The State of the World's Mangroves 2022. Global Mangrove Alliance.

Legg, C.J. and Nagy, L. 2006. Why most conservation monitoring is, but need not be, a waste of time. *Journal of Environmental Management* 78: 194–199, https://doi.org/10.1016/j.jenvman.2005.04.016.

Lee, S.Y., Hamilton, S., Barbier, E.B. et al. 2019. Better restoration policies are needed to conserve mangrove ecosystems. *Nature Ecology and Evolution* 3: 870–872, https://www.nature.com/articles/s41559-019-0861-y/.

Mascia, M. B., Pailler, S., Thieme, M. L. t al. 2014. Commonalities and complementarities among approaches to conservation monitoring and evaluation. *Biological Conservation* 169: 258–267, https://doi.org/10.1016/j.biocon.2013.11.017.

McDonald-Madden, E., Baxter, P. W., Fuller, R. A. et al. 2010. Monitoring does not always count. *Trends in Ecology & Evolution* 25, 547–550, https://doi.org/10.1016/j.tree.2010.07.002.

Newell, D., 2019. A test of the use of artificial nest forms in common swift *apus apus* nest boxes in southern England. *Conservation Evidence* 16: 24–26, https://www.conservationevidence.com/reference/pdf/6956

Nosek, B.A., Ebersole, C.R., DeHaven, A.C., et al. 2019. National differences in gender–science stereotypes predict national sex differences in science and math achievement. *Proceedings of the National Academy of Sciences* 115: 2600–2606. https://doi.org/10.1073/pnas.0809921106.

Ockendon, N., Amano, T., Cadotte, M. et al. W.J., 2021. Effectively integrating experiments into conservation practice. *Ecological Solutions and Evidence* 2: e12069. https://doi.org/10.1002/2688-8319.12069.

Pynegar, E., Gibbons, J., Asquith, N et al. 2021. What role should randomized control trials play in providing the evidence base for conservation? *Oryx 55*: 235–244. https://doi:10.1017/S0030605319000188

Rey Benayas, J.M., Newton, A. C., Diaz, A. et al. 2009. Enhancement of biodiversity and ecosystem services by ecological restoration: A meta-analysis. *Science* 325: 1121–1124. https://doi:10.1126/science.1172460

Roy, H.E.; Pocock, M.J.O.; Preston, C.D. et al. 2012. Understanding citizen science and environmental monitoring: final report on behalf of UK Environmental Observation Framework. Wallingford, NERC/Centre for Ecology & Hydrology.

Schleicher, J., Eklund, J., D. Barnes, M. et al. 2020 Statistical matching for conservation science. *Conservation Biology* 34: 538–549, https://doi.org/10.1111/cobi.13448.

Smith R.K., Dicks L.V., Mitchell R. et al. 2014 Comparative effectiveness research: the missing link in conservation. *Conservation Evidence* 11: 2–6, https://conservationevidencejournal.com/reference/pdf/5475.

Stewart-Oaten A., Murdoch W.W. and Parker K.R. 1986 Environmental impact assessment:' Pseudoreplication' in time? *Ecology* 67: 929–940, https://doi.org/10.2307/1939815.

Sutherland, W.J., Armstrong, D., Butchart, S.H.M. et al. 2010. Standards for documenting and monitoring bird reintroduction projects. *Conservation Letters* 3: 229–235, https://doi.org/10.1111/j.1755-263X.2010.00113.x.

Sutherland W.J., Robinson J.M., Aldridge D.C. et al. 2022 Creating testable questions in practical conservation: a process and 100 questions. *Conservation Evidence Journal,* 19, 1–7. https://doi.org/10.52201/CEJ19XIFF2753.

Tinsley-Marshall, P. Downey, H., Adum, G. et al. 2022. Funding and delivering the routine testing of management interventions to improve conservation effectiveness. *Journal for Nature Conservation* 67: 126184, https://doi.org/10.1016/j.jnc.2022.126184.

Vickery, J. A., et al. 2004. The role of agri-environment schemes and farm management practices in reversing the decline of farmland birds in England. *Biological Conservation* 119: 19–39, https://doi.org/10.1016/j.biocon.2003.06.004.

Walters, C. J., and Holling, C. S. 1990. Large-scale management experiments and learning by doing. *Ecosphere* 71: 2060–2068, https://doi.org/10.2307/1938620.

Wardropper, C.B., Esman, L.A., Harden, S.C. et al. 2022. Applying a 'fail-fast' approach to conservation in US agriculture. *Conservation Science and Practice* 4: e619, https://doi.org/10.1111/csp2.619.

White, H. 2010. A contribution to current debates in impact evaluation. *Evaluation* 16: 153–164, https://doi.org/10.1177/1356389010361562.

Wood, K.A. 2020 Negative results provide valuable evidence for conservation. *Perspectives in Ecology and Conservation* 18: 235–237, https://doi.org/10.1016/j.pecon.2020.10.007.

PART IV
TRANSFORMING SOCIETY

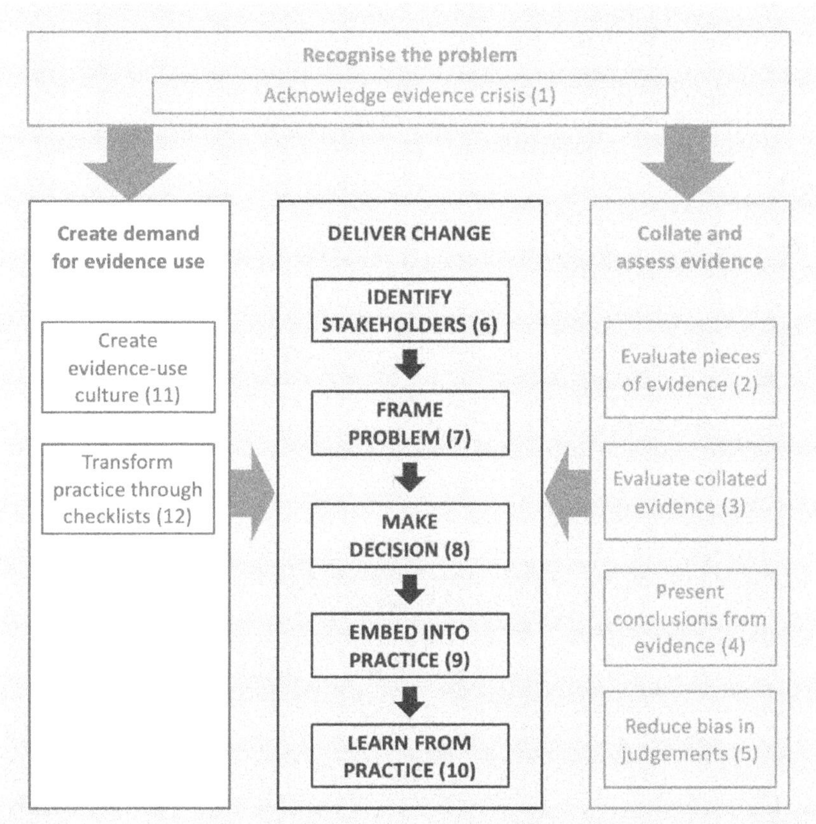

11. Creating a Culture of Evidence Use

Winifred F. Frick[1,2], Steven J. Cooke[3], Iain Dickson[4], Hazel A. Jackson[5], Kaylene E. Keller[6], Angelita De la Luz[7], Danni Parks[8], Tom McPherson[9], Ashley T. Simkins[4], Iroro Tanshi[10], Des B.A. Thompson[11], Paul Tinsley-Marshall[12], Lindsay C. Young[13]

Evidence is a prerequisite for effective conservation decisions, yet its use is not ubiquitous. This can lead to wasted resources and inadequate conservation decisions. Creating a culture of evidence use within the conservation and environmental management communities is key to transforming conservation. At present, there are a range of ways in which organisations can change so that evidence use becomes routinely adopted as part of institutional processes. Auditing existing use is a useful first stage followed by creating an evidence-use plan. A wide range of possible actions should encourage evidence use and ensure the availability of resources needed. Seven case studies show how very different organisations, from funders to businesses to conservation organisations, have reworked their processes so that evidence has become fundamental to their effective practice.

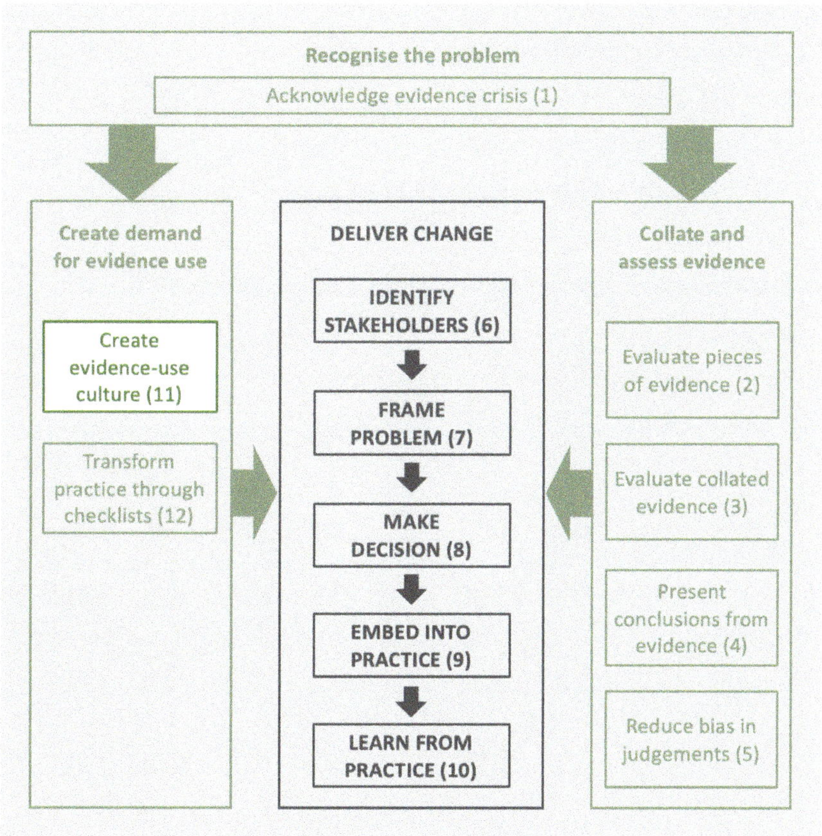

https://doi.org/10.11647/OBP.0321.11

[1] Bat Conservation International, 1012 14th Street NW, Suite 905, Washington, D.C. 20005, USA
[2] Department of Ecology and Evolutionary Biology, University of California Santa Cruz, 130 McAllister Way, Santa Cruz, CA 95060, USA
[3] Canadian Centre for Evidence-Based Conservation, Department of Biology and Institute of Environmental and Interdisciplinary Science, Carleton University, Ottawa, ON, Canada, K1S 5B6
[4] Birdlife International, The David Attenborough Building, Pembroke Street, Cambridge, UK
[5] Woodland Trust, Kempton Way, Grantham, Lincolnshire, UK
[6] US Fish and Wildlife Service, National Wildlife Refuge System 2800 Cottage Way, Suite W-2606 Sacramento, CA USA 95825
[7] Beeflow, E Los Angeles Ave., Somis CA, USA 93066
[8] The Whitley Fund for Nature, 110 Princedale Road London, UK
[9] Ingleby Farms and Forests, Slotsgade 1A, 4600 Køge, Denmark
[10] Small Mammal Conservation Organization (SMACON), Plot 8 Christina Imudia Street, Benin City, Nigeria
[11] NatureScot, Silvan House, 231 Corstorphine Road, Edinburgh, UK
[12] Kent Wildlife Trust, Tyland Barn, Chatham Road, Sandling, Maidstone, UK; current address: Butterfly Conservation, Manor Yard, East Lulworth, Wareham, Dorset, UK
[13] Pacific Rim Conservation, PO Box 61827, Honolulu, HI, USA 96839

Contents

11.1 Why Changing Cultures is Critical

Much of this book is devoted to detailing methods that could transform practice and increase effectiveness. However, creating a culture of evidence use in conservation practice has to overcome existing barriers to shift behaviour, priorities, and norms among potential evidence users and overcome the common disconnect between conservation scientists and the practitioners on the ground.

Shifting the field of conservation practice towards a *de facto* norm of evidence use requires an incisive and frank analysis of how conservation organisations operate, and how they can adapt and improve. It requires turning our analytic attention inward, being willing to admit failure, and ultimately adopting a mindset that results in adaptive practice. Changing attitudes and work practices can be challenging; most of us lack the necessary training required to help foster such changes.

As discussed in Chapter 1, shifting to evidence use as the norm has been successful in other fields, including medicine (Shortell et al., 2001), and is becoming increasingly common in other areas such as education (Slavin, 2002). Medical practice has standard processes for converting evidence into practice and has embedded evidence use into educational programmes (Ilic and Maloney, 2014). Yet, that revolution happened relatively quickly within a few decades (Guyatt et al., 1992). In medical practice, the cost of failure is human lives. The cost of failure of conservation practice is similarly high, with biodiversity and sometimes human lives at stake (Díaz et al., 2006), although the repercussions of the failure of conservation are not as immediate and therefore often discounted. However, we can learn from fields that have adopted evidence use as the cultural norm as well as from professional fields that have achieved cultural change.

Evidence use can be complicated, so there is a need for increasing conservation practitioners' familiarity and skills with evidence use as well as providing further training. In this chapter, we explore some of the challenges and offer ideas for how organisations can lead the conservation movement by shifting norms toward evidence use. All the authors of this chapter work for organisations that have increasingly adopted evidence-based processes.

11.2 Auditing Current Evidence Use

When organisations commit to using evidence, a useful starting point is to consider how evidence is currently used by the organisation as well as how it is used and valued by their peer network. Organisations can start with a self-assessment by asking questions about how often evidence is considered in decision making, what training is currently provided or available to staff, and how often tests of actions are carried out. This self-assessment can happen as part of strategic planning, priority setting, annual goal-setting, or as a stand-alone initiative. Several resources exist to help organisations get started with self-assessments and auditing their evidence use, including this book and the checklists in Chapter 12. Other online resources, such as the Conservation Standards (www.conservationstandards.org), offer a range of resources for developing new practices or training staff.

An initial audit of current evidence use may uncover areas where evidence use is already underway, even if not explicitly recognised. For example, Bat Conservation International was already publishing scientific papers testing actions to reduce threats to bats, but this was not recognised as part of their commitment to evidence use until they incorporated it as part of their contract with Conservation Evidence (www.conservationevidence.com) to become 'Evidence Champions' (see Section 11.4). Formalising this evidence creation as part of their contract as an Evidence Champion provided new recognition within the organisation of the value of staff time generating scientific products, and helped showcase how these products serve to aid conservation practice rather than simply as academic outputs.

An initial audit should ideally identify areas to improve or create new practices toward achieving routine evidence use. There is no single formula for success since organisations will vary in their structure and existing practices. The first step is to recognise the limits of current practices and identify ways in which new practices can become organisational habits. Studies of successful habit formation (Clear, 2018; Wood, 2019) can provide useful insights. For example, Clear (2019) makes a case for breaking big goals into smaller steps. Organisations could thus not just set strategic goals around evidence use but also create specific incremental tasks. Even small changes and modest accomplishments build success and lay a foundation for organisational habit formation for evidence use, as discussed further in Section 11.4 on Creating Expectations and Opportunities for Evidence Use.

An initial audit of evidence use should end with, 1) an understanding of the current practices of evidence use within an organisation, 2) some strategic goals toward changing practices, and 3) some specific targets for adopting new practices with plans for how to achieve such targets. If the audit reveals that the organisation has ample room for improvement (e.g. a low score on Checklist 11.1 below), it may be advisable to start with just one area of the organisation or try a pilot programme to build success and apply adaptive learning.

Checklist 11.1 provides a checklist that can serve as a starting point for auditing current practice and can also be adapted for periodic assessments to measure improvement, measure change over time or compare different branches within an organisation. The checklist can easily be adapted based on organisational structure and priorities.

In addition to doing a general audit of current evidence use by an organisation, each project/area of work could be audited to ensure staff are following the best available evidence or guidance and, if not, provide justification or reasoning behind decision making. An example from the medical field is the practice of regular audits by the National Health Service England (https://www.england.nhs.uk/clinaudit). If a doctor offers healthcare advice that differs from what is recommended in the National Institute for Health Care Excellence guidelines (www.nice.org.uk), then the doctor must justify their decision. These guidelines are not mandatory; however, their use is incentivised in part by the pressure of the audit to ensure reasonable justification for deviating from them.

Could a similar audit process be set up in conservation? Internal audits by organisations are valuable, but they may be inconsistent among organisations or may not be sufficiently self-critical. External audits that are part of project evaluations or reporting to funders, stakeholders, or collaborators may provide useful visibility and accountability. These external audits could

11.1 Checklist for assessing the extent of evidence use by organisations

- ☐ Do job descriptions for posts that involve making conservation decisions mention effective use of evidence?
- ☐ Do interviews for these posts include questions on evidence use?
- ☐ Do those who advise or manage plans and decisions routinely ask about the underpinning evidence?
- ☐ Is the key evidence underpinning plans and decisions made clear?
- ☐ Do processes exist to ensure decision makers routinely reflect on the documented evidence (such as from these sources: https://www.conservationevidence.com/content/page/127), for at least one decision?
- ☐ Do those deciding on actions routinely reflect on means of improving effectiveness through alternative management options?
- ☐ Routinely, say at least annually, does someone within the organisation establish, and subsequently document, a standard experimental test (replicated with control or comparisons)?
- ☐ Is training and professional development available on evidenceuse and testing?
- ☐ Do senior managers routinely (at least monthly) promote the use of evidence?
- ☐ Is there an organisational evidence delivery strategy?
- ☐ Are there processes for learning from challenges in project delivery?
- ☐ Are the impacts of the actions documented, regardless of the outcome, and made available to others?

Score /12

This checklist can be downloaded from https://doi.org/10.11647/OBP.0321#resources. It can be modified and tailored for specific uses.

include encouraging conservationists to explain how they reached their decision and what evidence they used. The Evidence-to-Decision tool (https://www.evidence2decisiontool.com/) could be adapted or used for this kind of audit (Christie et al., 2022). Many grant applications now require metrics of success, which could be used as a starting point for what evidence could be collected on a particular programme (Chapter 9). Similar audits could also take place on businesses/developers/consultants in their work to avoid, minimise, and offset biodiversity impacts, to ensure they are also based on the best available evidence.

11.3 Creating an Evidence-Use Plan

Once an organisation has the results of their initial audit of evidence use, it can begin to craft a plan for integrating evidence into its conservation practice. This will depend on how the group is organised. Is it a small team or part of a large organisation? Has leadership ensured that using evidence will be fundamental to the organisation's business (top down), or is this an effort where individuals and small teams are working to integrate evidence on a project-by-project basis (bottom-up)? It is worth the time to think through the desired achievements and how they will be delivered. At the organisational change scale, it will be important to develop a plan that works within the existing structure. Does it first need leadership buy-in, or is leadership directing the change? Will developing pilot projects to demonstrate the benefits be important as a first step, or is it possible to implement it across all projects at the same time (Keller and Schaninger, 2019)?

For example, a Conservation Measures Partnership (CMP) team is working on a project to provide guidance for the adoption of Conservation Standards by organisations (CMP 2021). Their approach involves breaking down the problem and identifying strategies and outcomes within the organisation. Successfully introducing new methods or practices for adoption by an organisation may require a plan outlining how the change will be implemented including a timeline. Organisations and teams can then initiate changes using high-priority decisions and evaluating the level of evidence needed (Sutherland et al., 2021). A decision process for prioritising time and effort is important as these practices become norms.

Box 11.1 lays out questions that can be used while developing the organisational plan or for individual projects. Not every project will create evidence or use the same level of evidence. At the start of a project, an assessment can be made to determine if evidence use is necessary for the project and the time and resources that can be dedicated to that part of the project. This can help to focus again on the highest priority evidence needs.

Box 11.1 Possible elements of an evidence-use plan

1. What are the goals of the organisation, programme, or project?
2. How can improved evidence use enhance these goals?
3. What is an appropriate commitment of time and resources?
4. How will an expectation of evidence use be achieved?
5. What opportunities will be created?
6. What capacity is needed?
7. What training is needed?
8. What will be the approach to failure?
9. How will results be disseminated?

11.4 Creating Expectations and Opportunities for Evidence Use

Creating expectations of evidence use needs to happen within and among organisations to drive the cultural sea-change toward new norms. Organisations need ways to publicly share their values, commitments, and approaches, but also ways to create internal processes and set expectations within their organisational culture. Changing organisational practices is akin to building new (and better) habits. The science of habit formation suggests that the level of 'friction' associated with a habit can be a strong determinant of whether it gets adopted (Wood, 2019). Behavioural change revolves around three elements: a cue, a routine, and a reward. Frictionless habits are those where the cues are obvious and attractive, and adoption happens with ease (Clear, 2018). These ideas can be used to help organisations find ways to make practices around evidence use as 'frictionless' as possible so that evidence use becomes routine and ultimately habitual.

Creating an expectation of evidence use within organisations requires participation by all staff. Senior leaders must lead by setting expectations for using evidence as desirable and routine. Without leadership engagement there may be less incentive, and thus slower adoption, for middle management or junior staff to allocate time and attention to practices that incorporate evidence use. To set the expectations, supervisors and leaders should set strategic priorities but also create structures and incentives for staff time spent on activities related to evidence use. Leaders also need to provide training for all staff so the practice of evidence use is clearly defined and staff can meet expectations. Creating expectations without accompanying support to implement changes is unlikely to succeed.

Ideally, staff across all levels in an organisation know how to use evidence in decision making and value its use. However, this may take time and investment to achieve. Organisational leaders may need to help direct staff to use available tools, such as the Evidence-to-Decision tool (Section 9.10.3; Christie et al., 2022) or the Conservation Standards (CMP, 2020), to identify what evidence is needed for different activities and decisions. All staff can contribute to setting expectations by identifying processes and workflows where evidence is identified, gathered, stored, and made available to others in the organisation. Expectations are reinforced among staff when evidence is presented or mentioned during briefings or other types of information exchange within organisations (e.g. if junior staff are briefing senior staff on a programme). An indicator that can be used to determine if an evidence practice is routine is to ask this question: Does your manager routinely ask about the underlying evidence for proposed projects? If this is a standard question during project briefings and annual reviews, it can help set expectations and determine if evidence practice is becoming routine. Likewise, external reviews and exchanges between organisations can help organisations innovate and adapt. Checklists (Chapter 12) are useful tools for adopting new practices that can be embedded at different stages of workflows for reinforcing best practices.

Time seems the most limited commodity in every organisation. Therefore, creating specific times for staff to complete tasks related to evidence use on a regular and recurring basis is effective. This can take place in the form of recurring meetings where staff review a checklist together or assign action items with deadlines during different stages of a workflow so that evidence is evaluated before decisions are made. Here, the concept of habit-stacking may be helpful. With habit-stacking, an existing habit is used as the cue for a new habit to make it more obvious and easier to adopt. Organisations may want to identify the triggers for using new tools such as evidence-use checklists or the Evidence-to-Decision tool by attaching them to routine tasks, such as at the start of re-occurring check-in meetings, before a proposal is submitted, or during the review of reports.

When organisations conduct goal-setting or performance evaluations for staff, review forms and meetings can include questions about work activities involving evidence use. Incorporating questions about evidence practice in such reviews can reinforce organisational values and also makes it explicit to staff that time spent on evidence use is valued and reviewed. Including questions about staff activities related to evidence practice in annual performance reviews or goal-setting provides an opportunity for feedback and for managers to learn if strategies to change practices are working. Feedback may also discover if sufficient time and resources have been allocated to achieve desired goals. Likewise, organisational practices and values around evidence use should be shared during the onboarding process for new employees. Specific practices and terminology may not be clear to new employees, and offering training materials and sharing expectations at the outset of employment may accelerate adoption and organisational progress. Setting expectations for evidence use can also be explicit in the hiring process. Evidence use can be mentioned in job descriptions, and questions on evidence use are incorporated into job interviews and the selection of candidates. How to structure this will vary depending on the type of position being offered and core duties, but statements about expectations of evidence use as an organisational commitment may be a way to signal an organisation's commitment and values related to evidence use.

Committing publicly to evidence use is an important part of creating professional norms of including evidence use (Sutherland and Wordley, 2017). The Evidence Champion programme by Conservation Evidence recognises those organisations that have adopted practices to help deliver evidence-based approaches. To earn recognition as an Evidence Champion, organisations must commit to at least one of a range of practices related to evidence use in conservation. Evidence Champions can then use the badge or logo on websites or branded materials to signal their commitment. Public statements of commitments to evidence use through logos, mission statements, or other branded outreach make explicit that an organisation takes evidence seriously.

Beyond formal processes and institutions involved in building capacity for evidence use or engaging in evidence application, there are also informal institutions that can be effective. Of particular note is the growing movement focused on developing 'communities of practice'. Communities of practice tend to be bottom-up informal collaboratives where a wide range of individuals with common interests support each other through sharing of practical experience (Lave and Wenger, 1991). A community of practice is already evident in the environmental evidence realm (see Cooke et al., 2017), which is particularly promising and bodes well for the continued development of the capacity for conducting evidence synthesis and incorporating it into decision-making processes. The Miradi Share platform available from the Conservation Measures Partnership is an example of a community of practices to help develop evidence-use approaches in conservation (https://www.miradishare.org).

11.5 Providing the Capacity to Deliver Evidence Use

There are now numerous tools available to help organisations develop better practices around evidence use, but creating capacity will require investments in staff time and expertise to establish and reinforce such practices. Creating or identifying staff positions with core responsibilities and duties centred around evidence use is ideal. For example, the United States government requires a named personnel position, Evaluation Officer, at major departments and bureaus who is responsible for an evidence-building plan (or 'learning agenda'). Not all organisations may be able to hire a new position specifically dedicated to evidence use, so finding ways to build evidence responsibility into existing roles is critical. As described above, there are many ways to create and reinforce both expectations and opportunities by adapting existing workflows within organisations.

Whether a new position is needed may depend on existing expertise within an organisation and the ability to shift workloads to accommodate new duties. Ultimately, the capacity needed within organisations includes staff members who have the time and expertise to design projects to use and create evidence, evaluate progress and identify success or failure, lead training, collaborate with and mentor co-workers, and answer questions. Organisations can use the Strategic Evidence Assessment Framework (Sutherland et al., 2021) to allocate available capacity most effectively. As evidence use becomes the adopted norm, hopefully, there will be a rise in organisations creating posts with responsibility for evidence use. Positions with evidence in the title have recently become much more common (see Chapter 13 for a list).

Ultimately, evidence use requires evidence to be available. Therefore, addressing the lack of availability of evidence is an important part of providing the capacity to deliver evidence use. Positions with a primary focus to create or contribute evidence (e.g. scientists) are essential to effective conservation delivery. Routine checks of what evidence is available and the quality of that evidence (using the Conservation Evidence database or other tools) and having plans for how to disseminate or share results of research that tests the efficacy of actions is key. Coordination among staff who are trained in scientific study design, with staff responsible for carrying out conservation projects, can help assure that conservation practice is adaptive and effective.

11.6 Training, Capacity Building, and Certification

Building a community of practice around evidence use requires creating and using training and capacity-building opportunities. The Conservation Measures Partnership has resources available for training and offers training courses (www.conservationstandards.org). The

IUCN and others offer various training modules that provide examples of online resources for training materials, such as those available through the Conservation Training website (https://www.conservationtraining.org). The Evidence in Conservation Teaching Initiative (Downey et al., 2021; https://www.britishecologicalsociety.org/applied-ecology-resources/about-aer/additional-resources/evidence-in-conservation-teaching) provides a range of open-access teaching material about evidence use, including lectures in nine different languages. There is a need for more resources and opportunities for training at all levels, including integration with University programmes, but also professional training courses. Foundations of Success (https://fosonline.org) offers courses on using the Conservation Standards that can be bespoke for organisations. Training courses could also lead to certification, which reinforces visibility and recognition of the importance of evidence use. The Conservation Evidence programme offers its Evidence Champion certification — certification based on commitments developed for each organisation.

On a larger scale, the US Cooperative Extension Service shows that it is possible to create a model for long-term and substantial government investment in transferring knowledge from scientific research to practitioners (Franz and Townson, 2008). This community-based education entails providing effort and resources toward educating, training, and motivating practitioners (Warner and Christenson, 2019).

There are myriad opportunities for providing training and certifications through professional or academic societies to bolster the conservation evidence movement. Like many themes in this chapter, the structure and nature of such training may depend on the target audiences and organisations involved, given the diversity of organisations involved in conservation practice around the world. Knowledge sharing is critical, as global approaches to conservation are developed, and historic and current inequities to access knowledge must also be acknowledged. Too often, certifications and training are costly and therefore inaccessible to many. Creating open-access tools and training (and making them accessible in multiple languages) are necessary to advance equitable and just approaches to conservation.

11.7 Learning from Failure

Failure in conservation can be defined as 'a lack of success in meeting stated outcomes and objectives' (Catalano et al., 2019). In practice, failure and success are usually not binary categorisations but it is often more useful to imagine projects along a spectrum, ranging from 100% failure to 100% success.

In business, learning from failure is recognised as part of innovating toward success. However, most people avoid the discomfort of acknowledging and discussing failure, which then limits opportunities for learning and improving. This point is illustrated by quotes from aeroscientist and former Indian president A.P. J. Abdul Kalam, 'if you fail, never give up because F.A.I.L. means "First Attempt In Learning"', and Bill Gates, 'it's fine to celebrate success but more important to heed the lessons of failure'. Chambers et al. (2022) express the concern that embracing, or even celebrating, failure may mean unsuccessful projects can be reframed as successes; they emphasise the need to ensure the lessons lead to change.

In some cases, the failure that occurs is a result of experimentally testing actions, which is considered 'intelligent failure' because it advances the knowledge frontier (Edmondson, 2011). Some other failures may be due to events that could not have reasonably been predicted, such as the outbreak of a novel disease. This section focuses on the remaining majority of failures where an action seemed as if it should have worked but did not, like pilots crashing fully functioning planes by confusing levers, as described in Chapter 1, or the various examples of failures to engage with communities described in Chapter 6.

How failure is best framed may vary with the project stage. During planning — where the aim is typically to identify factors that could negatively influence the project's results, assess their potential impact and develop appropriate mitigation strategies — potential failures are considered as 'risks'. During implementation, the terms 'challenges' or 'issues' might be more appropriate to ensure the gathering of information that can inform practice. After project completion, where the aim is to document learning to inform future practice, the term 'lessons learnt' may be most effective.

11.7.1 Why does failure occur?

Conditions are ripe for intelligent failure in conservation. Firstly, conservation action takes place in natural systems with high complexity and uncertainty. Secondly, many conservation actions involve actions whose effectiveness is uncertain. Thirdly, many conservation projects are dependent on a range of individuals and so may fail due to challenges in execution or ineffective processes.

When a plane crashes, it is obvious that failure has occurred and it is often relatively straightforward to determine how it happened and why the problem developed. For many conservation efforts, establishing these elements is considerably more challenging, with perceptions differing between individuals alongside the complexity and variability in ecological systems. Some may consider a project a success because it met all its short-term objectives, such as the successful training courses, while others might consider the same project a failure because it ultimately failed to contribute to a wider long-term goal, such as the species declining at the same rate. Similarly, some stakeholders may consider a project as an overall success because only a single component of a wider initiative failed to meet its aims, while others may perceive the entire initiative as a failure because the component that failed was the only one they considered important.

A group of conservation practitioners developed a taxonomy of 'root causes' of failure in conservation projects, as listed in Table 11.1 (Dickson et al., 2022). This taxonomy can be used to help identify root causes of failure, summarise projects in a standard manner, collate experiences on common issues, and help causes.

In practice, individuals may have different perceptions of the root causes. For example, one might believe the main reason a project failed was the unexpected bad weather that caused a key piece of equipment to fail, another might attribute failure to inadequate planning in selecting weather-proof equipment, while another might state the whole project was overdependent on technology.

Table 11.1 A taxonomy of reasons for project failure. (*Source:* from Dickson et al., 2022)

Area	Reasons for project failure
Planning, design or knowledge	
Knowledge inputs to project design	• Lacking sufficient information on the ecology of the conservation target. • Lacking sufficient knowledge of the social, cultural or economic conditions. • Lacking sufficient knowledge of local contexts and conditions (other than ecological or socio economic). • Lacking sufficient evidence on the effectiveness of the proposed solution.
Project design	• Project design not based on the identification of a clearly defined conservation problem. • The mechanism for addressing the problem proved insufficient. • The mechanism for addressing the problem proved insufficient and/or inadequate for bringing about the desired change (i.e. the project's Theory of Change did not work in practice). • Ineffective systems for capturing information on progress, effectiveness and impact. • Not allocating enough funding during the design phase to achieve the desired outcome. • Setting goals/objectives beyond what could be realistically delivered. • Using inappropriate or inefficient methods, techniques or materials.
Sustainability planning or exit strategy	• Not planning for likely changes in personnel. • Lacking a clear plan for disengaging from or ensuring the sustainability of the project.
Consultation during design phase	• Insufficient engagement/input during design phase from relevant stakeholder groups.
Team dynamics	
Project management	• Project management not providing effective support, supervision or guidance. • Inadequate delegation of roles and responsibilities within the team. • Lacking necessary adaptation of approach/roles, etc., when required. • Lacking necessary support/buy-in from senior management to the project team. • Ineffective management of funds allocated to the project. • Ineffective planning, consultation and feedback between management and others. • Management too far removed from day to day running of the project.

Area	Reasons for project failure
Project delivery	• Staff lacking motivation to implement project activities effectively. • Poor communication between those involved in implementation. • Lack of understanding by those involved in the project on objectives, priorities and resources. • Corruption by staff.
Internal governance structures	
Project governance structures	• Elements of project management and/or governance structures either missing and/or not functioning effectively. • Roles and responsibilities of those involved in the project not clearly defined. • Legal structures set up to facilitate the functioning of the project not clearly defined. • Lacking effective communication between levels of project governance.
Systems and structures for identifying risk/mismanagement	• The project lacked the proper structures and procedures necessary to identify and deal with risk or mismanagement.
Systems and structures for learning	• Project governance did not include an effective process for capturing lessons and determining when to act on these.
Resources	
Funding	• Delay in signing grant agreement meant that key activities could not be carried out in time. • Funding reallocated to cover other areas of work within organisation. • Project did not receive/raise co-funding needed. • Funding received was insufficient to complete project. • Funding not secured beyond length of initial grant period.
Human capacity and expertise	• Insufficient staff numbers to carry out effective implementation. • Staff involved in implementation unable to work effectively due to overly high workload. • Burden of administration (e.g. reporting, financial management, recruitment) negatively affected implementation. • Lack of necessary knowledge/skills/experience. • Loss of essential knowledge/skills/experience, and inability to effectively replace this.
Physical resources	• Lack of the physical resources needed to implement the project. • Resources/materials used in the project not maintained to the level required.

Area	Reasons for project failure
Stakeholder relationships	
Funder support	• Loss of, change in, or disconnects in support or engagement by the project funder.
Support from key stakeholders	• Lack of support/buy-in from existing relevant government agencies/individuals. • Inability to ensure continuity of existing support resulting from a change in relevant government agencies/individuals. • Not enough support from local communities in and around project. • Unintended impacts resulting from the project negatively affected delivery. • Lack of support from stakeholders owning/controlling land relevant to the project. • Inability to build support from general public in relation to the project's conservation goals. • Dysfunctional/non-existent relationships with stakeholder organisations supportive of the project's aims or working to achieve similar outcomes. • Dysfunctional/non-existent relationships with stakeholder organisations not supportive of the project's aims and/or working to achieve opposing outcomes.
Stakeholder agendas	• Key stakeholder agendas not aligned or in opposition to each other.
Corruption and illegal activities	• Corruption carried out by individuals not directly working on the project. • Illegal activity carried out by individuals not directly working on the project.
External events that cannot be predicted or influenced by the project	
Environmental events	• Climate/weather. • Other natural disasters. • Wildlife disease.
Human events	• Conflict/insecurity. • Disease affecting humans or their domesticated animals or plants.

11.7.2 Learning from failure

Because failure is generally associated with blame, and is typically emotionally unpleasant, active leadership is needed to create an organisational culture that values failure as a learning opportunity, to devise processes that provide safe ways to discuss failure without blame (Edmondson, 2011), and encourage innovation and experimentation.

The key stages are to consider four questions. What was expected to happen? What actually happened? What went well and why? What can be improved and how? The learning process will often centre around gathering and analysing an individual's perceptions of failure, both in relation to whether something is considered a failure and in relation to how and why it occurred. These perceptions may differ considerably between individuals and stakeholder groups depending on their role, knowledge, attitudes or underlying motivations. Understanding divergent views is often key to diagnosing the problem.

Improvement then occurs as a result of reflecting on the cause of the problem, considering what could have been done differently, and then considering lessons from other projects and studies. Testing different possible options (see Chapter 10), such as restructuring how some teams operate or providing additional training, would help create more of an evidence base to improve delivery.

This process can be adopted when a project is underway by reflecting on the outcome or by examining individual cases of failure. The process may be an informal consideration or a formal process of review. Table 11.2 below outlines a range of methods that can be applied at various stages of the project cycle.

Workflow practices, such as Scrum or Agile workflows, are built on ideas of transparency, inspection, and adaptation (Schwaber and Sutherland, 2020). Work is conducted in defined sprints with each sprint including a retrospective that allows the team to reflect and individuals to evaluate what did and did not work, and what measures could be taken to improve in the next sprint. Even if not all work in conservation fits the sprint workflow, this approach of routine assessment and transparency to discuss failure and incremental progress are valuable approaches to creating an organisational culture that adopts evidence use (Catalano et al., 2021). The Objectives and Key Results (OKR) framework can also be used as a way of tracking success in this middle-ground by collaboratively setting goals and identifying key measurable results to track (Panchadsaram, 2020).

At all stages, trying to understand potential underlying motivations, agendas and relationships between those involved is particularly useful. Specific categorisations of failure (or success) are sometimes adopted to defend a specific viewpoint or further a particular agenda where emphasising supportive criteria, aligning with particular allies, or incorporating a story into a wider narrative is given greater emphasis than an objective assessment of whether, how, and why the failure occurred.

Seeking evidence for the how and the why, using a common framework for describing problems, testing options, collating lessons, developing a process for learning, and sharing experiences will help ensure that efforts to learn from failure in conservation help drive similar improvements to those seen in other areas of practice.

Table 11.2 Summary of learning from failure methods.

Method	Description	Useful for	Links/guidance
Pause & Reflect Session	Exercise where participants are asked to review the following statements: • What should be happening? • What is actually happening? • What action (if any) do we need to take?	Analysing progress on a continual basis, for example, as a means of periodically reviewing previously identified risks, and potential new ones, and deciding on whether any adaptations need to be made.	USAID — Facilitating Pause & Reflect (https://usaidlearninglab.org/resources/facilitating-pause-reflect)
After-Action Review	Addressing similar questions to Pause & Reflect but carried out after a project or wider initiative has finished, or in response to a specific incident.	Assessing and analysing root reasons/root causes of failure concerning efforts that have finished, or after a specific incident/case of failure has occurred	USAID — After Action Review Factsheet (https://usaidlearninglab.org/resources/after-action-review-aar-guidance)
Risk Assessment	Prior assessment of potential risks and underlying assumptions.	Assessing, before work begins, assumptions, potential risks/causes of failure and developing potential mitigation strategies. The resulting information then forms a useful basis for ongoing review and assessment of progress and likelihood of failure and success.	Ecosystem risk assessment science for ecosystem restoration (https://www.iucn.org/news/ecosystem-management/202112/using-ecosystem-risk-assessment-science-ecosystem-restoration)
Pre-mortem	A similar exercise to a risk assessment where the emphasis is specifically on identifying potential reasons/root causes of failure.	Useful supplement to a wider risk assessment. Gathers input in an interactive, participatory way that may work better with some audiences than a more formal risk assessment.	Performing a project pre-mortem (https://www.atlassian.com/team-playbook/plays/pre-mortem)

11.8 Case Studies: Organisations who Shifted to Embrace Evidence Use

A wide range of organisations, from NGOs to governments, have shifted their working so that evidence use is increasingly embedded in practice. The following case studies show the diversity of routes adopted by eight, very different, organisations to achieve a culture of evidence use.

11.8.1 Bat Conservation International

Bat Conservation International (BCI) is a non-profit conservation organisation dedicated to ending bat extinctions worldwide. Founded 40 years ago, BCI has a long-standing history of working for bat conservation globally. In 2020, BCI launched a new 5-year strategic plan that focuses on programmes that deliver conservation outcomes (www.batcon.org). The strategic plan identified a portfolio of work focused on four core missions: implementing endangered species interventions, protecting and restoring habitats, conducting priority research to develop scalable solutions, and inspiring through experience (BCI, 2020).

The strategic planning process was a multi-year effort that recalibrated BCI's approach to focus more explicitly on strategies and activities that could achieve measurable outcomes (Salafsky et al., 2002). To start the strategic planning process, the organisation invested in a training course led by Nick Salafsky at Foundations of Success to teach staff the Open Standards for the Practice of Conservation (https://conservationstandards.org). This helped identify projects and activities with theories of change that could result in desired outcomes, as described in Chapter 7 (CMP, 2020). The course also helped BCI develop internal processes for evaluating projects, which included attention to assessing evidence. Perhaps more important than the course content itself, the course participation allowed staff to engage in shared learning and explore together whether existing work was evidence-based or not. Participating staff formed cross-functional teams that mixed across organisational hierarchy and departments, which helped reset organisational culture toward a more collaborative and growth-mindset community of practice.

A challenge for bat conservation practice is a paucity of evidence to support actions (Frick et al., 2020; Berthinussen et al., 2021). In the first edition of the synopsis of conservation evidence for bats published in 2014, there were only 78 actions identified and most of those had no evidence or unknown effectiveness (Berthinussen et al., 2014). In the latest edition of the synopsis (Berthinussen et al., 2021), there are now 200 actions identified. Yet, of those 200 actions, 60% (n = 119/200) have no evidence and another 22% (n = 44/200) are ranked with unknown effectiveness due to a limited number of studies (in most cases, there is a single study; www.conservationevidence.com). In sum, 81% of currently identified actions for bat conservation have no evidence or unknown effectiveness based on the evaluation standards set by Conservation Evidence. This lack of evidence limits the toolbox for implementing evidence-based strategies and indicates the need for integration of research to test strategies and report on efficacy.

One of the ways that BCI has responded to the need to create evidence is to commit to publishing results in scientific outlets on research testing actions, which is formally part of their Evidence Champion agreement with Conservation Evidence. In addition, BCI actively supports collaboration between its science and conservation departments and seeks opportunities for cross-functional teams and engagement. The commitment to do both conservation science and practice has proven successful. In the past five years, the number of staff with PhDs increased from 2 to 11. Organisational leaders also socialise the value of scientific products to advance the mission, and set expectations that work should be conducted in ways that can lead to scientific products, so that results are shared to advance conservation practice broadly. The agreement with Conservation Evidence to become Evidence Champions served to formalise a cultural shift that was already taking place within the organisation. The organisation created and hired a new Director of Conservation Evidence position in 2022.

The effort to integrate scientific practice into delivering conservation takes consistent attention and diligence. While growth at BCI has resulted in increased staff and number of projects, it also creates challenges to maintain consistency in processes. Even before the covid pandemic, BCI functioned as a distributed organisation with remotely located staff. Collaboration and team-based work happen almost entirely in a virtual space. Creating opportunities and time for cross-functional processes can be difficult as team sizes increase, especially in virtual environments. In reality, there are some gaps between organisational aspirations toward standardised evaluation of evidence use for projects and how all projects are actually developed. Some projects happen opportunistically or organically and, because most staff carry heavy workloads, sometimes due diligence toward standardised evaluations of evidence is skipped or overlooked. The key to continued success seems to be consistent leadership, to value and socialise the process and the need for evidence use, and to reflect those values in work management strategies. Much like adaptive management itself is an iterative learning process, the process of incorporating evidence use into organisational culture is itself an iterative and adaptive cycle.

11.8.2 Fisheries and Oceans Canada

Fisheries and Oceans Canada (DFO) is a federal government agency in Canada responsible for the protection and management of aquatic ecosystems and aquatic biodiversity to maintain ecosystem services (see https://www.dfo-mpo.gc.ca/index-eng.html). DFO is regarded as a 'science-based' organisation and has its own science unit to support management and decision-making. At present, there is no standard approach to evidence use within the Canadian federal government aside from general statements in Ministerial Mandate Letters about how evidence should inform and guide decisions. DFO has one of the most long-standing and formalised approaches to evidence use via a science advice process called the Canadian Science Advisory Secretariat (CSAS), which was founded in 1996/97. CSAS is the mechanism by which DFO provides peer-reviewed science advice used by DFO and made available to the public. Efforts focus on ensuring that CSAS outputs yield advice that is credible, relevant and legitimate and therefore provides the best possible advice to the Minister (who holds the ultimate decision-making authority), managers, rightsholders, stakeholders, and the public through peer review that is evidence-based, objective, impartial and respectful (CSAS, 2011). Evidence of various forms is synthesised and vetted by internal and external experts (spanning knowledge generators to evidence users from relevant sectors including NGOs, industry, rightsholders, etc.). There is no standard means of synthesis but it can range from a narrative review to a full systematic review with meta-analysis, often supplemented with expert advice.

Revisions to the fisheries protection provisions (largely about fish habitat) of the Fisheries Act, and its ongoing implementation, provide a good example of how CSAS is effective in making the consideration of evidence part of decision making. Early efforts included the development of a science framework (Rice et al., 2015) and, after a change in government and refinements to the Act, there began a series of CSAS exercises focused on different topics needed to inform its implementation. One particular topic explored in CSAS was the effectiveness of different off-setting strategies for fish habitat creation/restoration for substrate spawning fish. A full systematic review was commissioned that revealed that the evidence base was large but the evidence was generally of poor quality (Taylor et al., 2019). A second review was conducted that relaxed the criteria for inclusion (i.e. including studies that lacked replication or comparators) to assess the lower-quality evidence (Rytwinski et al., 2019). The CSAS was convened and at the workshop both evidence syntheses were discussed alongside expert input from habitat managers and scientists. Given that decisions will be made about habitat restoration with or without evidence, it was apparent that any information that nudged the practitioner into being able to make a 'better' decision would be beneficial. This approach highlighted the reality that evidence quality will vary and different synthesis methods can yield different outcomes, but, with an appropriate understanding of biases and limitations, all evidence has the potential to enable better decisions (CSAS, 2020).

The CSAS process is not perfect, especially as it relates to decisions for stock-specific fisheries' management that lack the expediency and transparency desired (see Archibald et al., 2021). However, for files that are less time sensitive, the process seems to be useful for equipping decision makers, managers, and practitioners with science-based management advice. Social science research showed that DFO managers placed the highest credibility on products generated by the CSAS process, showing that evidence is both valued and used (Young et al., 2016). Young et al. (2016) found that the CSAS process was regarded as a validator of knowledge for government employees given its emphasis on the critical evaluation and synthesis of evidence, thereby yielding institutionally-endorsed knowledge. CSAS has embedded a culture of evidence synthesis and use within DFO and has also led to improvements in how science is conducted to ensure that it is done in a manner that contributes meaningfully to the evidence base through generating high quality science that can be used in evidence synthesis (CSAS, 2020). As the Canadian government moves towards understanding evidence synthesis and use ecosystem in federal agencies, it is likely that more formal processes, such as CSAS, will be adopted by other agencies with science-based portfolios.

11.8.3 Ingleby Farms

Ingleby Farms owns 101,000 hectares of farmland and forest across nine countries and specialises in the production of high-quality food through sustainable agriculture and environmental improvement by farming with nature, not against it. It is a Conservation Evidence Champion and, as part of this commitment, is continuously reviewing how to integrate evidence use into actions and decisions.

Ingleby uses monitoring, especially birds (conducted by local ornithologists) and earthworm surveys (total count in 20 x 20 cm cube) to broadly determine overall ecosystem health and detect positive and negative changes. This is used to identify areas of concern and opportunities as well as forming the basis for looking at the success of measures.

For each farm, Ingleby has identified features of significance for designation as Privately Protected Areas (PPAs). Priorities are identified through the Significant Species database (Table 11.3), which classifies species according to global priorities alongside their occurrence on the farms. The other categories of features of significance are cultural (historic or important sites for local traditions, such as communal grazing) or recreational, such as fishing or picnicking. Globally, Ingleby has 2,769 hectares of formally protected land, all with management plans.

Table 11.3 Identifying significant species within Ingleby Farms using IUCN Red List, National lists and the status on farms.

Conservation status	Resident/breeding	Visitor/migrant/ possibly breeding	Occasionally seen
Critical	Top priority	High priority	Medium priority
Endangered	High priority	High priority	Low priority
Nationally critical	High priority	Medium priority	Low priority
Vulnerable	Medium priority	Low priority	Monitor
Threatened	Low priority	Low priority	Monitor
Near threatened	Low priority	Monitor	Monitor
Gradual decline	Monitor	Monitor	Monitor
Rare	Monitor	Monitor	Monitor
Special concern	Monitor	Monitor	Monitor
Least concern	Monitor	Monitor	Monitor
No special status	Monitor	Monitor	Monitor
Not listed	Monitor	Monitor	Monitor

Ingleby has used the Evidence Assessment Hierarchy (Sutherland et al., 2021) to identify an appropriate and realistic evidence strategy. This strategy is to assess the evidence for actions when faced with a problem or when updating farm management plans. With the time available, the plan is that, unless the decision was obvious or trivial, issues will be assessed by checking likely overall effectiveness (e.g. the effectiveness criteria of Conservation Evidence). If that appears contradictory to plans, or the issue is more critical, then this action will be considered in greater depth.

A few subjects require detailed assessment of evidence due to serious problems (such as the fly, *Enallodiplosis discordis,* affecting the highly important tropical dry forest in Peru) or global responsibilities for species, such as the golden sun moth (Victoria), Peruvian plantcutter (Peru) or rufous flycatcher (Peru). In these cases, Ingleby then works with the farm managers to consider the challenges (including as shown by monitoring) and options and uses the Evidence-to-Decision tool (Section 9.10.3; Christie et al., 2022) to systematically examine the issue, evaluate evidence, and make an informed decision. The evidence is balanced with local experience, knowledge, and values to assist the making of a decision. Ingleby's agreed strategy for using evidence and conducting tests on-farm comprises

1. Adopt effective actions, focusing on what is known to work from documented evidence or experience.
2. Routinely test where the likely gain in knowledge exceeds the cost of testing. These will be small individual actions rather than complex projects, such as different crop varieties or different planting options.
3. Look for occasional (ideally annually) opportunities for simple, but well-designed, tests to contribute to global knowledge.
4. Increase experimental rigour where possible and appropriate (controls, replication, randomisation).
5. Use annual bird and earthworm surveys on each farm to gauge overall progress and identify challenges.

Thus, when considering a problem, Ingleby identifies a range of options by consulting industry best practices, the available evidence (e.g. using the Conservation Evidence database), and what has worked in practice in the past. If there is uncertainty then experimental tests may be created. In Romania there was a decision to restore an area of meadow, but then debate as to whether to just leave it alone or spread meadow hay from an adjacent farm. This resulted in a simple split-plot experiment (Figure 11.1) currently underway. Another experiment recently created in New Zealand is to add bird perches in an attempt to increase natural tree regeneration.

Figure 11.1 Monitoring a test of adding hay (right) against a control (left) of no hay added. (*Source:* Tom McPherson, CC-BY-4.0)

Ingleby ensures the results from all trials and tests are made available to all within Ingleby, creating a database for collating and sharing the observed outcomes of all trials (production and conservation) on their intranet. Trials are added to the database as they commence to ensure all trial outcomes are documented, reducing the risk of documenting only the successes. Major projects are also communicated via Ingleby news and through social media.

Ingleby is also committed to supporting peer-reviewed environmental science and research, allowing interested parties to conduct unimpeded research on Ingleby's farms and publish their findings, regardless of results; the research must be published open access.

11.8.4 Kent Wildlife Trust

Kent Wildlife Trust is based in the county of Kent, in South East England. The Trust manages around 80 nature reserves, and works to develop and deliver nature-based solutions to enable the restoration of biodiversity and bioabundance, and enhance the carbon sequestration potential and resilience of landscapes. It engages with a wide range of stakeholders, from politicians and business leaders to landowners and local communities, to undertake advocacy work, advise, or work in partnership to deliver conservation outcomes and reconnect people with nature.

The development of an evidence culture at Kent Wildlife Trust was catalysed in 2015 with the creation of a new role of Conservation Evidence Ecologist, tasked with creating, coordinating and delivering a programme of monitoring and evidence to support the Trust's work. This was born out of a realisation that, while the organisation had always done monitoring, it had not always done so consistently, using standardised methods, or with methods appropriate to assessing target outcomes. Evidence did not inform adaptive management, interventions were used without consulting evidence for their effectiveness, and there was no culture of, or commitment to, testing effectiveness or publishing.

The new incumbent in the role of Conservation Evidence Ecologist attended a conference in 2017 at which Conservation Evidence was presenting, and recognised the significant opportunity provided by the summaries of evidence collated in its database and in becoming an Evidence Champion. The senior leadership of the Trust were persuaded of the benefits of pursuing a path towards becoming an Evidence Champion, though there was a challenge in resourcing the work needed, which delayed the ambition to progress. This was overcome when funding was secured from the National Lottery Heritage Fund as part of a project grant to resource the work required.

In support of becoming an Evidence Champion, a review was conducted in 2019 of 544 interventions that Kent Wildlife Trust carried out as part of its land management activities, for which Conservation Evidence had summarised the evidence. Each of these was assessed using the 'effectiveness categories' in the database to determine the likelihood of effectiveness based on the collated evidence. Close attention was paid to interventions that had been assessed as likely to be ineffective or harmful, unlikely to be beneficial, or as having a trade-off between benefits and harms. It was enlightening to discover how frequently interventions had been assessed in these categories (Figure 11.2). Interventions were then reviewed and a position on the continued use of each was determined, including a commitment to reviewing the use of ineffective interventions and regularly reviewing new evidence. The review of the 544 interventions took approximately one month for two staff members to carry out and was conducted by a Project Officer, supported by a volunteer, and overseen by the evidence lead.

Through a partnership agreement with Conservation Evidence, Kent Wildlife Trust has committed to periodically reviewing evidence for the interventions it uses, identifying opportunities to test the effectiveness of interventions and to publish results, contributing to the development of best practice in evidence use, and to producing an organisational conservation evidence strategy. At the time of writing, three tests of interventions are underway (bracken control, vegetated shingle restoration, and an experimental test of different wilding grazing regimes). Kent Wildlife Trust benefits from being able to publicly demonstrate an evidence-based approach with mechanisms to improve effectiveness that help to maximise project outcomes and lead the way in showing how organisations could adopt evidence-based practice. The Conservation Evidence function is now fully embedded and consists of a multidisciplinary team of ecologists, GIS specialists and data scientists overseen by a Conservation Evidence Manager, working to ensure that evidence informs adaptive management across all strands of the Trust's work.

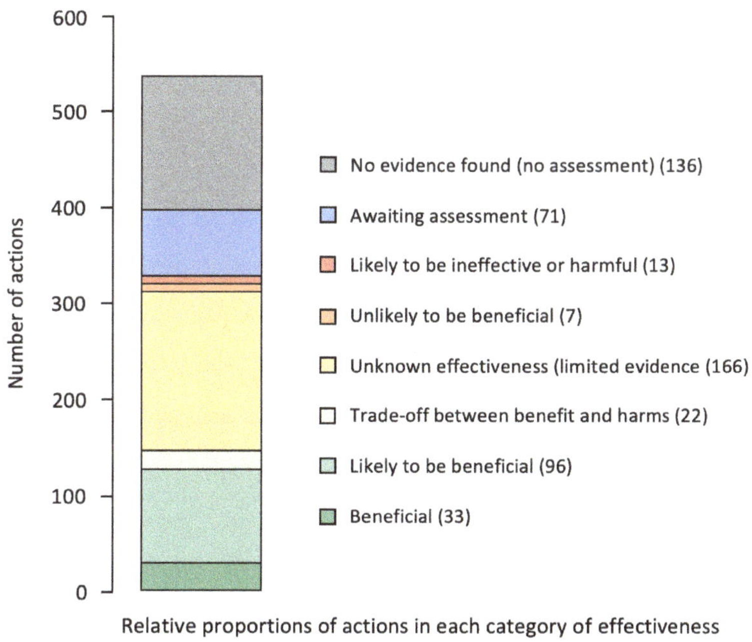

Figure 11.2 The number of interventions carried out by Kent Wildlife Trust classified by categories of effectiveness as then summarised by Conservation Evidence. (*Source:* Paul Tinsley-Marshall, CC-BY-4.0)

11.8.5 Pacific Rim Conservation

Pacific Rim Conservation (PRC) is a small US-based non-profit organisation whose mission is to maintain and restore native bird diversity, populations, and ecosystems in Hawaii and the Pacific Region. Pacific Rim Conservation was founded in 2006 out of a need for research-based management on native species, particularly birds, throughout Hawaii and the Pacific. Island species, particularly those in Hawaii, are some of the most imperilled on earth and, with so few individuals of some species, research was sorely needed to inform and deliver management actions. Pacific Rim Conservation works together with local communities, government agencies, and other conservation organisations to achieve their goals through direct conservation action. The organisation conducts research to understand avian biology, as well as changes and benefits to the ecosystem, in order to inform future conservation actions. It has published over 135 peer-reviewed papers on these outcomes.

For the first nine years of its existence, PRC operated as a small business before converting to a 501(c)3 non-profit organisation. That same year, PRC undertook strategic planning to better define their goals, and provide direct, measurable outcomes for whether they were achieving them. The result was a roadmap for how to design and implement projects in a step-wise, evidence-based manner. At a high level, this process entails

1. Conducting prioritisation and structured decision making to select projects on the front end ensuring that the species and ecosystems most at risk are those benefiting from conservation actions.
2. Once a project has been selected, establishing species and project-specific benchmarks that are tied to timelines.
3. Post-action (at each project phase) and post-mortem (if wildlife mortality is experienced during a project) debrief through a combination of meetings and anonymous surveys of both staff and immediate stakeholders.
4. Upon completion of a project or field season, immediate post-season dissemination of information to stakeholders and land managers through direct presentations and annual project reports.
5. Publication of the final results in the peer-reviewed literature.

Due to the organisation's small size (11 or fewer employees) and existing structure, they do not employ a single person responsible for evidence use, but have instead created a culture of evidence that is incorporated into every project and which ensures each staff member is accountable for collecting evidence. What follows is an example of how each of these steps has been put into practice with the organisation's largest project.

The current flagship initiative of PRC is the No Net Loss initiative (https://www.islandarks.org), which has the goal of creating new habitat for seabirds in the US Tropical Pacific, on high islands, matching acre for acre that are currently being lost to sea level rise on low lying atolls. They accomplish this by creating 'mainland islands' through predator exclusion fencing on high islands, and then translocating or attracting birds into the area to create new breeding colonies safe from both predators and projected sea level rise. The first step of this project was to select species that were most vulnerable to sea level rise through a prioritisation exercise, and then to select the appropriate source and recipient colonies to move them to. To prioritise species, PRC compiled a list of all the breeding seabirds in the region, scored each species for 10 criteria that reflected their extinction risk and vulnerability to climate change and invasive predators, and then summed the scores of all criteria to obtain an overall score and ranked the species in terms of overall conservation need. The top 10 candidates then had species profiles compiled to determine if they were suitable for translocation.

A similar process was used to select source and recipient colonies. Factors that were considered in assessing suitability as a source or restoration site included elevation, presence of predators, ability to exclude or eradicate predators, and several other anthropogenic risk factors. Pacific Rim Conservation also considered a colony to be a suitable source if it was: 1) at risk of inundation from sea level rise and storm surge such that the long-term persistence of the colony is in jeopardy; 2) subject to predation by invasive species that has not been effectively managed and would be difficult to manage; and 3) large enough to sustain removal of the desired number of individuals for several years. They considered a site to be suitable for restoration if: 1) it was not at risk of inundation; 2) predators and other anthropogenic threats were absent, had been eliminated or effectively managed, or could be effectively managed on a long term scale; 3) there were no serious logistical constraints that could limit the ability to safely move birds to them in a timely manner, and sufficient facilities to carry out the action or the facilities could be reasonably constructed without damaging the integrity of the site. The result of these three prioritisation activities was a science-based repeatable approach that maximised potential success and minimised risk from an outcomes, safety and financial standpoint. The results of this exercise were then published in the peer-reviewed literature so that other organisations could use the exercise to potentially prioritise other actions (Young and VanderWerf, 2022).

Once the project parameters (species and sites) had been selected, project-specific benchmarks were set along with associated timelines. Since this project involved both restoration of habitat (i.e. building predator-free reserves) and the translocation of species, there were multiple categories used. The overall project benchmarks were:

1. Habitat outcomes:
    a. The number of sites established
    b. The total number of acres of predator-free habitat on high islands that have been created
    c. The number of native plants out-planted
    d. The number or acreage of invasive plants removed
2. Translocation outcomes:
    a. % of chicks that survive capture and transfer to release site
    b. Body size of fledged chicks

c. % chicks that fledge from the new colony

d. % translocated chicks that return to the release site (by age four)

e. Number birds fledged from other colonies that visit the translocation site

f. Number birds fledged from other sites that recruit to the new colony

g. Reproductive performance of birds breeding in the new colony.

3. The number of people reached:

   a. Number of people brought to see the site in person

   b. Number of public presentations

   c. Number of peer-reviewed papers published

   d. Number of media articles and associated reach

The project has been ongoing since 2015, and has involved four species across three sites. To date, all habitat specific goals have been met, and two of the four species have met their final long-term outcomes with the other two having met the stage specific outcomes (i.e. not enough time has passed to achieve the long term objectives). While there have been stage and time specific failures (usually mortality events if associated with the translocation or predator breaches in the case of the enclosures), none of these was considered overall project failures.

What has been key to using evidence-based conservation within PRC is not only prioritising projects and setting stage specific benchmarks, but conducting regular evaluations with staff and project partners. This is typically done after each major project stage (i.e. completion of a predator fence, or completion of a translocation event for the year), and then at the end of each season. Feedback is gathered on the process as well as the metrics being used, and adjustments are made, if needed. Considerable effort is put into keeping stakeholders informed by presenting data at the end of each season both in written format (i.e. annual reports) and visually through presentations. Efforts are made to present both the challenges and successes in order to ensure a holistic picture is being put forward. Finally, at the end of each project, at least one, and often multiple peer-reviewed publications are distributed on the project to ensure the greater conservation community may benefit.

The system employed by PRC is one that started top down during a critical inflexion point in the organisation and has been incorporated into every staff position and project plan so much so that it is not treated as a stand-alone organisational unit. As such, it continues to have seamless integration into the existing work and culture of the organisation, ensuring its long-term persistence.

11.8.6 US Fish and Wildlife Service National Wildlife Refuges System

The US Fish and Wildlife Service's National Wildlife Refuge System (NWRS) manages a network of 560 protected areas across the United States. Founded in 1903, NWRS has a long history of conservation of species and ecosystems. In 2010 the NWRS invested in an Inventory and

Monitoring (I&M) programme designed to address information needs and evaluate the effectiveness of conservation strategies implemented on NWRS lands.

The NWRS is making progress toward supporting a culture of evidence use. Currently, the NWRS is engaged in a process to focus monitoring data collection on specific species or ecosystem at each refuge, develop SMART (specific, measurable, achievable, relevant, and time-bound) objectives for each species or ecosystem and identify what indicators will be used to evaluate the objectives that were set. In the NWRS Pacific Southwest Region, the I&M team used the Conservation Standards (CMP, 2020) to set up the structure for connecting conservation targets, conservation strategies and indicators. This has been designed as an iterative process where refuge staff use professional judgement in most cases to set conservation targets and evaluation criteria. As data for the indicators are either collected or existing data are analysed these initial target values will be re-evaluated and updated. The process of developing objectives, having data to evaluate the objectives and the ability to update targets, when moving from professional judgement to an evidence base, provides a flexible structure that can be scaled to the staff time and funding available. A subset of refuges identified strategies and outcomes of their strategies that can be used to increase learning and to better understand how the organisation's conservation strategies impact its conservation targets. In addition to developing methods and processes for integrating evidence into decisions, there is a need for easy methods for capturing the data, results, and lessons learned. As part of the process of developing evidence, the I&M team has been developing tools, data standards, and visualisations of the data used to set targets, as well as the lessons learned. The change from only publishing information in summary reports to capturing information in standardised data structures has increased the I&M team's ability to look across all the refuge conservation targets to identify common challenges, what evidence is available for different strategies, and where to focus efforts.

Trying to change the culture in a large organisation responsible for managing a wide variety of ecosystems, with varying levels of funding and staffing, takes time. Nevertheless, they are starting to see a change in how some individual staff approach projects to increase their learning and ensure that others can build on the knowledge they have gained. The changes are dependent on additional support for staff to learn how to use and develop evidence, and for leadership to support and ask for evidence. Now that they are seeing the change with early adopters, they are taking the next steps to scale it up to be an organisation-wide approach.

11.8.7 Whitley Fund for Nature

The Whitley Fund for Nature (WFN) is a fundraising and grant-giving nature conservation charity offering recognition, training, and grants to support the work of proven grassroots conservation leaders across the Global South. Since its founding in 1993 as a UK charity, WFN has channelled £20 million to 200 conservation leaders in 80 countries, primarily across Asia, Africa, and South America.

WFN's grants programmes offer laddered support to those spearheading local solutions to the global biodiversity and climate crises. Through its flagship Whitley Awards Ceremony and Continuation Funding programme, WFN supports work rooted in science and community involvement that benefits wildlife, landscapes, and people. Award recipients gain funding, skills, and increased visibility, resulting in international profile boost, media attention, additional investment opportunities, and improved access to decision makers.

WFN supports a culture of evidence use and continues to fine tune its application process to take evidence into account. WFN started working with Conservation Evidence several years ago because, as funders, the organisation recognises the importance of not trying to reinvent the wheel, sharing knowledge, and learning from failure. As a grant-maker, WFN also wants to encourage openness among conservationists (including the sharing of both positive and negative results), avoid unnecessary duplication, and support the scale-up of effective environmental solutions. WFN therefore also recognises the value of adopting innovative approaches and sharing these results.

The use of Conservation Evidence was integrated into WFN's application screening process, at first by checking the What Works in Conservation literature themselves and then by integrating it into their application guidelines for efficiency. WFN currently recommend all applicants check the Conservation Evidence website to reference examples of conservation interventions and their effectiveness captured in the literature. This will help to inform project design and monitoring.

For shortlisted applications to the Whitley Awards and Continuation Funding programme, WFN requires candidates to provide evidence of success by asking Principle Investigators what makes them confident the proposed activities will succeed and be effective in achieving the desired project outcomes. Applicants must provide evidence that their proposed methods will be effective, drawing from either their experience to-date, peer-reviewed publications, grey literature (non-published data), and relevant examples from other projects.

In this way, WFN is mindful that most published evidence originates from the Global North and is found behind paywalls, acting as a barrier to many, and thereby recognises the need to take valuable unpublished evidence and experience into account. This type of evidence is more accessible to NGOs and practitioners in the Global South, where conservation is underfunded yet much needed, and where effective work is being delivered by committed teams.

WFN asks shortlisted applicants to complete a form (see Table 11.4), focusing on their objectives and assumptions rather than providing a long list of every planned intervention or activity. This exercise aims to encourage applicants to scrutinise their project design, and base actions on evidence to increase the chance of successful outcomes. The main aspiration of this approach is to ensure that applicants have gone through a rigorous process as part of their decision making, even if the formal scientific evidence is sparse.

The WFN grant application process is, of course, an iterative learning process. It continues to evolve, as WFN takes on board feedback from grantees and practitioners on the ground.

Among WFN's network of Whitley Award winners across the globe, several have become Evidence Champions, acting as ambassadors for Conservation Evidence and putting the approach of evidence use into practice. WFN's vision is that by taking evidence into account and using this to assess applications as a standard part of its review process, they will help practitioners improve the likelihood of success and the delivery of effective conservation outcomes that can be brought to scale. This mode of application assessment is already seen

Table 11.4 The Whitley Fund for Nature evidence summary table used for shortlisted projects.

Intervention:	Removing ghost fishing gear				
Evidence source[a]	Type of evidence[b]	Direction and strength of results[c]	Relevance (Low/ Medium/ High)[d]	Evidence quality (L/M/H)[e]	Overall confidence (L/M/H)[f]
Personal experience	Observation	150 kg discarded nets removed from a small area of benthic habitat. Tube sponges quickly established.	High	Medium	High
Peer-reviewed publication: Melli et al. (2017) assessing marine debris in a Site of Community Importance in the north-western Adriatic Sea	Observation	Litter-fauna interactions were high, with most of the debris (65.7%) entangling or covering benthic organisms.	High	Medium	Medium
https://www.conservationevidence.com/actions/2206	Synthesis — no evidence	none	High	High	Medium

[a] E.g. Peer-reviewed publication, expert opinion, grey literature report, personal experience

[b] E.g. synthesis, experimental, observational, anecdotal, theoretical/modelling.

[c] Was the result strongly positive, weakly positive, mixed, or no effect?

[d] How relevant is the evidence in terms of geography, taxa or habitat? Does the evidence relate directly to the intervention effectiveness?

[e] Depends on the type of evidence, but also sample size and experimental design.

[f] From a combination of the direction/strength of results, relevance and quality across all evidence types.

in the way many funders (including WFN) request a budget or Theory of Change. The result of including evidence is expected to not only be to aid the recovery of the natural world, but to also attract increased funding to the environment sector.

11.8.8 Woodland Trust

The Woodland Trust is a UK charitable organisation committed to tackling the climate and nature emergencies through trees and woods. Being 'evidence-led' is of critical importance for informing where and how to prioritise and deploy limited resources, in a cost-effective way to achieve maximum impact. Evidence is key to understanding the challenges our natural world is facing, and where the organisation's expertise, advocacy, and practical delivery can best be applied to make impactful and meaningful change in the short and long term. By harnessing the power of evidence to inform and underpin our interactions and communications with a wide range of audiences, from decision makers, politicians, scientists, woodland managers, landowners and indeed its supporters, the Woodland Trust can act and speak with confidence and credibility.

The transformation to an effective culture of using evidence has required clarity on what being evidence-led means for the variety of different functions of the organisation and how it can help us target our limited resources for greatest value and impact. This transformation was given a real boost with the creation of a dedicated conservation team over 15 years ago, and in 2018 by greater focus and investment in a Conservation Evidence and Outcomes team. This team of scientists has a range of specific skills and experience and able to lead the development of improved engagement with science and research through a growing evidence toolkit, enhanced communications across the organisation and externally, including webinars with external guests, quarterly newsletters, and regular site visits with the organisation's delivery teams. There has also been further development of systems, processes, and training with this shift.

This has led to an improved understanding of why and how the power of evidence should be harnessed, with evidence now becoming embedded within all functions of the organisation and clear visibility in our organisational strategy. Utility and application of evidence are being enabled through the synthesis of evidence into review papers and briefings, through funding and commissioning research to fill key evidence gaps, through the development of evidence-based best practice guides for the creation and restoration of trees and woods and collecting and harnessing data to understand where and how action for change should be delivered. The Woodland Trust published the first ever State of the UK's Native Woods and Trees report in April 2021 (Reid et al., 2021). This provided a comprehensive review of the evidence, and this will continue to shape the Trust's policy influencing goals, improved its reputation as trusted leaders with policy makers and landowners, identified evidence gaps, and significantly informed its own strategic direction, increasing the organisation's ambitions for protection and restoration of existing woods and trees. Woodland Trust evidence reviews, practical guides, reports, and citizen science datasets are all freely shared for others to use. For example, evidence reviews are published in Applied Ecological Resources, where the Woodland Trust is a silver member.

Meanwhile, the culture and commitment to using evidence within the Woodland Trust continues to grow with new initiatives such as a new monitoring, evaluation and learning framework. This framework is currently in development using the open conservation standards and expert conservation training 366 References for practitioners and landowners. This training allows them to develop skills and capability in line with evidence-based guidance. Integrating evidence is the key to success. As Evidence Champions with the Conservation Evidence initiative, the Woodland Trust continues to build important collaborations and relationships with partners, to bring together and share knowledge, tools and evidence to galvanise action, and enhance the quality and longevity of conservation outcomes on the ground.

References

Archibald, D.W., McIver, R. and Rangeley, R. 2021. Untimely publications: Delayed Canadian fisheries science advice limits transparency of decision-making. *Marine Policy* 132: 104690, https://doi.org/10.1016/j.marpol.2021.104690.

Bat Conservation International (BCI). 2020. *Our Mission to End Bat Extinctions Worldwide: Strategic Plan 2020–2025* (Austin, TX: Bat Conservation International), https://www.batcon.org/wp-content/uploads/2020/10/BCI_Strategic_Plan_2020_2025_web.pdf.

Berthinussen, A., Richardson, O.C. and Altringham, J.D. 2014. *Bat Conservation: Global Evidence for the Effects of Interventions* (Exeter, UK: Pelagic Publishing Ltd.).

Berthinussen, A., Richardson, O.C. and Altringham, J.D. 2021. *Bat Conservation: Global Evidence for the Effects of Interventions (2021 Edition)* (Cambridge: University of Cambridge), https://www.conservationevidence.com/synopsis/pdf/32.

Catalano, A.S., Jimmieson, N.L. and Knight, A.T. 2021. Building better teams by identifying conservation professionals willing to learn from failure. *Biological Conservation* 256: 109069, https://doi.org/10.1016/j.biocon.2021.109069.

Catalano, A.S., Lyons-White, J., Mills, M.M., et al. 2019. Learning from published project failures in conservation. *Biological Conservation* 238: 108223, https://doi.org/10.1016/j.biocon.2019.108223.

Chambers, J.M., Massarella, K. and Fletcher, R. 2022. The right to fail? Problematizing failure discourse in international conservation. *World Development* 150: 105723, https://doi.org/10.1016/j.worlddev.2021.105723.

Christie, A.P., Downey, H., Frick, W.F., et al. 2022. A practical conservation tool to combine diverse types of evidence for transparent evidence-based decision-making. *Conservation Science and Practice* 4: e579, https://doi.org/10.1111/csp2.579.

Clear, J. 2018. *Atomic Habits: Tiny Changes, Remarkable Results* (New York: Penguin Random House).

CMP. 2020. *Open Standards for the Practice of Conservation* (Conservation Measures Partnership), https://conservationstandards.org/wp-content/uploads/sites/3/2020/10/CMP-Open-Standards-for-the-Practice-of-Conservation-v4.0.pdf.

Conservation Measures Partnership (CMP). 2021. *Conservation Standards Adoption — A Change Management Strategy* (Conservation Measures Partnership), available at: https://conservationstandards.org/wp-content/uploads/sites/3/2022/02/Conservation-Standards-Adoption_Change-management-TOC_CMP-Learning-Initiative.pdf.

Cooke, S.J., Johansson, S., Andersson, K., et al. 2017. Better evidence, better decisions, better environment: Emergent themes from the first environmental evidence conference. *Environmental Evidence* 6: Article 15, https://doi.org/10.1186/s13750-017-0092-0.

CSAS. 2011. *Policy on Observers* (Canadian Science Advisory Secretariat), http://www.dfo-mpo.gc.ca/csas-sccs/process-processus/observers-observateurs-eng.html.

CSAS. 2020. *Science Advice on the Effectiveness of Spawning Habitat Creation for Substrate Spawning Temperate Fish,* Science Advisory Report 2020/013 (Canadian Science Advisory Secretariat), https://www.dfo-mpo.gc.ca/csas-sccs/Publications/SAR-AS/2020/2020_013-eng.html.

Díaz, S., Fargione, J., Chapin, F.S., et al. 2006. Biodiversity loss threatens human well-being. *PLoS Biology* 4: e277, https://doi.org/10.1371/journal.pbio.0040277.

Dickson, I., Butchart, S.H.M., Catalano, A., et al. 2022. Introducing a common taxonomy to support learning from failure in conservation. *Conservation Biology*, https://doi.org/10.1111/cobi.13967.

Downey, H., Amano, T., Cadotte, M., et al. 2021 Training future generations to deliver evidence-based conservation and ecosystem management. *Ecological Solutions and Evidence* 2: e12032, https://doi.org/10.1002/2688-8319.12032.

Edmondson, A.C. 2011. Strategies for learning from failure. *Harvard Business Review*, https://hbr.org/2011/04/strategies-for-learning-from-failure.

Franz, N.K. and Townson, L. 2008. The nature of complex organizations: The case of Cooperative Extension. *New Directions for Evaluation* 2008: 5–14, https://doi.org/10.1002/ev.272.

Frick, W.F., Kingston, T. and Flanders, J. 2020. A review of the major threats and challenges to global bat conservation. *Annals of the New York Academy of Sciences* 1469: 5–25, https://doi.org/10.1111/nyas.14045.

Guyatt, G., Cairns, J., Churchill, D., et al. 1992. Evidence-based medicine: A new approach to teaching the practice of medicine. *JAMA* 268: 2420–25, https://doi.org/10.1001/jama.1992.03490170092032.

Ilic, D. and Maloney, S. 2014. Methods of teaching medical trainees evidence-based medicine: A systematic review. *Medical Education* 48: 124–35, https://doi.org/10.1111/medu.12288.34.

Keller, S. and Schaninger, B. 2019. *Beyond Performance 2.0: A Proven Approach to Leading Large-Scale Change* (New York: John Wiley and Sons, Inc).

Lave, J. and Wenger, E. 1991. *Situated Learning: Legitimate Peripheral Participation* (Cambridge: Cambridge University Press).

Melli, V., Angiolillo, M., Ronchi, F., et al. 2017. The first assessment of marine debris in a site of community importance in the north-western Adriatic Sea (Mediterranean Sea). *Marine Pollution Bulletin* 114: 821–30, https://doi.org/10.1016/j.marpolbul.2016.11.012.

Panchadsaram, R. 2020. What is an OKR? Definition and examples. *What Matters: FAQs*, https://www.whatmatters.com/faqs/okr-meaning-definition-example.

Reid, C., Hornigold, K., McHenry, E., et al. 2021. *State of the UK's Woods and Trees 2021* (Woodland Trust), https://www.woodlandtrust.org.uk/state-of-uk-woods-and-trees/.

Rice, J., Bradford, M.J., Clarke, K.D., et al. 2015. The science framework for implementing the fisheries protection provisions of Canada's Fisheries Act. *Fisheries* 40: 268–75, https://doi.org/10.1080/03632415.2015.1038381.

Rytwinski, T., Elmer, L.K., Taylor, J.J., et al. 2019. How effective are spawning-habitat creation or enhancement measures for substrate-spawning fish? A synthesis. *Canadian Technical Report of Fisheries and Aquatic Sciences* 3333: 183pp, https://www.fecpl.ca/wp-content/uploads/2019/09/Rytwinski-et-al-2019_Offsets_CSAS.pdf

Salafsky, N., Margoluis, R., Redford, K.H., et al. 2002. Improving the practice of conservation: A conceptual framework and research agenda for conservation science. *Conservation Biology* 16: 1469–79, https://doi.org/10.1046/j.1523-1739.2002.01232.x.

Schwaber, K. and Sutherland, J. 2020. The 2020 Scrum Guide. *Scrum Guides*, https://scrumguides.org/scrum-guide.html.

Shortell, S.M., Zazzali, J.L., Burns, L.R., et al. 2001. Implementing evidence-based medicine: The role of market pressures, compensation incentives, and culture in physician organizations. *Medical Care* 39: I-62–I-78, https://doi.org/10.1097/00005650-200107001-00005.

Slavin, R.E. 2002. Evidence-based education policies: Transforming educational practice and research. *Educational Researcher* 31: 15–21, https://doi.org/10.3102/0013189X031007015.

Sutherland, W.J., Downey, H., Frick, W.F., et al. 2021. Planning practical evidence-based decision making in conservation within time constraints: The Strategic Evidence Assessment Framework. *Journal for Nature Conservation* 60: 125975, https://doi.org/10.1016/j.jnc.2021.125975.

Sutherland, W.J. and Wordley, C.F. 2017. Evidence complacency hampers conservation. *Nature Ecology and Evolution* 1: 1215–16, https://doi.org/10.1038/s41559-017-0244-1.

Taylor, J.J., Rytwinski, T., Bennett, J.R., et al. 2019. The effectiveness of spawning habitat creation or enhancement for substrate-spawning temperate fish: A systematic review. *Environmental Evidence* 8: Article 19, https://doi.org/10.1186/s13750-019-0162-6.

Warner, P. and Christenson, J.A. 2019. *The Cooperative Extension Service: A National Assessment* (New York: Taylor and Francis), https://doi.org/10.4324/9780429309809.

Wood, W. 2019. *Good Habits, Bad Habits: The Science of Making Positive Changes that Stick.* (London: Pan Macmillan).

Young, N., Corriveau, M., Nguyen, V.M., et al. 2016. How do potential knowledge users evaluate new claims about a contested resource? Problems of power and politics in knowledge exchange and mobilization. *Journal of Environmental Management* 184: 380–88, https://doi.org/10.1016/j.jenvman.2016.10.006.

Young, L.C. and VanderWerf, E.A. 2022. *Conservation of Marine Birds* (Cambridge, MA: Elsevier).

12. Transforming Practice

Checklists for Delivering Change

Tatsuya Amano[1], Longji Bako[2], Marina Best[3], Nicolas Boenisch[4], Philipp Boersch-Supan[5], Des Browne[6,7], Yvonne Buckley[8], Mark Burgman[9], Marc W. Cadotte[10], Stefano Canessa[11], Samantha H. Cheng[12], Alec P. Christie[7,13,14], Geoffroy Citegetse[15], Carly N. Cook[16], Steven J. Cooke[17], Gemma Cranston[18], Lynn V. Dicks[14], Angelita De la Luz[19], Iain Dickson[20], Harriet Downey[21], Johanna Eklund[22], Nafeesa Esmail[23,24], Paul J. Ferraro[25], Alison Field[26], Martin Fisher[27], Robert Freckleton[28], Winifred Frick[29,30], H. Charles J. Godfray[31], Matthew J. Grainger[32], Rhys Green[14], Mark Hancock[33,34], Victoria Hemming[35], Jonathan Hughes[36], Robyn Irvine[37], Hazel A. Jackson[38], Kaylene E. Keller[39], Julia Koricheva[40], Charles Latrémouille[41], Marissa McBride[9], Angela R. McLean[31], Tom McPherson[42], William H. Morgan[14], Matt Muir[43]*, Nibedita Mukherjee[44], David O'Brien[45], Nancy Ockendon[46], Danni Parks[47], Silviu Petrovan[7,14], Maud Quinzin[48], Nicola Randall[49], Ali Mohammad Rezaie[50], Dilys Roe[51], David Rose[52], Libby Rumpff[53], Ullrika Sahlin[54], Nick Salafsky[55], Stefan Schindler[56], Tom Sheldon[57], Ashley T. Simkins[20], Rebecca K. Smith[7,14], Jonathan Spencer[26], William J. Sutherland[7,14], Iroro Tanshi[58], Nigel G Taylor[14], Eleanor R. Tew[59], Des B.A. Thompson[60], Ann Thornton[14], Paul Tinsley-Marshall[61], Thomas B. White[14], Kate Willott[14] Jeremy D. Wilson[62], Bonnie C. Wintle[63], Thomas A. Worthington[14], Hiromi Yamashita[64], Lindsay C. Young[65]*

Delivering a revolution in evidence use requires a cultural change across society. For a wide range of groups (practitioners, knowledge brokers, organisations, organisational leaders, policy makers, funders, researchers, journal publishers, the wider conservation community, educators, writers, and journalists), options are described to facilitate a change in practice, and a series of downloadable checklists is provided.

 https://doi.org/10.11647/OBP.0321.12

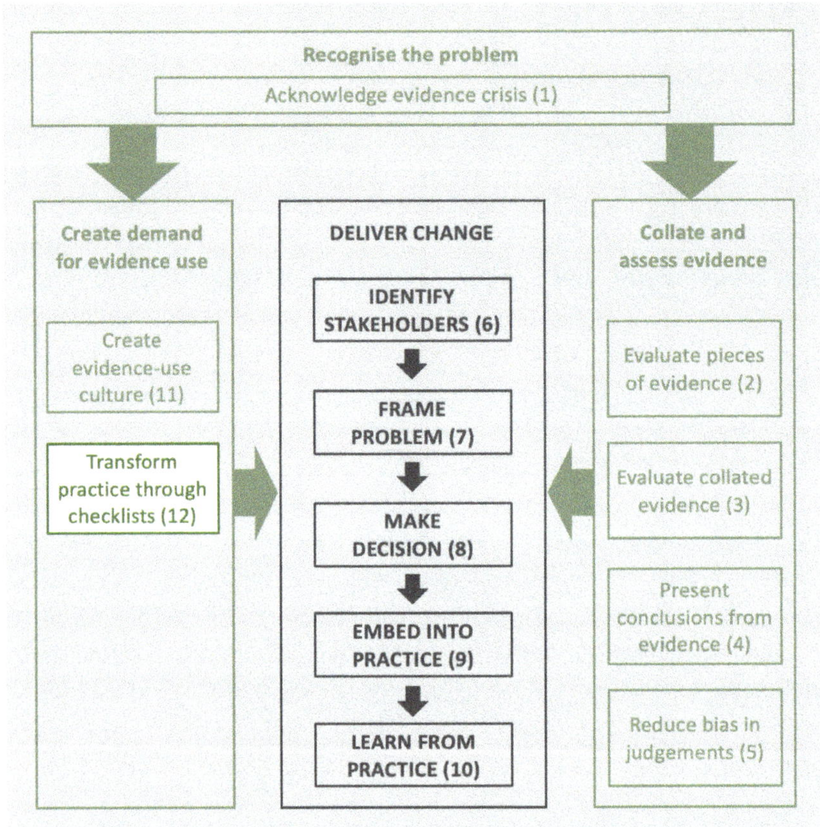

[1] School of Biological Sciences, University of Queensland, Brisbane, 4072 Queensland, Australia
[2] The AP Leventis Ornithological Research Institute (APLORI), University of Jos Biological Conservatory, PO Box 13404, Laminga, Jos-East LGA, Plateau State, 930001, Nigeria
[3] Parks Canada Agency, Indigenous Conservation, Indigenous Affairs and Cultural Heritage, 30 Rue Victoria, Gatineau, Quebec, Canada
[4] Foundations of Success Europe, 7211 AA Eefde, The Netherlands
[5] British Trust for Ornithology, The Nunnery Thetford Norfolk, UK; Department of Geography, University of Florida, Gainesville, FL 32611, USA
[6] Houses of Parliament, Parliament Square, London SW1A 0PW
[7] Biosecurity Research Initiative at St Catharine's College (BioRISC), St Catharine's College, Cambridge, UK
[8] School of Natural Sciences, Trinity College, Dublin, Ireland
[9] Centre for Environmental Policy, Imperial College London, London, UK
[10] Biological Sciences, University of Toronto-Scarborough, 1265 Military Trail, Toronto, ON, Canada
[11] Institute of Zoology, Zoological Society London, Regent's Park, London, UK
[12] Center for Biodiversity and Conservation, American Museum of Natural History, New York, NY 10026, USA
[13] Downing College, University of Cambridge, Regent Street, Cambridge, UK
[14] Department of Zoology, University of Cambridge, The David Attenborough Building, Pembroke Street, Cambridge, UK
[15] BirdLife International, Mermoz Pyrotechnie, Lot 23 Rue MZ 56, Dakar, Senegal
[16] School of Biological Sciences, Monash University, Clayton, VIC 3800, Australia
[17] Carleton University, Ottawa, ON, Canada
[18] University of Cambridge Institute for Sustainability Leadership, The Entopia Building, 1 Regent Street, Cambridge CB2 1GG, UK

[19] Beeflow, E Los Angeles Ave., Somis CA, USA 93066
[20] Birdlife International, The David Attenborough Building, Pembroke Street, Cambridge, UK
[21] Woodland Trust, Kempton Way, Grantham, Lincolnshire, UK
[22] Digital Geography Lab, Department of Geosciences and Geography, Faculty of Sciences, University of Helsinki, Helsinki, Finland
[23] The Wilder Institute, 1300 Zoo Road NE, Calgary, Alberta, Canada
[24] Women for the Environment Africa (WE Africa) Fellow
[25] The Center for Behavioral & Experimental Agri-Environmental Research, Johns Hopkins University, Baltimore, MD, USA
[26] Field and Forest, Land Management, Winchester, UK
[27] Fauna & Flora International, The David Attenborough Building, Pembroke Street, Cambridge, UK
[28] School of Biosciences, University of Sheffield, Sheffield, UK
[29] Bat Conservation International, 1012 14th Street NW, Suite 905, Washington, D.C. 20005, USA
[30] Department of Ecology and Evolutionary Biology, University of California Santa Cruz, 130 McAllister Way, Santa Cruz, CA 95060, US
[31] Oxford Martin School, University of Oxford, Oxford, UK
[32] Norwegian Institute for Nature Research (NINA), PO Box 5685 Torgarden, 7485 Trondheim, Norway
[33] RSPB Centre for Conservation Science, RSPB Scotland, North Scotland Office, Etive House, Beechwood Park, Inverness, UK
[34] Cairngorms Connect, Achantoul, Aviemore, UK
[35] Department of Forest and Conservation Sciences, The University of British Columbia, Forest Sciences Centre, 3041 - 2424 Main Mall, Vancouver BC, V6T 1Z4 Canada
[36] UN Environment World Conservation Monitoring Centre, 219 Huntingdon Rd, Cambridge, UK
[37] Parks Canada Agency, Ecosystem Conservation, Protected Areas Establishment and Conservation, 30 Rue Victoria, Gatineau, Quebec, Canada
[38] Woodland Trust, Kempton Way, Grantham, Lincolnshire, UK
[39] US Fish and Wildlife Service, National Wildlife Refuge System 2800 Cottage Way, Suite W-2606
[40] Department of Biological Sciences, Royal Holloway University of London, Egham, Surrey, UK
[41] Partner, Conservation Measures Partnership (CMP), Montreal, Canada,
[42] Ingleby Farms and Forests, Slotsgade 1A, 4600 Køge, Denmark
[43] United States Fish and Wildlife Service International Affairs, 5275 Leesburg Pike, Falls Church, VA 22041, USA
[44] College of Business, Arts and Social Sciences, Brunel University, London, UK
[45] NatureScot, Great Glen House, Leachkin Road, Inverness, UK
[46] Endangered Landscapes Programme, Cambridge Conservation Initiative, The David Attenborough Building, Pembroke Street, Cambridge, UK
[47] The Whitley Fund for Nature, 110 Princedale Road, London, UK
[48] Ocean Discovery League, P.O. Box 182, Saunderstown, RI 02874, USA
[49] Centre for Evidence Based Agriculture, Harper Adams University, Newport, Shropshire, UK
[50] Department of Civil Engineering, University of Asia Pacific, Dhaka 1205, Bangladesh
[51] International Institute for Environment and Development, 235 High Holborn, London, UK
[52] School of Water, Energy and the Environment, Cranfield University, College Road, Cranfield, Bedford, UK
[53] School of Ecosystem and Forest Sciences, University of Melbourne, Parkville 3052, Melbourne, Australia
[54] Centre for Environmental and Climate Science, Lund University, Lund, Sweden
[55] Measures of Success, Bethesda, Maryland, USA
[56] Department Biodiversity & Nature Conservation, Environment Agency Austria (Umweltbundesamt), Vienna, Austria
[57] Science Media Centre, 183 Euston Road, London, UK
[58] Small Mammal Conservation Organisation SMACON, Plot 8, Christiana Imudia, Irhirhi, Benin City, Nigeria
[59] Forestry England, 620 Bristol Business Park, Coldharbour Lane, Bristol, UK
[60] NatureScot, Silvan House, 231 Corstorphine Road, Edinburgh, UK
[61] Kent Wildlife Trust, Tyland Barn, Chatham Road, Sandling, Maidstone, UK; current address: Butterfly Conservation, Manor Yard, East Lulworth, Wareham, Dorset, UK
[62] RSPB Centre for Conservation Science, 2 Lochside View, Edinburgh, UK

[63] School of Ecosystem and Forest Sciences, University of Melbourne, Victoria 3010 Australia
[64] Ristumeikan Asia Pacific University, Beppu, Oita, Japan; Nagoya University Graduate School of Environmental Studies, Nagoya, Aichi, Japan
[65] Pacific Rim Conservation, PO Box 61827, Honolulu, HI, USA 96839

* The findings and conclusions in this chapter are those of the authors and do not necessarily represent the views of the US Fish and Wildlife Service.

Contents

12.1 The Importance of Checklists

A checklist is a list of systematically ordered actions that enables users to consistently perform each action and record its completion while reducing any errors that may be caused by missing out crucial steps (Gawande, 2010). Checklists can be used to help organisations and individuals consistently measure and monitor outputs and outcomes, and hence develop a culture of continuous quality improvement. They have been shown to be effective in improving outcomes in other areas of practice, powerfully described in *The Checklist Manifesto* (Gawande, 2010). Aircraft safety has been transformed by the use of checklists. This was exemplified in the 'Miracle of the Hudson' landing of US Airways Flight 1549 on the River Hudson, where the pilots ran through a checklist before flying, worked through the engine failure checklists when attempting to restart the engines after colliding with a flock of geese, and then, after landing on the Hudson River, ran through the evacuation checklist.

Gawande (2010) organised the creation of a checklist for use in surgery: the World Health Organization's Surgical Safety Checklist (Figure 12.1). Some of the 19 questions appear obvious. Are you operating on the correct patient? Is it the correct operation? Are you at the correct site for operation? Are you operating on the correct side of the patient? Has the area been disinfected recently? Have all the tools and swabs been removed after the operation? A randomised controlled experiment to test if the checklist was actually effective (Haugen et al.,

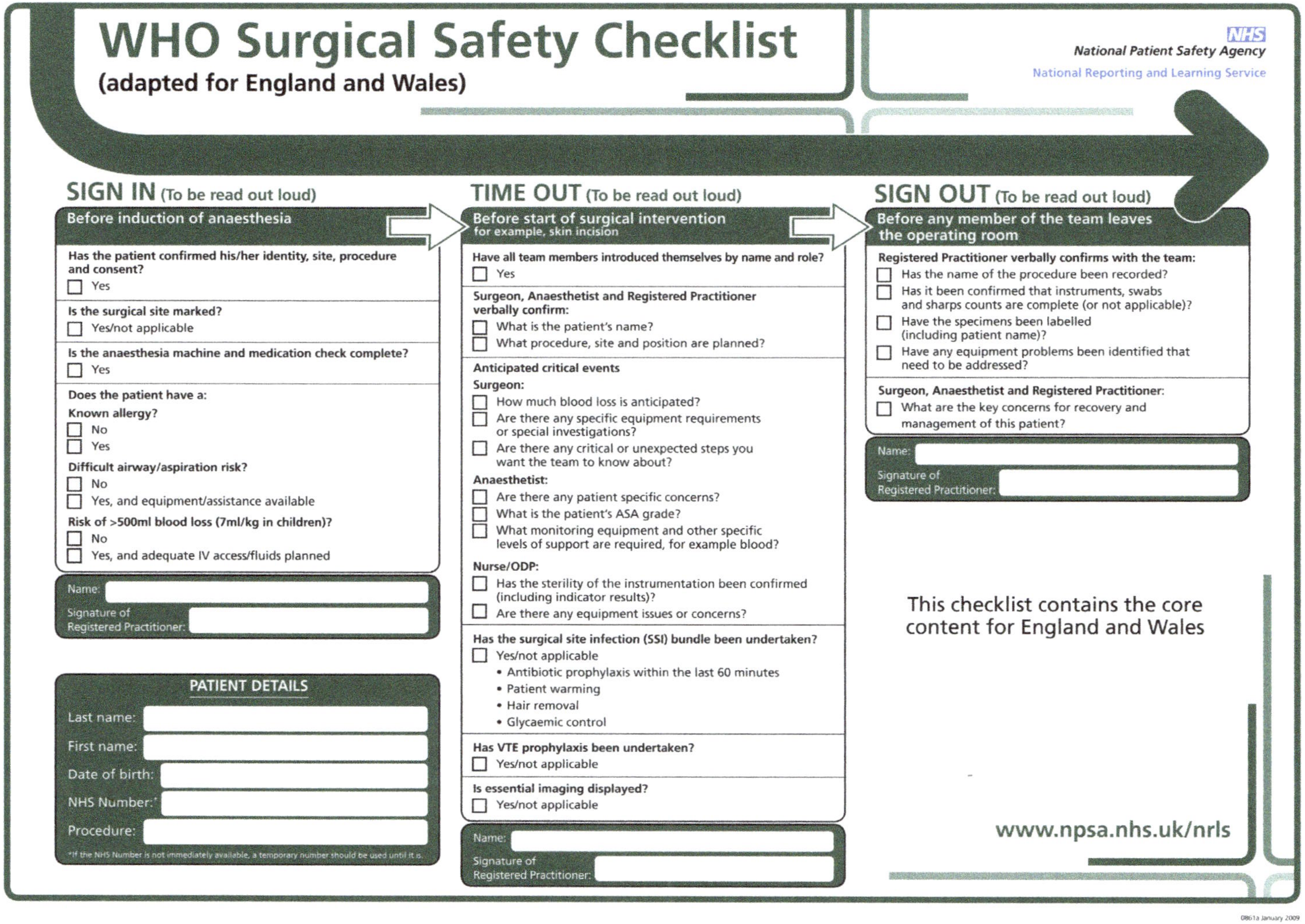

WHO Surgical Safety Checklist
(adapted for England and Wales)

NHS
National Patient Safety Agency
National Reporting and Learning Service

SIGN IN (To be read out loud)
Before induction of anaesthesia

Has the patient confirmed his/her identity, site, procedure and consent?
☐ Yes

Is the surgical site marked?
☐ Yes/not applicable

Is the anaesthesia machine and medication check complete?
☐ Yes

Does the patient have a:
Known allergy?
☐ No
☐ Yes
Difficult airway/aspiration risk?
☐ No
☐ Yes, and equipment/assistance available
Risk of >500ml blood loss (7ml/kg in children)?
☐ No
☐ Yes, and adequate IV access/fluids planned

Name:
Signature of Registered Practitioner:

PATIENT DETAILS
Last name:
First name:
Date of birth:
NHS Number:*
Procedure:
*If the NHS Number is not immediately available, a temporary number should be used until it is.

TIME OUT (To be read out loud)
Before start of surgical intervention
for example, skin incision

Have all team members introduced themselves by name and role?
☐ Yes

Surgeon, Anaesthetist and Registered Practitioner verbally confirm:
☐ What is the patient's name?
☐ What procedure, site and position are planned?

Anticipated critical events
Surgeon:
☐ How much blood loss is anticipated?
☐ Are there any specific equipment requirements or special investigations?
☐ Are there any critical or unexpected steps you want the team to know about?
Anaesthetist:
☐ Are there any patient specific concerns?
☐ What is the patient's ASA grade?
☐ What monitoring equipment and other specific levels of support are required, for example blood?
Nurse/ODP:
☐ Has the sterility of the instrumentation been confirmed (including indicator results)?
☐ Are there any equipment issues or concerns?

Has the surgical site infection (SSI) bundle been undertaken?
☐ Yes/not applicable
- Antibiotic prophylaxis within the last 60 minutes
- Patient warming
- Hair removal
- Glycaemic control

Has VTE prophylaxis been undertaken?
☐ Yes/not applicable

Is essential imaging displayed?
☐ Yes/not applicable

Name:
Signature of Registered Practitioner:

SIGN OUT (To be read out loud)
Before any member of the team leaves the operating room

Registered Practitioner verbally confirms with the team:
☐ Has the name of the procedure been recorded?
☐ Has it been confirmed that instruments, swabs and sharps counts are complete (or not applicable)?
☐ Have the specimens been labelled (including patient name)?
☐ Have any equipment problems been identified that need to be addressed?

Surgeon, Anaesthetist and Registered Practitioner:
☐ What are the key concerns for recovery and management of this patient?

Name:
Signature of Registered Practitioner:

This checklist contains the core content for England and Wales

www.npsa.nhs.uk/nrls

0861a January 2009

Figure 12.1 The World Health Organization checklist for surgeries (adapted for England and Wales). (*Source:* Panesar et al., 2010, © 2010, with permission from Elsevier)

2015) found that when the checklist was used, rates of complications during surgery decreased from 19.9% to 11.5%, and the mean length of a hospital stay was reduced by 0.8 days. The reduction in mortality from 1.6% to 1.0% was not statistically significant.

This chapter provides a series of checklists, each aimed at a different group of users, to help improve processes in using evidence and making decisions. Whilst the narrative of this chapter focuses on conservation, the checklists are equally applicable to other disciplines. These checklists are available on the associated website at https://doi.org/10.11647/OBP.0321#resources so can be modified for individual use. They are open access (Creative Commons CC-BY-4.0). Whilst attribution is welcome, it is not essential.

The checklists provided here differ from those in surgery or aircraft safety in which the user has to check every applicable box. These are checklists of what ideal evidence-based practice looks like and thus are targets, providing the opportunity to reflect on actions that are not checked. The checklists are inevitably somewhat repetitive across groups.

12.2 The Decision-Making Process

The heart of this book is a set of processes for delivering more effective conservation through improving the use of evidence in policy and practice. The process for making decisions based on evidence is summarised in Checklist 12.1, as a way to enable a simple check of whether good practice is being adopted.

12.3 Organisations

As the importance of using evidence becomes accepted, organisations will need a shift in emphasis such that evidence use is routinely embedded in decision making, as described in Chapter 9.

Embedding evidence use will require organisations to build capacity, train and mentor staff, and commit to employing new recruits skilled in using evidence. Recruitment and training should ensure conservation practitioners develop the motivation and necessary skills to embed evidence in practice. Checklist 12.2 describes the activities that can be considered by organisations interested in becoming evidence based.

There is also a need to add to the existing evidence base by testing actions and ensuring the results are made available to others, for example, by publishing results, adding them online and documenting methods, results and data on open science platforms (such as the Center for Open Science). Some journals have a specific objective of attracting articles from practitioners such as *Ecological Solutions and Evidence*, the associated Applied Ecology Resources platform (https://www.britishecologicalsociety.org/applied-ecology-resources/) and the *Conservation Evidence Journal*, which also encourages the publication of short papers by practitioners.

Documenting actions that are not successful, or which are only partially successful, is as informative as documenting those that *do* work. This allows practitioners to learn from the problems faced or mistakes made by others and thus avoid potentially costly errors. In a review

12.1 Checklist of components of evidence-based decision making

Note that numbers refer to the relevant book chapter or section.

- ☐ Have relevant stakeholders and local partners been identified, consulted, and involved in formulating goals and strategies appropriately? (6)
- ☐ Have the features of interest been identified? (7.2)
- ☐ Have the problems been identified? (7.3,7.4)
- ☐ Is the aim of the project clear? (7.5)
- ☐ Have the range of possible solutions to those problems been identified? (7.6)
- ☐ Have the knowledge needs been identified? (7.7)
- ☐ Has the evidence been gathered and assessed? (2,3,4)
- ☐ Have the costs and acceptability of these options been evaluated in a standardised and inclusive manner? (2.4)
- ☐ Have decisions been made, and expert advice solicited, in ways that reduce bias? (5)
- ☐ Has an appropriate structured decision-making process been adopted? (8)
- ☐ Have the decisions been embedded in the organisation's processes? (9)
- ☐ Has the testing of options been embedded in the proposed project? (10)
- ☐ Has an evidence-based organisational culture been created? (11)

This checklist can be downloaded from https://doi.org/10.11647/OBP.0321#resources. It can be modified and tailored for specific uses.

of the papers in the *Conservation Evidence Journal*, Spooner et al. (2015) showed that 31% of tested interventions were considered as unsuccessful.

Ideally, at least annually, organisations should establish a test (which can be very simple) to examine the effectiveness of an action. This could be done by comparing the action to no treatment, or by directly comparing different treatments or different means of carrying out a technique in one study (Smith et al., 2014).

12.3.1 Easy wins for organisation leaders

A key role of leaders is to create a culture where evidence use is encouraged and to enable the activities described in Checklist 12.1. This can be achieved through a combination of processes to ensure evidence use is embedded, such as those in Chapter 9, and routinely asking about the basis of decisions.

12.2 Checklist of evidence use by organisations

- ☐ Do job descriptions for posts that involve making decisions mention effective use of evidence?
- ☐ Do interviews for posts that involve making decisions include questions about evidence use and structured processes for making decisions (do interviewees use evidence and how)?
- ☐ Are audits routinely (e.g. annually) completed on the extent and effectiveness of evidence use in the organisation?
- ☐ Is there a plan outlining how evidence use will be conducted and improved in the organisation?
- ☐ Do decision makers have access to relevant evidence in an accessible form?
- ☐ Is it routine for staff to reflect on the existing evidence before carrying out actions?
- ☐ Do managers routinely ask about the use of evidence?
- ☐ Do those who overview plans and decisions routinely ask why the choice was made and how evidence was used?
- ☐ Do processes exist so that decision makers routinely (e.g. monthly) reflect on the documented evidence underpinning their work?
- ☐ Do those deciding on management actions routinely reflect on alternative management options and means of improving effectiveness?
- ☐ Is the evidence underpinning statements and decisions made clear in guidance documents?
- ☐ Is it made explicit whether the guidance used was evidence-based (e.g. Downey et al., 2022)?
- ☐ Is it routine to combine (evidence-based) guidance document recommendations with additional up-to-date evidence to justify management actions?
- ☐ Do goal-setting or performance evaluations for staff include questions about evidence use?
- ☐ Is training available, and taken up, on the use of evidence and testing actions?
- ☐ Is training available to help test actions and publish results?
- ☐ Are there processes by which, as an organisation, gaps in the available evidence are highlighted and filled by testing?

- ☐ Is it routine (at least annually) to create an experimental test to compare the effectiveness of an action against a control, or to compare different treatments?
- ☐ Are results of tests documented and made available regardless of the outcome?

This checklist can be downloaded from https://doi.org/10.11647/OBP.0321#resources. It can be modified and tailored for specific uses.

While some of the proposals in this chapter will take time to deliver, there are easy wins that the leaders of organisations could put into action very simply, listed in Checklist 12.3.

12.3 Checklist of eight 'easy wins' for leaders to consider enacting

- ☐ Ensure that job advertisements for decision makers specify the need to understand evidence-based practice.
- ☐ Make someone responsible for creating and delivering a strategy for evidence use.
- ☐ Establish a process of providing training on the principles of evidence use.
- ☐ State that reporting on evidence use (e.g. an outline of how evidence was incorporated) is expected in plans and reports produced by the organisation.
- ☐ Establish a process so that contracted reports require a statement on evidence use.
- ☐ Include the standard question, 'Does your manager routinely ask about the underlying evidence?', in annual reviews of practitioners and decision makers.
- ☐ Create a process that ensures applications for funding include reflections on the underlying evidence.
- ☐ Make someone responsible for ensuring that tests of an action are regularly initiated, for example at least annually, and results published.

This checklist can be downloaded from https://doi.org/10.11647/OBP.0321#resources. It can be modified and used.

12.4 Knowledge Brokers

The fundamental aim of this book is to facilitate the process of embedding evidence into practice, in order to make conservation actions more effective and efficient. Many of the anticipated changes will occur through practitioners and decision makers being more involved in research, by researchers being more involved in practice, and by organisations encouraging such collaborations.

Knowledge brokers can be useful in creating these cross-disciplinary communities (Checklist 12.4 suggests possible actions for knowledge brokers). Evidence-based medicine, for example, has a network of knowledge brokers bridging the gap between the researcher and frontline doctors and nurses. These serve to translate research findings into implications for clinical practice. Similarly, there is a need to create a cohort of evidence-based conservation knowledge brokers. These advisors could either be employed within organisations or within established external entities providing services to a range of organisations, for example, research institutes such as universities (Cook et al., 2013). Kadykalo et al. (2021) call such individuals and organizations 'evidence bridges' that would help foster research-practitioner connections and synthesise and distribute research. Knowledge brokers would have three main roles:

1. **Aiding the interpretation of evidence for decision-making:** As evidence-based conservation becomes increasingly accepted, the evaluation and interpretation of evidence will likely become more important. This consists of both evaluating the strength of evidence from different sources and relating the relevance of research to local conditions. For example, we expect that sometimes this interpretation will be provided in response to questions about an individual case (such as an advisor answering a question from a land manager) and sometimes provided as generic advice (such as an organisation providing advice to their staff about how they can treat invasive plants in their region). Interpreting the evidence for local conditions should be done in close consultation with practitioners.

2. **Producing accessible information that incorporates evidence:** Knowledge brokers can also help the creation of evidence-based guidance. As described in Chapter 9, this is key, in particular as an accepted standard for actions undertaken by ecological consultants.

3. **Assisting with the design of tests to assess effectiveness:** It is unreasonable to expect that most practitioners have the training and skills to design experiments, collect and analyse data or disseminate the results. There is a need for knowledge brokers who can assist with these tasks.

The Woodland Trust (UK) is one example of an organisation fulfilling this knowledge broker role through their Outcomes and Evidence team, which consists of 21 staff members (at the time of writing) whose expertise includes woodland ecology, citizen science, evidence use, land management, monitoring and evaluation, tree health, carbon and soil science, data science, and more. The team funds, co-designs, and co-delivers research with policy makers and practitioners, provides evidence syntheses and summaries, produces evidence-based guidance, provides bespoke advice to landowners, helps in the design of projects and experiments, disseminates results, and engages with internal and external stakeholders to enable evidence-based delivery of conservation outcomes in a wide variety of accessible formats.

12.4 Checklist for evidence use by knowledge brokers

- ☐ Have the knowledge needs of the target organisation been determined?
- ☐ Has there been an exploration of the range of possible actions that could be taken, including variations in implementation?
- ☐ Has the evidence been collected, shared, and stored in a manner that aligns with the community's best interests and rights?
- ☐ Is evidence made available to practitioners in a usable manner?
- ☐ Are there means for informing users of the full range of possible actions?
- ☐ Can evidence-based guidance be created for repeated activities?
- ☐ Have processes been adopted to ensure the evidence used is as up-to-date as possible?
- ☐ Is the range of evidence used appropriate to the questions being asked?
- ☐ Is help provided, where necessary, to ensure tests of actions are designed effectively?
- ☐ Is assistance provided, where necessary, to enable analysis and documenting of tests of effectiveness?
- ☐ Have processes been adopted to reflect on practice, especially where there were problems?

This checklist can be downloaded from https://doi.org/10.11647/OBP.0321#resources. It can be modified and tailored for specific uses.

12.5 Practitioners and Decision Makers

Those who make decisions and those who convert these into actions are central to the use of evidence in practice (see Checklist 12.5). These include a diverse range of individuals (e.g. nature reserve managers, reserve wardens, farmers, rangers, environmental site managers, and foresters). As a result of changes in expectations and requirements, and as evidence-use skills become essential for employment, over time we expect that evidence use will become standard practice.

12.6 Commissioners of Reports and Advice

Checklist 12.6 looks at means of establishing whether reports are evidence-based. These may be reports from consultants or internal reports. There is research indicating that conservation guidance documents, and actions then commonly recommended, are weakly based on

12.5 Checklist for evidence use by practitioners and decision makers

- ☐ Is it routine to consider whether other management options may be more effective and/or cost-effective?
- ☐ Is it routine to consider whether modifications to actions put in place might improve efficiency?
- ☐ Is it routine to consult a wide variety of evidence sources including documented evidence, professional experience, practitioner knowledge and local and indigenous knowledge when deciding upon actions?
- ☐ Are there processes for making the evidence used in plans and decisions transparent?
- ☐ Are there processes by which gaps in the evidence are highlighted within the community and filled by testing (and sharing of results)?
- ☐ Is there an evidence use plan?
- ☐ Are tests of effectiveness routinely embedded within conservation action?
- ☐ Are the results of tests of actions documented and shared regardless of the outcome?
- ☐ Are practitioners/decision makers involved in work to help apply and expand the evidence base?
- ☐ Is it routine to search for evidence when planning new projects?
- ☐ Are practitioners/decision makers appropriately trained in evidence use (e.g. how to find and synthesise evidence)?

This checklist can be downloaded from https://doi.org/10.11647/OBP.0321#resources. It can be modified and tailored for specific uses.

evidence. Hunter et al. (2021) showed that ecological mitigation and compensation measures recommended in practice often have a limited evidence base backing them up, and most actions were justified by using guidance documents. However, Downey et al. (2022) showed that the majority of guidance documents for the mitigation and management of species and habitats in the UK and Ireland do not cite scientific evidence to justify the recommended actions and are often outdated. This risks carrying out actions that are ineffective, or not carrying out the most cost-effective action. As a result of ineffective actions, limited funds could be wasted, staff or volunteer morale could be damaged, and there is a risk of bad publicity or reputational effects.

12.6 Checklist for ensuring reports are evidence-based

- ☐ Are sufficient resources allocated to finding and synthesising evidence?
- ☐ Does the report include a transparent process for evidence use?
- ☐ Does the evidence used in the report appear appropriately comprehensive and up-to-date?
- ☐ Does the report clarify any knowledge gaps?
- ☐ Does the report appropriately reflect on the uncertainties of the evidence?
- ☐ Have any conflicts of interest that might influence evidence use in the report been declared?
- ☐ Does the guidance look into the costs and acceptability of different actions, and do statements of financial costs specify what is included?
- ☐ Do consultants have a statement about their approach to evidence use on their website?
- ☐ Do consultants have robust processes for evidence use and testing?

This checklist can be downloaded from https://doi.org/10.11647/OBP.0321#resources. It can be modified and tailored for specific uses.

12.7 Funders and Philanthropists

Funders may be in the best position to deliver a fundamental change in evidence use. Conservation practice would be transformed if funders expected transparent evidence use, and agreed to fund work that improves the evidence base (see Checklist 12.7). Furthermore, the expected efficiency gains, as outlined in Chapter 1, should result in the delivery of substantially enhanced outcomes for the same funding. In turn, such increased efficiency may make funding conservation more attractive.

The most important questions for funders to ask applicants are: Does the applicant have a history of using evidence? What processes are in place for using evidence? How well are the proposed actions evidenced? The main ways in which funders can help increase and embed the use of evidence in practice are listed below. They can:

1. *Ask applicants why they believe their proposal will work*: There is a range of ways (see Chapter 9) in which funders can ask why the applicant believes their proposal will be effective and ensure that evidence has not been cherry-picked to support a particular course of action.

2. *Encourage rigorous testing of actions*: Funders can encourage experimental testing, analysis and publication of the effects of interventions by allowing a) a percentage

of grant money to be spent assessing the effectiveness and b) sufficient time to complete this evaluation. This could be an expected component of large projects and an optional element of small grants focussing on those opportunities where testing is appropriate (Section 10.3).

3. *Fund learning organisations*: Funders can favour organisations that are learning by evaluating and testing interventions and ideas; they can rate applicants on their past record of project evaluation and their record of making results publicly available. Funders should, however, make it clear that the criterion relates to whether their evaluation is rigorous and the organisation is learning, rather than to claims about how wonderful the programme is.
4. *Promote open data*: Funders can make it a requirement that the data collected, such as species recorded or responses to interventions, is stored in standard open access databases (using adequate geospatial data protection where necessary in order to protect vulnerable species and habitats), and they can provide funds for the necessary data curation work.
5. *Encourage submission to journals that are making progress toward open science*, for example by following the Transparency and Openness Promotion Guidelines (https://www.cos.io/initiatives/topguidelines).
6. *Ensure evidence base and infrastructures are funded*: Effective delivery of evidence-based practice requires that the evidence is collated, synthesised and made available for use, for example, through guidance or decision support tools. There is also a need for funding of adequate training/capacity building and materials.
7. *Accept publication as part of the reporting process*: A major barrier to dissemination is that many organisations do not have the time to both write a report for the funders and to publish a scientific paper. Accepting a final report in the form of a draft paper, or papers, reduces this problem.
8. *Establish a common register of projects*: Funders can establish an open register of funded projects (preferably a joint register across many funders), which also raises the expectation that robust data will be generated and subsequently made available to practitioners and scientists. This is also useful for testing for publication bias (e.g. on the success of a given type of project) and reducing duplication of projects or tests of interventions.
9. *Fund organisations committed to evidence use*: Funders can show a preference for funding organisations with a clear, demonstrated commitment to evidence-based practice (e.g. Conservation Evidence 'Evidence Champions'). There is a need for evidence certification such as through professional bodies.

12.7 Checklist for philanthropists and funders to encourage evidence use

- ☐ Does the funding call include a requirement to reflect on the evidence underpinning the proposal?
- ☐ Is the applicant asked to outline their history of evidence use?
- ☐ Are there processes for reviewing the use of evidence in proposals?
- ☐ Is there a process for encouraging and supporting projects to embed tests within actions?
- ☐ Is there a process to encourage data collected to be made openly available, when it does not cause ethical issues or conservation risks?
- ☐ Is there a plan to pre-register the work, and to analyse the sample effort required to detect important changes?
- ☐ Are project funder reports detailing outcomes made freely available online? Can draft publications, for submission to peer-reviewed journals, be accepted as a replacement for reports?
- ☐ Is there a stated expectation for an honest appraisal, including when the project is much less successful than expected?
- ☐ Does the funding call favour teams that include practitioners or decision makers alongside researchers?
- ☐ Is sufficient funding available, across the discipline or subject area, for evidence synthesis, guidance creation, training and capacity building?

This checklist can be downloaded from https://doi.org/10.11647/OBP.0321#resources. It can be modified and tailored for specific uses.

12.8 The Research and Education Community

Evidence-based medicine depends on an enormous pool of scientists who conduct basic research to underpin medical practice. It also relies on entire institutions that conduct syntheses of evidence, including members of the Cochrane Collaboration. Researchers and the education community have a large part to play in the practice of evidence-based conservation. Their roles are to:

1. *Identify priorities for evidence synthesis*: Engage with stakeholders, practitioners, managers, and policy makers to identify gaps in synthesis; to decide whether an evidence map, subject-wide evidence synthesis, systematic review, or dynamic synthesis is the most appropriate approach; and to create a process for delivering.

2. *Conduct research that fills in gaps in the evidence base*: Questions suitable for research might be a) identified by policy makers or practitioners (Sutherland et al., 2006, 2009), b) identified as gaps in policy (Sutherland et al., 2010a), c) identified as gaps in systematic reviews or synopses of evidence (e.g. evidence synthesis for primate conservation identified numerous evidence gaps, Junker et al., 2020). Building relationships with policy makers and practitioners assists in identifying knowledge gaps and facilitating the dissemination of research findings.
3. *Facilitate timely publication, and dissemination, of research findings*: The average length of time between the end of data collection and the final date of publication for studies of conservation interventions has been estimated at 3.2 years (Christie et al., 2021); this delay is further hindering practitioners from accessing up-to-date evidence. Such delays can be reduced by disseminating findings to relevant stakeholders, and timely publishing of results, including preprints where appropriate, in open access sources if possible.
4. *Encourage a culture of evidence-based practice*: Routinely develop questions to test the evidence base, and disseminate the need for evidence synthesis or primary research. Setting an agenda for priority conservation activities would direct projects to areas where learning could provide the greatest benefits.
5. *Teach effective decision making*: We envisage a major growth area to be the provision of training in effective decision-making skills (as described in this book) in a range of languages and at different levels. Chapter 10 gives further details on how this could be achieved.
6. *Teach approaches to quantifying the effectiveness of interventions*: Opportunities for generating convincing evidence are often missed or ineffective experimental designs are employed. There is thus a need for improved knowledge of experimental design, data collation, data analysis and the publication process.
7. *Introduce, by journals, reporting standards*: It is often difficult or impossible to extract key data from papers. This situation would be improved by introducing reporting standards, in a machine-readable manner, for research papers to facilitate subsequent use of data in synthesis and meta-analysis; similar advances have been made by editors of medical journals (Pullin and Salafsky, 2010). One solution is to form an International Committee, similar to that of the Medical Journal Editors (https://www.icmje.org/), and publish 'uniform requirements for manuscripts' such as reference styles.
8. *Introduce, by journals, standards for evidence synthesis methodology and reporting*: Review articles vary considerably in their rigour and reliability. Editors should consider using published checklists to improve the standards of review articles that are attempting to synthesise existing evidence (Pullin et al., 2022; O'Dea et al., 2021).

9. *Request, by journals, that authors report the costs of interventions*: Encourage authors to report the cost of an intervention in a manner that allows policy makers to translate this information to their management context and to determine the most cost-effective alternative. Consistent methods for reporting costs would enable economic return-on-investment evaluations as part of evidence synthesis.
10. *Introduce, by journals, measures to reduce barriers to the publication of evidence:* Publishers can take a number of measures to reduce hindrance and delays in publication. For example, they could introduce less strict formatting requirements for the initial screening of articles, or conduct preliminary peer review before submission, or move to new publication models where articles are published before peer assessment.
11. *Standardise methods*: A major problem with reviewing environmental evidence is that a wide range of methods is used, which makes comparisons difficult. Some of this variation is for sensible reasons, but much appears not to be. Standardising methods and outcome measurements, as has been done for reintroductions (Sutherland et al., 2010b) is, therefore, a useful step that could be applied more widely.
12. *Standardise terminology*: When comparing studies, standard terminology is required. Salafsky et al. (2008) suggest standard terms for the major threats, actions and habitats. Similarly, Mascia et al. (2014) provide standardised terminology for approaches to monitoring and evaluation. A useful analogy is to consider medicine before common names were agreed for diseases and potential cures, greatly hindering systematic science.
13. *Agree upon and use standard information repositories*: Create effective, attractive information systems for storing the outcomes of evidence-based planning and decision making, and monitoring and testing. Applied Ecology Resources (https://www.britishecologicalsociety.org/applied-ecology-resources/), Conservation Evidence (https://www.conservationevidence.com/), Panorama (https://panorama.solutions/en) and GBIF (https://www.gbif.org/conservation) are some stores for such material.
14. *Proactively bridge gaps between practice and research* (see Kadykalo et al. 2021): Facilitate conversations and networking between researchers and practitioners to help identify evidence needs, disseminate evidence produced to the wider community. Consult practitioner groups frequently to help mould research plans, and identify practical effective solutions.

12.8 Checklist for researchers and educators to support evidence use

- ☐ Do researchers, practitioners and policy makers in the field regularly interact and identify knowledge gaps?
- ☐ Does the research question fill a gap in the existing evidence base for policy and practice? If not, can the question be adapted to do so?
- ☐ Is the experimental design as rigorous as practically possible?
- ☐ Can the research project (including the methods and questions to be asked) be pre-registered?
- ☐ Are methodological and reporting standards (for either primary studies or evidence syntheses, as appropriate to the research question) available and followed?
- ☐ Do publications testing actions routinely report the costs of those actions?
- ☐ Have any conflicts of interest that might influence interpretation of the research been declared?
- ☐ Does the research clarify any remaining knowledge gaps?
- ☐ Is there a plan in place for timely publication of research results?
- ☐ Is it possible to publish results as a preprint?
- ☐ Are the results documented and shared regardless of the outcome?
- ☐ Are the conclusions of research made available to relevant practitioners and policy makers in a manner that is useful to them?
- ☐ Is data added to widely used information repositories, where appropriate?
- ☐ Is training available, and taken up, on the use of evidence?
- ☐ Is training available on effective decision making?
- ☐ Is training available, and taken up, on the testing of actions?

This checklist can be downloaded from https://doi.org/10.11647/OBP.0321#resources. It can be modified and tailored for specific uses.

References

Christie, A.P., White, T.B., Martin, P.A., et al. 2021. Reducing publication delay to improve the efficiency and impact of conservation science. *PeerJ* 9: e12245, https://doi.org/10.7717/peerj.12245.

Cook, C.N., Mascia, M.B., Schwartz, M.W., et al. 2013. Achieving conservation science that bridges the knowledge-action boundary. *Conservation Biology* 27: 669–78, https://doi.org/10.1111/cobi.12050.

Downey, H., Bretagnolle, V., Brick, C., et al. 2022. Principles for the production of evidence-based guidance for conservation actions. *Conservation Science and Practice* 4: e12663, https://doi.org/10.1111/csp2.12663.

Gawande, A. 2010. *The Checklist Manifesto: How to Get Things Right* (London: Profile Books).

Haugen, A.S., Søfteland, E., Almeland, S.K., et al. 2015. Effect of the World Health Organisation checklist on patient outcomes. *Annals of Surgery* 261: 821–28, https://doi.org/10.1097/SLA.0000000000000716.

Hunter, S.B., zu Ermgassen, S.O.S.E., Downey, H., et al. 2021. Evidence shortfalls in the recommendations and guidance underpinning ecological mitigation for infrastructure developments. *Ecological Solutions and Evidence* 2: e12089, https://doi.org/10.1002/2688-8319.12089.

Junker, J., Petrovan, S.O., Arroyo-Rodríguez, V., et al. 2020. A severe lack of evidence limits effective conservation of the world's primates. *BioScience* 70: 794–803, https://doi.org/10.1093/biosci/biaa082.

Kadykalo, A.N., Buxton, R.T., Morrison, P. et al. 2021. Bridging research and practice in conservation. *Conservation Biology* 35:1725–1737, https://doi.org/10.1111/cobi.13732.

Mascia, M.B., Pailler, S., Thieme, M.L., et al. 2014. Commonalities and complementarities among approaches to conservation monitoring and evaluation. *Biological Conservation* 169: 258–67, https://doi.org/10.1016/j.biocon.2013.11.017.

O'Dea, R.E., Lagisz, M., Jennions, M.D. et al. 2021. Preferred reporting items for systematic reviews and meta-analyses in ecology and evolutionary biology: a PRISMA extension. *Biological Review*, 96:1695–1722, doi:10.1111/brv.12721.

Panesar, S.S., Carson-Stevens, A., Fitzgerald, J.E., et al. 2010. The WHO surgical safety checklist — junior doctors as agents for change. *International Journal of Surgery* 8: 414–6, https://doi.org/10.1016/j.ijsu.2010.06.004.

Pullin, A.S., Cheng, S.H., Jackson, J.D., et al. 2022. Standards of conduct and reporting in evidence syntheses that could inform environmental policy and management decisions. *Environmental Evidence* 11: 16, https://doi.org/10.1186/s13750-022-00269-9.

Pullin, A.S. and Salafsky, N. 2010. Save the whales? Save the rainforest? Save the data! *Conservation Biology* 24: 915–17, https://doi.org/10.1111/j.1523-1739.2010.01537.x.

Salafsky, N., Salzer, D., Stattersfield, A.J., et al. 2008. A standard lexicon for biodiversity conservation: Unified classifications of threats and actions. *Conservation Biology* 22: 897–911, https://doi.org/10.1111/j.1523-1739.2008.00937.x.

Smith, R.K., Dicks, L.V., Mitchell, R., et al. 2014. Comparative effectiveness research: The missing link in conservation. *Conservation Evidence* 11: 2–6, https://conservationevidencejournal.com/reference/pdf/5475.

Spooner, F., Smith, R.K. and Sutherland, W.J. 2015. Trends, biases and effectiveness in reported conservation interventions. *Conservation Evidence* 12: 2–7, https://conservationevidencejournal.com/reference/pdf/5494.

Sutherland, W.J., Adams, W.M., Aronson, R.B. et al. 2009. One hundred questions of importance to the conservation of global biological diversity. *Conservation Biology* 23: 557–67, https://doi.org/10.1111/j.1523-1739.2009.01212.x.

Sutherland, W.J., Albon, S.D., Allison, H., et al. 2010a. The identification of priority policy options for UK nature conservation. *Journal of Applied Ecology* 47: 955–65, https://doi.org/10.1111/j.1365-2664.2010.01863.x.

Sutherland, W.J., Armstrong-Brown, S., Armsworth, P.R., et al. 2006. The identification of 100 ecological questions of high policy relevance in the UK. *Journal of Applied Ecology* 43: 617–27, https://doi.org/10.1111/j.1365-2664.2006.01188.x.

Sutherland, W.J., Armstrong, D., Butchart, S.H.M., et al. 2010b. Standards for documenting and monitoring bird reintroduction projects. *Conservation Letters* 3: 229–35, https://doi.org/10.1111/j.1755-263X.2010.00113.x.

13. Supplementary Material from Online Resources

This chapter provides links to sources of evidence, to material that aids teaching evidence use, and to tools for delivering change. It also lists 1,100 collaborators in the Conservation Evidence project that underpins much of this book.

Contents

 https://doi.org/10.11647/OBP.0321.13

13.1 Sources of Evidence

13.1.1 Catalogue of evidence sources in conservation

https://www.conservationevidence.com/content/page/127

13.1.2 Non-English evidence

The 53 members of the translatE (https://translatesciences.com/) project identified 365 relevant journals (https://www.conservationevidence.com/journalsearcher/nonenglish) in 16 languages (other than English) from 30 countries, screened 423,840 papers and identified 1,356 relevant papers (https://www.conservationevidence.com/data/nonenglishstudies) that describe tests of conservation actions.

13.1.3 Grey literature catalogue

The criteria for inclusion are that the reports must be online, must give details of the action and consequences and must be organised, for example, if the reports are numbered or ordered by the date completed. Report/grey literature series which have already been searched by Conservation Evidence can be found here, https://www.conservationevidence.com/journalsearcher/consrep, and individual reports that tested actions and met the inclusion criteria can be found here, https://www.conservationevidence.com/data/consrep.

13.2 Teaching Evidence Use

13.2.1 Evidence-use teaching materials

The Evidence in Conservation Teaching Initiative (Downey et al., 2021; https://www.britishecologicalsociety.org/applied-ecology-resources/about-aer/additional-resources/evidence-in-conservation-teaching) provides a range of open-access teaching material about evidence use, including lectures in nine different languages, with more underway.

13.2.2 Courses teaching evidence use

145 undergraduate, postgraduate or professional development courses teach evidence-based conservation (https://doi.org/10.1002/2688-8319.12032 — see supplementary material) — the courses are taught by a group of 117 educators from 23 countries (Downey et al., 2021).

13.3 Building the Evidence Base

13.3.1 Assessing impact

The PRISM tool (Practical Impact Assessment Methods Small and Medium-sized Conservation Projects, https://www.cambridgeconservation.org/resource/prism-evaluation-toolkit) provides users with guidance and methods for evaluating outcomes/impacts resulting from five different kinds of conservation action: Awareness and Attitudes, Capacity Development, Livelihoods and Governance, Policy, and Species and Habitat Management.

13.3.2 Material to aid publication of results

Oryx provides a library of recommended open access software for research, analysis and writing (https://www.oryxthejournal.org/authors/tools-and-resources/), the guide *Writing for Conservation* (https://www.oryxthejournal.org/writing-for-conservation-guide/) and suggestions for promoting the research conclusions (https://www.oryxthejournal.org/authors/promoting-your-research/). Lövei's (2019) open access coursebook, *Writing and Publishing Scientific Papers: A Primer for the Non-English Speaker)*

Templates make it easier to write a journal publication. These are available for *Conservation Evidence Journal* (https://conservationevidencejournal.com/collection/journaldetails) and *Oryx* (https://www.oryxthejournal.org/authors/guidelines-for-authors/).

13.3.3 Tools for creating systematic maps and reviews

The Collaboration for Environmental Evidence has a set of tools for creating systematic maps and reviews (https://environmentalevidence.org).

13.3.4 Journals providing assistance to authors not writing in their first language

A number of academic journals have committed to tackling language barriers. Tatsuya Amano and Martin A. Nuñez created a list outlining which conservation journals provide assistance for authors not writing in their first language (https://docs.google.com/spreadsheets/d/1n24A3O2EvuTk1WH5PHSHr6nWlvdCPuUaA6D6CADt6gs/edit#gid=0).

13.4 Delivering Change

13.4.1 Conservation organisations that have evidence roles

The best indication that evidence use has become mainstream will be organisations creating posts with responsibility for evidence use. Positions with evidence in the title have recently become much more common (some are listed here, https://docs.google.com/document/d/1tV32besT3IinUBvqTDF_SxsgbhkgyVbPu1PajazuXhs/edit), which indicates how evidence-based practice is becoming mainstream.

13.4.2 Funders require use of evidence

Funders have identified ten different ways in which the evidence underpinning a project proposal can be established (https://conbio.onlinelibrary.wiley.com/doi/full/10.1111/cobi.13991). To aid grant applicants, a group co-created some key principles for including evidence in proposals (https://conbio.onlinelibrary.wiley.com/doi/full/10.1111/cobi.13991). As more funders adopt this principle we expect it to transform evidence use.

13.4.3 Journals with processes to place research in context

As described in Section 12.8, research papers often, problematically, did not place the results in the context of the existing literature (too much 'standing on the toes of giants', rather than on their shoulders). To help overcome this problem, 40 conservation-focused journals (Sutherland et al., 2020; https://conbio.onlinelibrary.wiley.com/doi/full/10.1111/cobi.13555) established processes to ensure that authors outline how they have placed the literature in context.

13.4.4 Databases link to evidence of effective actions

As part of collaboration with other organisations, Conservation Evidence searches are embedded within species pages of other conservation data resources including the IUCN Red List website (https://www.iucnredlist.org/), the British Trust for Ornithology BirdFacts (https://www.bto.org/understanding-birds/birdfacts), the Agreement on the Conservation of African-Eurasian Migratory Waterbirds (https://www.unep-aewa.org/en/species/conservation-evidence) and the National Biodiversity Network (https://species.nbnatlas.org/) so that the page describing each species links to any relevant evidence on actions for that species (e.g. skylark https://species.nbnatlas.org/species/NHMSYS0000530139#literature).

13.5 Collaborators

13.5.1 The 1,100 named contributors to the Conservation Evidence project

At least 1,100 named collaborators created the work of the Conservation Evidence project that underpins this book (https://conservationevidenceblog.wordpress.com/2022/02/14/the-conservation-evidence-project-a-truly-collaborative-and-international-effort).

References

Downey, H., Amano, T., Cadotte, M., et al. 2021. Training future generations to deliver evidence-based conservation and ecosystem management. *Ecological Solutions and Evidence* 2: e12032, https://doi.org/10.1002/2688-8319.12032.

Lövei, G.L. 2019. *Writing and Publishing Scientific Papers: A Primer for the Non-English Speaker* (Cambridge: OpenBook Publishers), https://doi.org/10.11647/OBP.0235).

Sutherland, W.J., Alvarez-Castañeda, S.T., Amano, T., et al. 2020. Ensuring tests of conservation interventions build on existing literature. *Conservation Biology* 34: 781–83, https://doi.org/10.1111/cobi.13555.

Sutherland, W.J., Dicks, L.V., Petrovan, S.O., et al. 2021. *What Works in Conservation 2021* (Cambridge: Open Book Publishers), https://doi.org/10.11647/OBP.0267.

Tinsley-Marshall, P., Downey, H., Adum, G., et al. 2022. Funding and delivering the routine testing of management interventions to improve conservation effectiveness. *Journal for Nature Conservation* 67: 126184, https://doi.org/10.1016/j.jnc.2022.126184.

Checklists, Boxes and Tables

Checklists

Boxes

Tables

Figures

Index

About the Open Book Publishers Team

Alessandra Tosi was the managing editor for this book.

Gisele Parnall performed the copy-editing and proofreading.

Lucy Barnes indexed the volume.

Katy Saunders designed the cover. The cover was produced in InDesign using the Fontin font.

Luca Baffa typeset the book in InDesign and produced the paperback and hardback editions. The text font is Source Serif Pro; the heading font is Source Sans Pro. Luca Baffa produced the EPUB, AZW3, PDF, HTML, and XML editions — the conversion is performed with open source software such as pandoc (https://pandoc.org/) created by John MacFarlane and other tools freely available on our GitHub page (https://github.com/OpenBookPublishers).

For information on the environmental impact of Open Book Publishers, see https://blogs.openbookpublishers.com/environmental-impact/

This book need not end here...

Share

All our books — including the one you have just read — are free to access online so that students, researchers and members of the public who can't afford a printed edition will have access to the same ideas. This title will be accessed online by hundreds of readers each month across the globe: why not share the link so that someone you know is one of them?

This book and additional content is available at:

https://doi.org/10.11647/OBP.0321

Donate

Open Book Publishers is an award-winning, scholar-led, not-for-profit press making knowledge freely available one book at a time. We don't charge authors to publish with us: instead, our work is supported by our library members and by donations from people who believe that research shouldn't be locked behind paywalls.

Why not join them in freeing knowledge by supporting us: https://www.openbookpublishers.com/support-us

You may also be interested in:

What Works in Conservation

2021

William J. Sutherland, Lynn V. Dicks, Silviu O. Petrovan, Rebecca K. Smith (eds.)

https://doi.org/10.11647/OBP.0267

Terrestrial Mammal Conservation

Global Evidence for the Effects of Interventions for Terrestrial Mammals Excluding Bats and Primates

Nick A. Littlewood, Ricardo Rocha, Rebecca K. Smith, Philip A. Martin, Sarah L. Lockhart, Rebecca F. Schoonover, Elspeth Wilman, Andrew J. Bladon, Katie A. Sainsbury, Stuart Pimm, William J. Sutherland

https://doi.org/10.11647/OBP.0234

Negotiating Climate Change in Crisis

Steffen Böhm, Sian Sullivan (eds.)

https://doi.org/10.11647/OBP.0265

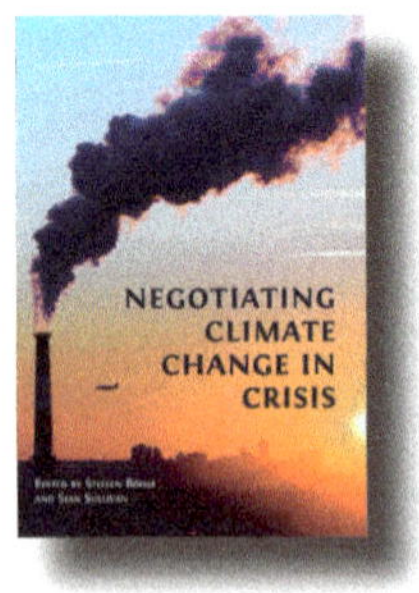

www.ingramcontent.com/pod-product-compliance
Ingram Content Group UK Ltd.
Pitfield, Milton Keynes, MK11 3LW, UK
UKHW062010041026
15591UKWH00017B/1256

9 781800 648562